Handbook of Pollution and Hazardous Materials Compliance

Environmental Science and Pollution Control Series

Additional Volumes in Preparation

Handbook of Pollution and Hazardous Materials Compliance

A Sourcebook for Environmental Managers

Nicholas P. Cheremisinoff
Madelyn Graffia

**National Association of Safety and Health Professionals, Inc.
Morganville, New Jersey**

CRC Press
Taylor & Francis Group
Boca Raton London New York

CRC Press is an imprint of the
Taylor & Francis Group, an **informa** business

First published 1996 by Marcel Dekker

Published 2020 by CRC Press
Taylor & Francis Group
6000 Broken Sound Parkway NW, Suite 300 Boca
Raton, FL 3487-2742

First issued in paperback 2020

ISBN-13: 978-0-367-57964-7 (pbk)
ISBN-13: 978-0-8247-9704-1 (hbk)

Visit the Taylor & Francis Web site at
http://www.taylorandfrancis.com

and the CRC Press Web site at
http://www.crcpress.com

PREFACE

This book has been written as a guide for environmental health and safety managers. Many companies today combine the management of environmental and occupational safety and health issues under one division, group, or even focal individual. Indeed this makes a great deal of sense because of direct as well as subtle interrelations between the environmental statutes and OSHA (Occupational Safety and Health Act) regulations. This is particularly true when implementing RCRA, CERCLA/SARA, or even TSCA, since specific OSHA safety standards are applicable.

This volume provides an overall working reference for environmental managers so that they can gain a working knowledge of the environmental/safety statutes and therefore a rapid understanding of what their organization/company must do in order to comply. There are ten chapters to this book. Chapter 1 covers the Federal Hazard Communication Act and elements common to individual state Right-To-Know legislation. Chapter 2 provides an overview of the OSHA regulations, with emphasis given to hazardous waste handling operations and emergency response. Chapter 3 provides a concise overview of the Clean Air Act and its amendments. Chapter 4 covers the Clean Water Act, and Chapter 5 covers the Safe Drinking Water Act. Chapters 6 through 10 cover statutes and subject matter which are closely linked regarding solid waste management and landfill disposal practices. Chapter 6 provides a discussion of the Comprehensive Environmental Responsibility, Compensation, and Liability Act (CERCLA), and Chapter 7 covers SARA (Superfund Amendments and Reauthorization Act). Both CERCLA and SARA have commonly been referred to by the general public as Superfund Laws and are often linked with RCRA (Resource Conservation and Recovery Act), which is covered in Chapter 8. Chapter 9 provides an overview of the technical requirements for managing facilities and issues associated with facility and property transfers. Finally, Chapter 10 covers overall management skills and concepts for multimedia facilities.

The authors have written this book both as a textbook for environmental management students and as a working reference for the environmental and health and safety manager. Its intent is to provide working knowledge of the compliance issues that must be met by all companies.

A special thanks is extended to the many students who have been instructed by the authors at New Jersey Institute of Technology and the Center for Environmental Management at Fairleigh Dickinson University, as they have helped shape ideas impacting on the organization of content. Additionally, gratitude is extended to the publisher for the fine production of this volume.

Nicholas P. Cheremisinoff
Madelyn Graffia

CONTENTS

1 THE HAZARD COMMUNICATION ACT

WHAT IS THE HAZARD COMMUNICATION ACT?

The Hazard Communication Act is a key OSHA rule that affects facilities that manage regulated hazardous materials. This federal regulation is known as the 29 CFR 1910.1200 standard. Included are requirements for training employees on the hazards associated with chemicals used in the workplace. Training programs must be established and maintained by the facility. Additionally, Material Safety Data Sheets must be available at the workplace and accessible to employees at all times. This standard also addresses the labeling requirements for all types of containers. The *Hazard Communication Act*, or more commonly known as *Right-To-Know*, is a fundamental piece of legislation aimed at educating workers on the hazards associated with chemicals handled and encountered in the workplace.

PURPOSES OF THE HAZARD COMMUNICATION ACT

The purpose of the Hazard Communication Act is to ensure that hazards of all chemicals produced or imported by chemical manufacturers are evaluated and that information on their hazards is transmitted to employers and employees within the manufacturing sector.

Employers in the manufacturing sector are required to provide information to their employees about hazardous chemicals by means of a Hazard Communication Program, labels and other forms of warning, Material Safety Data Sheets, and information and training. The purpose of the Hazard Communication Standard is threefold:

1. To ensure that the hazards of all chemicals produced or imported are evaluated and that information concerning their hazards is transmitted to employers and employees. This transmittal of information is to be accomplished by means of comprehensive hazard communication programs, which are to include container labeling and other forms of warning, Material Safety Data Sheets (MSDS), and employee training.

2. To comprehensively address the issue of evaluating the potential hazard of chemicals and communicating information concerning hazards and appropriate protective measures to employees.

3. To preempt any legal requirements of a state or political subdivision of a state, pertaining to the subject. Evaluating the potential hazards of chemicals, hazard communication, appropriating protective measures for employees, and developing and maintaining a written hazard communication program, which should include an inventory of all chemicals. There are also important requirements on labeling and on the preparation, distribution, and use of Material Safety Data Sheets.

The standard applies to chemical manufacturers, importers, distributors, and employers in the manufacturing sector in SIC Codes 20 through 39. In addition, the standard applies to any chemicals known to be present in the workplace in such a manner that employees may be exposed under normal conditions of use or in a foreseeable emergency.

Five important factors to be considered in applying the Hazard Communication Standard are:

- *Who is responsible for information dissemination?*
- *Who is to be informed?*
- *What is the information?*
- *How is the information to be transmitted?*
- *How can the information be standardized?*

Who informs employees is specified by OSHA. Manufacturers and importers have the responsibilities to evaluate, generate, and transmit information on hazardous materials not only to their employees, but the product end user. All employers have the responsibility to transmit this information to workers. However, not all business services or industries are covered by the standard.

The *target persons* to be informed are workers who handle or are potentially exposed to the material and are in the industrial manufacturing sectors. Who is to be informed is a concern spelled out in the standard. There are also others who need this information, including doctors, nurses, health and safety professionals, and supervisors.

What kind of information is needed is specified in the standard. Six basic types of information required are:

- Material identification.
- Company identification.
- Material properties.
- Hazard information.
- Protective information.
- Emergency information.

Criteria for each type is specified in the standard to varying degrees in 29 CFR 1910.1200, Section (g)(2)(i to xii).

How information is to be transmitted is specified for three modes of communication:

- Labeling.
- Material Safety Data Sheets.
- Training.

Containers of hazardous materials must be labeled by manufacturers or importers. MSDSs which are cross referenced to the label are intended to detail the information on the material's properties, its hazards, and safe handling practices. Employee training is essential and must cover at a minimum information necessary to understand the labeling requirements and the use of MSDSs.

Two aspects of this law are intended to promote the standardization of information:

1. The generic performance criteria, which includes:

   a. Hazard determination: The hazards of chemicals must be evaluated by chemical manufacturers and importers. The information must be passed on to employers in the manufacturing sectors who purchase hazardous chemicals.

  b. Written hazardous communication programs: Employers covered by the regulation must develop a hazard communication program to transmit information on hazardous chemicals to their employees.

  c. Labeling and other forms of warning must be placed on containers of hazardous chemicals.

  d. Material Safety Data Sheets must be developed to transmit hazard information to manufacturing employees and employers.

  e. Employee information and training must be provided: This includes identifying work operations where hazardous chemicals are present as well as the means that employees can take to protect themselves.

  f. Release of trade secret information: There are provisions for the release and protection of the secret information.

2. Preemption of state laws which are not consistent with the OSHA standards.

HAZARD COMMUNICATION: A HISTORICAL PERSPECTIVE

Prior to the passage of the federal Occupational Safety and Health Act of 1970, the communication of information to workers about the hazards of materials they were using was primarily a voluntary responsibility of industry. Amendments to the Longshoremen's Act of 1969 required the use of Material Safety Data Sheets (MSDS) to convey hazard information to workers. When transfer of hazard information occurred, it was influenced by several factors, including:

1. Market forces.
2. Trade secrets.
3. Available toxicity data.
4. Emergency situations.
5. Potential for high hazards.
6. Warnings from health and safety professionals.

7. Worker demands.
8. Liabilities.

Industries that were relatively safety conscious requested health and safety information for materials they purchased and therefore a demand was placed on manufacturers to provide such data.

In a properly designed Hazard Communication Program, hazard information should emphasize the prevention of accidents or situations that could cause fires, explosions, acute poisonings, or personal injuries and disfigurements. Safety training of personnel should concentrate on these risks. Communication between health and safety professionals and workers concerning dangerous materials can be indirect, with information filtering through supervisors or management.

The reasons for the communication of chemical hazards information include reducing personnel absences, loss of equipment, and lost time, and of course the concern for the safety of people. Many companies may be motivated to provide employee safety training simply to avoid costly legal suits. However, the increased flow of information seems to have an opposite effect, resulting in increased tort liability cases by workers. The 1970s were characterized by a rapid growth of public consciousness about chemical hazards. The Occupational Safety and Health Act put forth legal requirements for protecting workers against unsafe work environments. Hazard Communication has become an internal part of the Occupational Safety and Health Administration's (OSHA) requirements and whatever an employer's motivation, the fact is that it is the law.

IMPACTS OF OTHER AGENCIES

Other federal and state agencies have incorporated their own hazard communication standards. The U.S. Department of Transportation (DOT) has labeling, placarding, and manifesting requirements for the shipping and transport of hazardous materials. The U.S. Environmental Protection Agency (EPA) enforces several regulations requiring various levels of hazard communication. These include FIFRA (Federal Insecticide, Fungicide, and Rodenticide Act), TSCA (Toxic Substances Control Act), and RCRA (Resource Conservation Recovery Act). The Food and Drug Administration (FDA) has regulations governing the labeling of food and pharmaceuticals. These laws, coupled with greater public and worker consciousness of chemical hazards, have had a dramatic effect on market forces which have promoted hazard communication in the past. These forces, along with tort liability suits, cause many industries to assume responsibility for assessment and communication of chemical hazards.

WHAT ARE HAZARDS?

Hazardous waste sites and manufacturing operations pose a multitude of health and safety concerns, any one of which could result in serious injury or death. These hazards are a function of the nature of the sites and operations, as well as the nature of the work being performed. They include:

- Fire.
- Explosion.
- Toxic vapor, fluid, or gaseous releases.
- Corrosive vapors or their release.
- Overexposure to noise.
- Electrical shock.
- Release of cryogenic fluids.
- Slips, trips, and falls of workers.
- Fragmentation of a rapidly spinning grinding wheel.
- Personnel eye exposure to arc welding operations.
- Sudden release of pressurized fluids (gases).
- Exposure to a mechanical power transmission apparatus.
- Exposure to other mechanical equipment.
- Violent chemical reactions.

Several factors distinguish a hazardous waste site environment from other occupational situations involving hazardous substances. One factor is the large variety and number of substances that may be present at such a site. Any individual location may contain hundreds or even thousands of chemical hazards. In addition, the identity of the substances on site are frequently unknown, particularly in the initial stages of an investigation. Workers are subject not only to the hazards of direct exposure, but also to dangers posed by the disorderly physical environment of hazardous waste sites and the stress of working in chemical protective clothing and/or using respirators.

The combination of all these conditions results in a working environment that is characterized by numerous and varied hazards which:

- May pose an immediate danger to life or health.
- May not be immediately obvious or identifiable.
- May vary according to the location on site and the task being performed.
- May change as site activities progress.

WHAT ARE HAZARDOUS CHEMICALS?

A hazardous chemical is a material that is potentially dangerous to human health or the environment. OSHA defines hazardous chemicals in terms of health hazards and physical hazards. Tables 1 and 2 illustrate several hazard characteristics.

Health Hazards

Acute health hazards--Refers to exposure to a chemical where the dose is a single, rapid event. The term "dose" refers to amount or concentration of a chemical the subject is exposed to over a period of time. While this type of exposure is usually short, the results or effects may be irreversible. An acute exposure is a single dose in which protective mechanisms of the body are overcome.

Chronic health hazards--Refers to an exposure or dose delivery that is administered over a period of time and frequency that may be continuous or intermittent, but generally the period is time. Chronic exposures are characterized by a long duration and low level concentration, where the body capacity for detoxification is exceeded over a period of time.

Subacute--Closely resembles acute exposures and results from 12 to 40 doses which may be frequent, repeated, or extended over hours or even days.

Carcinogenic health hazards--This terminology refers to a carcinogenic or cancer-causing agent, under the OSHA regulations, if it has been evaluated by the International Agency for Research on Cancer (IARC) and is listed by IARC as a carcinogen or potential carcinogen.

TABLE 1
SELECTED EXAMPLES OF CHEMICALS THAT CREATE HEALTH HAZARDS

Health Hazard	Chemicals That Create the Hazard
Carcinogen	Aldrin, formaldehyde, ethylene dichloride, methylene chloride dioxin
Toxic	Xylene, phenol, propylene oxide
Highly toxic	Hydrogen cyanide, methyl parathion allyl alcohol, sulfur dioxide
Reproductive toxic	Methyl cellosolve, lead
Corrosive	Sulfuric acid, sodium hydroxide, hydrofluoric acid
Irritant	Ammonium solutions, stannic chloride calcium hypochlorite, magnesium dust
Sensitizer	Epichlorohydrin, fiberglass dusts
Hepatotoxin	Vinyl chloride, malathion, dioxane, acetonitrile, carbon tetrachloride phenol, ethylenediamine
Neurotoxin	Hydrogen cyanide, endrin, mercury cresol, methylene chloride, carbon disulfide, xylene
Nephrotoxin	Ethylenediamine, chlorobenzene, hexachloronaphthalene, acetonitrile allyl alcohol, phenol, uranium

---

TABLE 1 (Continued)

SELECTED EXAMPLES OF CHEMICALS THAT CREATE HEALTH HAZARDS

Health Hazard	Chemicals That Create the Hazard
Agents that damage: Blood	Nitrotoluene, benzene, cyanide, carbon monoxide
Lungs	Asbestos, silica, tars, dusts
Eyes or skin	Sodium hydroxide, ethylbenzene, allyl alcohol, nitroethane
Ethanolamine	Sulfuric acid, liquid oxygen, phenol, propylene oxide, ethyl butyl ketone

---

TABLE 2

SELECTED EXAMPLES OF CHEMICALS THAT POSE PHYSICAL HAZARDS

Physical Hazard	Chemicals That Pose the Hazard
Combustible liquids	Fuel oil, crude oil, other heavy oils
Flammables	Gasoline, isopropyl alcohol, acetone spray cans that use butane propellants
Explosives, Pyrophorics	Dynamite, nitroglycerine, ammunition, yellow phosphores, white phosphorus, silane gas, lithium hydride
Water reactives	Potassium, phosphorus pentasulfide, sodium hydride
Organic peroxides	Methyl ethyl ketone peroxide, dibenzoyl peroxide, dibutyl peroxide
Oxidizers	Sodium nitrate, magnesium nitrate bromine, sodium permanganate, calcium hypochlorite, chronic acid

Toxic--Toxic chemicals are defined as either *highly toxic* or *toxic*. Highly toxic chemicals are defined as follows:

- The chemical has a median lethal dose (LD_{50}) of 50 milligrams or less per kilograms of a body weight when administered orally to albino rats weighing between 200 and 300 grams each.
- The chemical has a median lethal dose (LD_{50}) of 200 milligrams or less per kilograms of body weight when administered by continuous contact on the bare skin of albino rabbits weighing between 2 and 3 kilograms each for a period of 24 hours.
- The chemical has a median lethal concentration (LC_{50}) in air if 200 parts per million by volume or less of gas or vapor or 2 milligrams per liter or less of mist, fume, or dust, and administered through continuous inhalation for 1 hour to albino rats weighing between 200 and 300 grams each.

A chemical is *toxic* if it meets any of the following criteria:

- The chemical has a median lethal dose (LD_{50}) of more than 50 milligrams per kilogram but not more than 500 milligrams per kilogram of body weight when administered orally to albino rats weighing between 200 and 300 grams each.
- The chemical has a median lethal dose (LD_{50}) of more than 200 milligrams per kilogram but not more than 1000 milligrams per kilogram of body weight when administered by continuous contact for 24 hours onto the bare skin of albino rabbits weighing between 2 and 3 kilograms each.
- The chemical has a median lethal concentration (LC_{50}) in air if more than 200 parts per million by volume or gas or vapor, or more than 2 milligrams per liter but not more than 20 milligrams per liter of mist, fume or dust is administered by continuous inhalation for 1 hour to albino rats weighing between 200 and 300 grams each.

Corrosives--Corrosive chemicals cause a visible destruction of living tissue. The regulations are stated under the DOT safety standards in 49 CFR Part 173. The test used to establish corrosivity exposes albino rats to a chemical for 4 hours. If, after the exposure period, the chemical has destroyed or irreversibly changed the structure of the tissue at the site of contact, then the chemical is considered to be a corrosive material.

Irritants--Chemicals that are irritants cause a reversible inflammatory effect on skin or eyes. Tests for determining whether the chemical is an irritant are defined in 16 CFR 1500.41. Tests for eye irritants are defined in 16 CFR 1500.42.

Sensitizers--A chemical is defined as a sensitizer if a large number of an exposed population develops an allergic reaction on normal tissue after repeated exposure to the chemical. The effects typically are reversible once the exposure ceases.

Human organ hazards--Chemicals are also considered a health hazard if exposure causes damage to any of the human organs. This includes liver enlargement, kidney disease or malfunction, excessive nervousness or a decrease in motor functions, decreases in lung capacity, and other organ damage.

Physical Hazards

A chemical is defined as a physical hazard if it has properties or characteristics which characterize a combustible liquid, a flammable, explosive, pyrophoric, or unstable material. Also, a chemical is deemed hazardous if it is a compressed gas, an organic peroxide, or an oxidizer.

Compressed gases/flammables--Compressed gases and flammable liquids represent special classes of hazards and risks. Laboratories, manufacturing operations, gas welding, and other areas using flammable materials, where such gases and liquids are stored are considered hazardous environments.

Compressed gases and flammable liquids must be handled and stored properly in order to prevent spills or vapors from causing fires or explosions. Relatively small amounts of gases or vapor can ignite from an arc caused by an electric switch, cigarette, open flame, or other source of ignition. Vapors can travel considerable distances to ignition sources and flash back to the original source, causing explosions and fire. These areas where compressed gases and flammable liquids are used must be properly designed and well maintained. These areas must also have specific safety features such as eyewashes, safety showers, and fire blankets. Fire protection features, such as fire resistant construction and fire extinguishing systems, will further reduce the hazard potentials.

SCOPE AND APPLICATION OF THE STANDARD

The Hazard Communication Act covers employers in the manufacturing sector. Chemical manufacturers have the primary obligation to evaluate chemicals for their hazards and to develop and transmit information found on Material Safety Data Sheets and labels. Employers have an obligation to develop a hazard communication program that includes MSDSs, labels, lists, and training.

The MSDS is the primary vehicle for transmitting information, but labels also provide hazard warnings. The trade secret provisions of the standard are, however, very broad. Chemical manufacturers or employers can claim any chemical they designate as a trade secret and withhold the identity of the ingredient from the exposed workers. Access to trade secret identities is only provided to health professionals and even then only under very limited circumstances and conditions. This does not mean, however, that the toxicity and hazard properties of the trade secret ingredient are withheld.

The standard's coverage is limited only to the manufacturing sector, SIC Codes 20-39. Included in these SIC codes are basic manufacturing industries such as chemical, electrical, rubber, steel, auto, textile, etc. There are, however, literally millions of workers in related industries that are not covered by the standard. The only requirement that does provide some indirect coverage to these excluded industries is that all chemical manufacturers must label their chemical containers before shipment from its manufacturing facility. Safety information must appear on labels of chemical containers, which includes the appropriate hazards warning as determined by the manufacturer, and the name and address of the chemical manufacturer. There is no requirement that MSDSs must be shipped to users outside SIC codes 20-39, nor is there a requirement that excludes industry employers leave labels intact.

Laboratories in the manufacturing sector are not subject to the standard's full requirements. For laboratories in covered industries, employers are required to leave labels intact, maintain and make available copies of MSDSs, and apprise laboratory workers of chemical hazards.

It is the AFL-CIO's position that all workers exposed to toxic chemicals in all industries should be covered by the standard. Exposure to toxic chemicals, not an arbitrary SIC code determination, should be the real basis for coverage under the standard.

Chemical manufacturers and importers of chemicals are required to evaluate all chemicals they produce or import to determine if the chemicals are hazardous as defined by the standard. Only those chemicals the manufacturer or importer determines to be hazardous are subject to the standards labeling, MSDSs, chemical inventory reporting, and training provisions.

Chemicals listed in 29 CFR 1910.1000 Subpart Z and by the American Conference of Governmental Industrial Hygienists (ACGIH) with Threshold Limit Values (TLV) are defined as hazardous by the standard and are subject to the standard's provision.

Chemicals that are regulated OSHA carcinogens or are listed as potential carcinogens in the National Toxicology Program (NTP) or in the International Agency for Research on Cancer (IARC) monographs, are legally defined as carcinogens and are subject to the standard's provisions.

From the standpoint of health hazards, chemicals which are known to harm animals or humans are specifically addressed by the OSHA standards. There is, however, some ambiguity as to which effect reported in animal studies triggers coverage of a chemical. OSHA's interpretation of this provision determines the extent of its coverage, which in general is very broad. The standard covers most chemicals for which well conducted animal tests show positive results, or is limited primarily to OSHA and ACGIH listed chemicals.

Pure chemicals and chemical mixtures are covered by the standard. For mixtures which have been tested as a whole, the results of the testing may be used to make a hazardous determination. For mixtures which have not been tested as a whole, the mixture is presumed to present the same health hazards as do hazardous components which comprise 1.0% or greater of the mixture, or 0.1% or greater concentrations for carcinogens.

Chemicals, food, drugs, cosmetics, consumer products, and hazardous wastes are subject to the labeling provisions of other federal statutes and are therefore exempt from the labeling provisions of the OSHA standard when labeled according to other statutes.

The Hazard Communication Act is a performance oriented standard with six major elements:

- Hazard determination.
- Hazard communication program.
- Labels or other forms of warning.
- Material Safety Data Sheets (MSDS).
- Employee information and training.
- Trade secret provisions.

Each of these elements are described below.

HAZARD DETERMINATION

Chemical manufacturers and importers are required to evaluate the chemicals which they produce or import to determine if they are hazardous. Other employees covered by the standard may rely upon hazard determinations performed by the manufacturer or importer. Chemicals manufacturers, importers, or employers who evaluate chemicals are required to identify and consider the scientific evidence concerning the physical hazard and health hazards of such chemicals. Specific definitions of physical hazards covered by the standard are set fourth in the definition section of the standard, that is combustible liquid, compressed gas, explosive, etc.

For health hazards, evidence which is statistically significant and which is based on at least one positive study conducted in accordance with established scientific principles is considered to be sufficient to establish a hazard if the results meet the definitions of health hazards set forth in Appendix A of the standard. Appendix A, which is mandatory, sets forth the health effects covered by the standard. Appendix A includes definitions of what constitutes a carcinogen, corrosive agents, highly toxic and toxic substance, irritant and sensitizer, and lists of target organ effects to illustrate the kinds of additional effects that are covered by the standard. Appendix B, which is mandatory, sets forth the hazard determination procedures which must be utilized in evaluating chemicals. The hazard determination requirement is performance oriented. There are mandatory sources of information that are listed for consultation. Certain criteria which must be followed in all hazard determinations include:

1. Determinations made by NTP, IARC, or OSHA that a chemical is a carcinogen or potential carcinogen.

2. Epidemiological studies and case reports on the adverse health effects must be considered in the evaluation.

3. The results of animal testing must be used to predict the health effects that may be experienced by exposed workers.

4. The results of any studies which are designed and conducted according to established scientific principles and which report statistically significant conclusions regarding the health effects of a chemical are considered a sufficient basis for a hazard determination and must be reported on the MSDS. For acute health hazards, the definitions of what constitutes an adverse health effect in animal studies are set forth in Appendix A. For chronic health effects, the manufacturer appears to have considerable flexibility in determining which results of animal tests constitute an adverse health effect and trigger coverage under the standard. Manufacturers and importers also are permitted to report the results of other scientifically valid studies which tend to refute the finding of the hazard. Appendix C, which is nonmandatory, sets forth a list of information sources which may be consulted in making a hazard determination.

Chemical manufacturers, importers, or employers evaluating chemicals are required to describe in writing their hazard determination procedures and must make these written procedures available upon request to employees, employee representatives, OSHA, and the National Institute for Occupational Safety and Health (NIOSH).

It is important to note that there is a "Floor" of over 600 substances that are automatically considered to be hazardous for the purposes of this standard. The "Floor" consists of any chemicals contained in:

1. OSHA regulated substances in Subpart Z of 1910 regulation.

2. The Threshold Limit Values for Chemical Substances and Physical Agents, published annually by the American Conference of Governmental Industrial Hygienists.

3. The National Toxicology Program (NTP) Annual Report on Carcinogens.

4. Monographs published by the International Agency for Research on Cancer (IARC).

The regulation allows the hazard determination of mixtures to be treated differently than the hazard assessment for pure substances. In general, if a mixture of chemicals has been tested as a whole to determine its health hazards or physical hazards, then the evaluator may use the results of the tests to determine whether or not the mixture presents a health or physical hazard. If the product mixture does not show evidence of being hazardous on the whole, then components in the concentration range of 1% or more need to be evaluated for toxic and hazardous characteristics. If there is evidence that any component may present a health hazard at less than these percentages, then the mixture is considered to be a health hazard. If a mixture has not been tested as a whole to determine its hazards, then the evaluator may rely on any scientifically valid data to assess the physical hazards of the entire mixture.

THE WRITTEN HAZARD COMMUNICATION PROGRAM

Employers covered by SIC codes 20-39 are required to develop and implement a written hazard communication program for their workplace which sets forth the requirements for labeling, warning signs, MSDSs, and how training will be met. The written programs must be made available to employees, employee representatives, OSHA, and NIOSH upon request.

The written program should not be created just to be filed away to prove compliance. This important document needs to be kept up to date. An out-of-date compliance program can pose safety risks and lead to fines, if discovered by OSHA. Some facilities place their programs on one or two year cycles for review. The program should be reviewed or revised when there are organizational changes. This is particularly important if individuals' names or positions are documented within the program.

A list of hazardous chemicals known to be present in the workplace must be compiled. Chemicals may be listed by an identity, including trade names or code names, that is referenced on the MSDS, and lists may be compiled by workplace or work area. The list is for chemicals currently present, and there is no requirement to maintain lists of chemicals for any period of time.

The Hazard Communication Program must set forth the methods the employer will use to inform employees of non-routine tasks and the hazards associated with chemicals contained in unlabeled pipes in their work areas. The AFL-CIO recommends the labeling or placarding of pipes and valves with appropriate identity and hazard information.

Employers are required to develop methods to inform any contractor working in the facility of the hazardous chemicals present and of appropriate protective measures. Employers may rely on existing hazard communication programs which meet the criteria set forth.

Program Elements

Key elements to be included and explained in a written hazard communication program are:

- Labels and other forms of warning.
- Material Safety Data Sheets (MSDS).
- List of hazardous chemicals.
- Hazards of non-routine tasks.
- Training and information.
- Multi-employer workplace hazards and SOPs.

Labels and Other Forms of Warning

In this section of the written hazard communication program, the description and explanation of the labeling system used on chemical containers must be described. Keep in mind that the labeling system used must convey the hazards of the container's contents. The requirement is primarily intended to cover containers leaving the facility, or for secondary containers not in control of the employee who filled them. Of course this assumes that the primary container had appropriate warning labels. If the container is to leave the facility, the labels must include the name and address of the facility of origination. Commercially available hazard warning labels, such as the Hazardous Material Information System (HMIS) and the National Fire Protection Association (NFPA) labels, may be used as in-house labeling systems. Some of these systems use numerical and alphabetical codes to convey hazards. Although these labeling systems do not convey target organ effects, their use is permitted, as long as the written hazard communication program addresses the issue, and simulates needed information from the Material Safety Data Sheet (MSDS).

Material Safety Data Sheets

All employers know that every hazardous chemical present at the facility must have an updated MSDS to warn employees about the hazards of the chemicals they work with. But employers also have the responsibility to describe how MSDSs will be made available to employees, and how MSDSs will be updated and maintained.

Hazardous Chemicals List

Another requirement of the Written Hazard Communication Standard is the hazardous chemicals list. This list should contain all materials and chemicals in the facility that are considered hazardous. Just as MSDSs, this list must be kept up to date and available to employees. This list was originally intended to be included in the written communication program, however, some facilities may find that their lists are, within the written program itself, exceedingly long and difficult to manage. For example, some facilities may find it easier to keep this list in an electronic data base, accessible to all employees. In a case such as this, instruction on how to access the database would be necessary within the written program.

Non-Routine Tasks

Within the written hazard communication program, employees must be informed of the hazards of non-routine tasks. Whether the information about the specific tasks is provided in a scheduled training session, or just prior to undertaking a non-routine task, it must be spelled out in the program. Applicable examples are tasks such as tank cleaning or line-flushing, in preparation for maintenance work. If other documents already exist for typical, non-routine tasks, the employer may reference these documents in the written program as well.

Training and Information

Within the written program, employers must describe the method and timing of training and information dissemination associated with the hazardous chemicals located at the facility. Among the requirements of the information to be provided are:

- The requirement of the Hazard Communication Standard.
- Any operations in the work area where hazardous chemicals are present.
- The location and availability of the written hazard communication program, including the list of hazardous chemicals and MSDSs.

Elements of training that need to be included are:

- Methods and observations used to detect the presence or release of hazardous chemicals in the workplace.
- Physical and health hazards of those chemicals.
- Measures employees can take to protect themselves from those hazards.
- Details of the hazard communication program developed by the employer.

Training should include an explanation of the labeling system, how to read an MSDS, and how the employees can obtain and use the appropriate hazard information to keep themselves safe.

Multi-Employer Workplaces

Also to be included in the written program is a description of the methods by which an employer will inform the employees of other employers located at the same facility, of the hazards at the workplace.

In addition to this, the program should address the possible hazards that the contract employer, or his employees, may pose to the facility's employees. For example, a painting contractor should be informed of the hazards presented by the chemicals and processes in a room that he or she is hired to paint. Employees need to be informed of the potential hazards of any paint fumes generated while the contractor is at the facility. Employers who produce, use, or store hazardous chemicals at a workplace in such a way that the employees of other employers may be exposed shall additionally ensure that the hazard communication programs developed and implemented under this paragraph include the following:

1. The methods the employer will use to provide other employers with a copy of the Material Safety Data Sheet, or to make it available at a central location in the workplace for each hazardous chemical the employers' employees may be exposed to while working.

2. The methods the employer will use to inform other employers of any precautionary measures that need to be taken to protect employees during the workplace's normal operating conditions and in foreseeable emergencies.

3. The methods the employer will use to inform the other employers of the labeling system used in the workplace.

The employer shall make the written hazard communication program available, upon request, to employees, their designated representative, the Assistant Secretary, and the Director, in accordance with the requirements of 29 CFR 1910.1200(e). Many employers may have the individual elements of a written hazard communication program outlined in already existing documents. It is not necessary to rewrite or reprint these entire documents again within the written program. It is acceptable to reference the existing documents within the written program, as long as the criteria and intent of the written hazard communication program is met.

LABELS AND PLACARDS

A label is intended to provide an immediate warning about hazardous materials in the workplace. It may be in written or graphics form and must be tied in with the training and MSDSs. The label must include the identity of the hazardous chemicals and appropriate warnings about the hazards. The label format is optional, and existing systems may be used. Some alternatives to labeling are allowed, such as large posters describing contents and hazards of a series of similar reactors. Chemical manufacturers, importers, and distributors must also include the name and address of the manufacturer of materials being shipped out of a plant. Substances covered by other federal laws are exempt from HCS's shipping container labeling requirements.

MSDSs complement the label and are more complex. They include:

1. The identity used on the label and the specific chemical's identity of the hazardous ingredients.

2. Physical and chemical characteristics of hazardous materials.

3. Fire, explosion, and reactivity hazards.

4. Health hazards, both short and long term.

5. Safe handling and use procedures.

6. Emergency and first aid procedures.

7. Date and name of the responsible party preparing the MSDS.

The Hazard Communication Standard requires chemical manufacturers to make MSDSs available to their employees. It also requires them to supply MSDSs to downstream manufacturing employees, who in turn must make them available to employees. Clearly then, the chemical manufacturers face the challenge of preparing complete MSDSs. Manufacturing employers must be sure that their workers do indeed get the message.

Since the standard is performance oriented, the format is optional rather than rigid. Different companies employ different systems. The new rule permits employers to continue to use effective systems so long as they contain the necessary information and effectively warn workers about hazardous chemicals. There is an advantage in following the ANSI Z129.1 standard to promote uniformity throughout industry.

Chemical manufactures, importers, and distributors are responsible for labeling on containers. For example, the hazards for a given material in an identified container may be posted in the workplace where it is readily accessible to employees.

In the workplace, manufacturers may use alternatives to actually putting labels on containers of hazardous chemicals shipped to employers in the manufacturing sector. Materials covered by other federal labeling laws are exempt from the shipping container labeling requirements.

The label itself must include the identity of the hazardous chemicals, which means their chemical or common names as keyed to the MSDS. The labels must also contain appropriate warnings about hazards, as determined by the employer. This does not mean the employer must test each substance in his workplace but may obtain the necessary information from the chemical manufacturer, importer, or distributor. In addition, the label for materials being shipped out of a plant must also contain the name and address of the manufacturer. In all cases, even where the label is in the form of a placard or other material not affixed to a moveable container, the information must be readily accessible to employees. Refer to Figure 1.

When identifying a hazardous material on a label, an employer may use the chemical or common name of the material, but then he must use the same name on the required list of hazardous chemicals, the label, and the MSDS. In any case, the MSDS must also include the specific chemical identity of the hazardous ingredients. The chemical name must be in accord with the Chemical Abstract Service (CAS) systems of the American Chemical Society or that of the International Union of Pure and Applied Chemistry (IUPAC).

Bona fide trade secrets are protected in the Hazard Communication Standard (HCS) while providing for the protection of exposed employees. This is done by providing for the limited trade secret disclosure to health professionals under prescribed conditions of need and confidentiality.

A hazard warning may be in the form of pictures or symbols, or words alone or in combination. The need for training to ensure that

Figure 1. Example of a complete label.

workers understand the hazard warnings is apparent. This need exists regardless of the method of written communication one chooses.

There are several exemptions to the labeling requirements. These exemptions apply to substances covered by other federal laws and regulations. Substances containing any chemicals that come under the labeling requirements of any of the following acts are not subject to the standard's shipping container labeling requirements:

1. Federal Insecticide, Fungicide, and Rodenticide Act.

2. Federal Food, Drug, and Cosmetic Act.

3. Federal Alcohol Administration Act.

4. Consumer Product Safety Act.

These acts are administered by the Environmental Protection Agency, the Food and Drug Administration, the Bureau of Alcohol, Tobacco, and Firearms, and the last two by the Consumer Product Safety Commission.

The standard has limited application to laboratories as follows:

1. Labels on incoming containers of hazardous chemicals must be left intact.

2. The MSDSs that accompany the hazardous chemicals must be kept and made readily accessible to laboratory employees.

3. Employees must be apprised of the hazards of the chemicals in their workplace in accordance with the employee information and training requirements of this standard.

In Figure 2, a summary of an acceptable graphics system for hazard identification is shown. This is an example developed by the National Paint and Coatings Association for a system for informing workers about the seriousness of a hazard and how to protect themselves from it. The system is called Hazardous Materials Identification System (HMIS). The severity index varies from zero, meaning minimal hazard, to four, meaning severe hazard. Readily identifiable are the symbols for safety glasses, gloves, aprons, respirators, and boots.

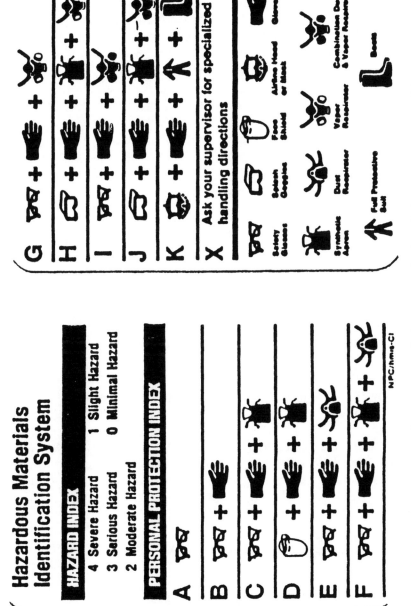

Figure 2. National Paint and Coatings Association system for identifying hazardous materials.

Alternatives to labeling each vessel or container with hazardous chemicals deal with representative posting, process container information, piping systems, and "immediate use" portable containers. For example, the warning for a row of identical reactors using the same process may consist of a large poster describing the materials contained, their hazards, and protective measures. Since the contents of pipelines will often vary with time, the pipes themselves need not be labeled. The hazards of the materials in these unlabeled pipes must nevertheless be addressed as part of the employer's hazard communication program. For example, the flow of materials from one vessel to another may be described in the standard operating procedure available to employees. Also portable containers into which hazardous chemicals are transferred need not be labeled.

It is the responsibility of the employers to be sure that:

1. Containers of hazardous chemicals are labeled. For example, the employer must leave the chemical manufacturer's incoming label in place if it is adequate. If the hazardous material was generated internally or for any other reason is unlabeled, the employer must leave the chemical manufacturer's incoming label in place if it is adequate. If the hazardous material was generated internally or for any other reason is unlabeled, the employer must provide one with appropriate information.

2. Labels must be legible, which means their format, type size, position, and other factors, although not specified, should be such that the label can be read easily by employees.

Chemical manufacturers, importers, and distributors shipping materials must label containers of hazardous chemicals and ensure that the labels do not conflict with the Hazardous Materials Transportation Act. Refer to Figures 3 through 5 for examples.

THE MATERIAL SAFETY DATA SHEET

All chemicals that are manufactured, imported, sold, or used in a manufacturing process must be accompanied by a Material Safety Data Sheet (MSDS), as defined in 29 CFR 1910.1200. Chemical

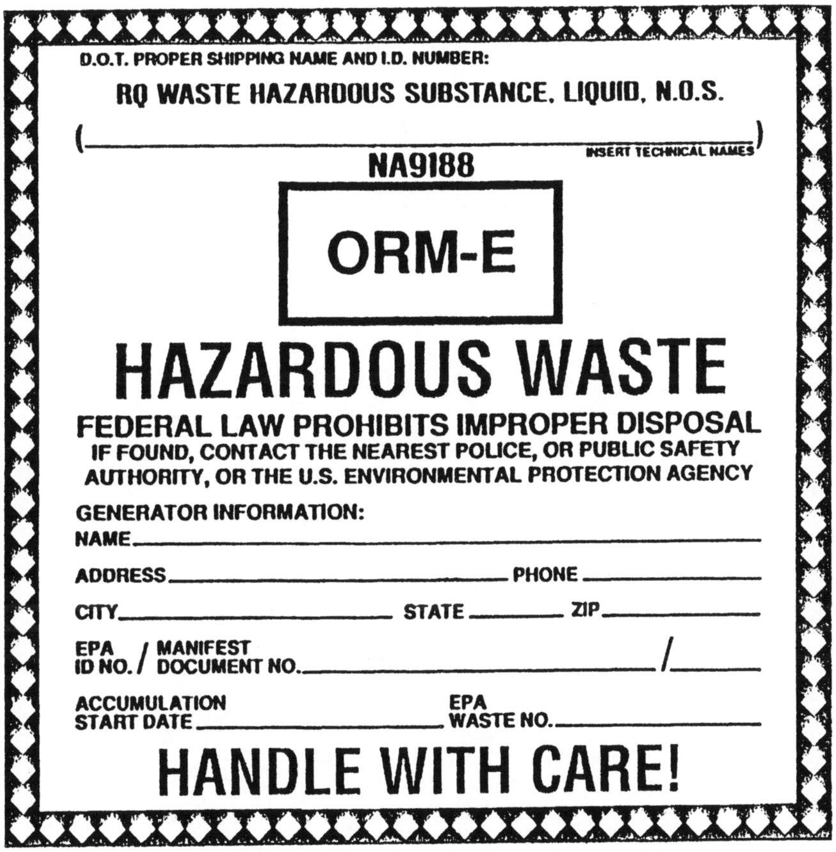

D.O.T. PROPER SHIPPING NAME AND I.D. NUMBER:

RQ WASTE HAZARDOUS SUBSTANCE, LIQUID, N.O.S.

(______________________________________)
INSERT TECHNICAL NAMES

NA9188

ORM-E

HAZARDOUS WASTE

FEDERAL LAW PROHIBITS IMPROPER DISPOSAL

IF FOUND, CONTACT THE NEAREST POLICE, OR PUBLIC SAFETY AUTHORITY, OR THE U.S. ENVIRONMENTAL PROTECTION AGENCY

GENERATOR INFORMATION:

NAME ______________________________

ADDRESS ______________________ PHONE __________

CITY ______________ STATE ________ ZIP __________

EPA / MANIFEST
ID NO. / DOCUMENT NO. ____________________ / ______

ACCUMULATION START DATE ______________ EPA WASTE NO. ______________

HANDLE WITH CARE!

Figure 3. Example of a hazardous waste label. From EMED Co., Buffalo, NY, with permission.

```
┌─────────────────────────────────────────────┐
│            NONHAZARDOUS WASTE                 │
├─────────────────────────────────────────────┤
│            PACKING MATERIAL                   │
├─────────────────────────────────────────────┤
│  NO HEALTH HAZARD                             │
│  COMBUSTIBLE                                  │
│  NON-REACTIVE                                 │
│  NO CONTACT HAZARD                            │
├─────────────────────────────────────────────┤
│  FIRE:   Use water-based extinguisher         │
│          or Class A dry chemical extinguisher.│
│                                               │
│  SPECIAL INFORMATION:                         │
│                                               │
│     This material is  RECYCLED.               │
│     Please segregate properly.                │
├─────────────────────────────────────────────┤
│  Accumulation Start Date:   /  /   Ship Date:  /  / │
│  Generator (Dept./Bldg./Name):                │
│  Name:                           Phone:_____│
│  Internal Shipper (Dept./Bldg./Name):         │
│  For Disposal, Contact:  Recyling Center at X6875. │
│                                               │
│  Comments: _____  │
└─────────────────────────────────────────────┘
```

Figure 4.   Example of an employer's waste label for a nonhazardous waste.

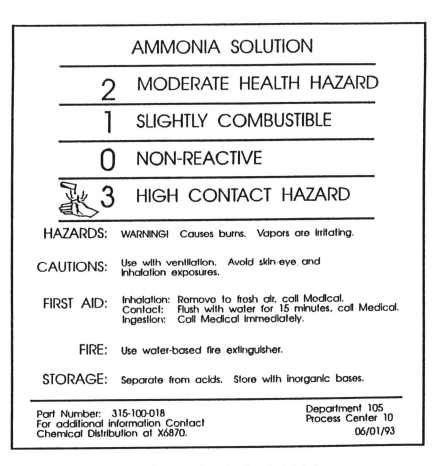

Figure 5.   Example of an employer's chemical label.

manufacturers or importers must provide the MSDS to distributors and employees with their initial shipment of chemicals and with the first shipment after data on the MSDS has been updated. Distributors who sell to others distributors or employers, must provide the MSDS and updates to their customers. An MSDS or equivalent must be available for review by employees in the workplace and must contain, at a minimum, the following information:

1. Chemical and Common Names--If the chemical is a single substance, the chemical and common names must be identified. If the hazardous chemical is a mixture and has been tested as a whole to determine its hazards, the chemical and common names of the ingredients that contribute to the known hazards and the common names of the mixture itself must be identified.

   If the hazardous chemical is a mixture that has not been tested as a whole, the following apply: chemical and common names of all ingredients that have been determined to be health hazards and that comprise 1% or greater of the composition, or that are carcinogens and comprise 0.1% or greater, must be identified; the chemical or common names of chemicals in the mixture could be released in amounts greater than the permissible exposure limit must be identified; the chemical and common names of all ingredients determined pose a physical hazard when present in the mixture must be identified.

2. Physical and Chemical Characteristics--such as vapor pressure, flash point, boiling point, melting point, specific gravity, solubility, and molecular weight must be included.

3. Physical Hazards--such as the potential for fire, explosion, and reactivity must be included. Incompatibilities and segregation practice may also be included in this part of the MSDS.

4. Health Hazards--The health hazards of the chemical must be included. Signs and symptoms of exposure, and any medical conditions that are generally recognized as being aggravated by exposure to the chemical must be delineated.

5. Primary Routes of Entry--Information pertaining to primary routes of entry such as inhalation, ingestion, skin absorption, and eye or skin contacts required.

6. Permissible Exposure Limit--The OSHA permissible exposure limit or American Conference of Governmental Industrial Hygienists Threshold Limit Value and any other exposure limit used or recommended by the chemical manufacturer, importer, or employer preparing the MSDS must be included.

7. Carcinogen or Potential Carcinogen--The MSDS also must include information as to whether the chemical is listed in the National Toxicology Program Annual Report on Carcinogens or has been published as a carcinogen in the International Agency for Research on Cancer Monographs or has been listed by OSHA.

8. Safe Handling Procedures--This section of the MSDS includes such items as appropriate industrial hygiene practices, protective measures for handling the chemical during repair and maintenance of contaminated equipment, and procedures for clean up of spills.

9. Control Measures--such as appropriate engineering controls, protective clothing and equipment, and work practices are detailed.

10. Emergency and First Aid Procedures--include emergency actions to be taken for exposure through all the primary routes.

11. Manufacturer's and Other Information--The MSDS also must include the name, address, and telephone number of the chemical manufacturer, importer, or other responsible party who prepares or distributes the MSDS. Additionally, the date of the MSDS preparation or the last change to it must be included.

The purpose of the MSDS is to serve as the primary vehicle for transmitting detailed hazard information to employees and manufacturing employers. Since the MSDS is such a key element in the hazard communication program, it should be prepared before labels are created.

The reason the MSDS is such a key part of the hazard communication program is that it contains all of the information on which other elements of the standard are based. Specifically, the labels and training will be derived from the information generated from the MSDSs.

Chemical manufacturers and importers who normally generate the MSDSs, must also obtain or develop an MSDS for each hazardous chemical they handle. The information in it must accurately reflect current scientific evidence which may require judgment.

Chemical manufacturers or importers who normally generate the MSDS must ensure that distributors and downstream customers are provided with an MSDS for each hazardous material. That is, MSDSs must be provided with the initial purchase, either attached to the container or sent to the purchaser prior to, or at the time of the next shipment. Distributors have the responsibility of ensuring that MSDSs are provided to other distributors and manufacturing purchasers.

A manufacturing employer must be sure that an MSDS for each hazardous chemical in the workplace is accessible to his employees. He may use the documents provided by his suppliers. In some cases, he may have to generate his own MSDSs.

The evaluation of a hazard depends on the available scientific data and the professional judgement applied to the interpretation of the information. The OSHA standard mandates certain criteria for hazard determination. For example, if OSHA has promulgated a permissible exposure limit for a given chemical or the American Conference of Governmental Industrial Hygienists (ACGIH) has published a Threshold Limit Value (TLV) for a substance, it must be listed as hazardous in the MSDS when present at 1% or greater. Similarly, if a material has been identified as a suspect or confirmed an animal or human carcinogen by OSHA or in the most recent report of the National Toxicology Program (NTP) or by the International Agency for National Cancer (IANC) present at 0.1% or greater, then this information must be indicated on the MSDS.

Other sections of the MSDS deal with precautions and procedures for safe handling, hygienic practices, and cleanup procedures for spills. The MSDS must also include any generally applicable control measures which are known to the chemical manufacturer, importer, or employer such as appropriate engineering controls, worker practices, or personal protective equipment.

The MSDS must contain emergency and first aid procedures, the date of preparation, and changes in the information given in the MSDS. For example, if the chemical manufacturer or importer becomes aware of any significant information regarding the hazards of a chemical or ways to protect workers against a hazard, this new information must be added to the MSDS within three months. To make additional information available, the MSDS must include the name, address, and telephone number of the party responsible for the document.

Miscellaneous requirements include that the MSDSs must be in English and contain no blanks. If the preparer can find no information for a given category on the MSDS, he must mark it to indicate that no applicable information was found. Where a series of mixtures have similar hazards and contents, a single MSDS may be prepared which applies to all of the similar mixtures. No universal format for the MSDS is specified. MSDSs are based on the adequacy and accuracy of the information, not format.

An example of a Material Safety Data Sheet is shown on the next page (Figure 6). In this case, the MSDS contains all the necessary information, but the arrangement of the sections is different than in the OSHA 20 form. For example, the health information is presented on the first page. Also, though not required, symbols and the hazard rating system are given.

Also the CHEMTREC number is given in the upper left hand corner of the Shell MSDS. CHEMTREC, a CMA (Chemical Manufacturers Association) operation, runs 24-hour, toll-free information hotline to assist in a chemical transportation emergency.

When the labels and MSDSs have been prepared, key things to remember are that they must be accessible to employees at their workplace, and that they must be tied in with the training program. CMA strongly supports the Hazard Communication Standard since it is an excellent way to help prevent confusion that differing state laws often bring. This position is a reflection of the chemical industry's safety commitment.

Material Safety Data Sheet
May be used to comply with
OSHA's Hazard Communication Standard.
29 CFR 1910.1200. Standard must be
consulted for specific requirements.

U.S. Department of Labor
Occupational Safety and Health Administration
(Non-Mandatory Form)
Form Approved
OMB No. 1218-0072

IDENTITY *(As Used on Label and List)*

Note: Blank spaces are not permitted. If any item is not applicable, or no information is available, the space must be marked to indicate that.

Section I

Manufacturer's Name	Emergency Telephone Number
Address *(Number, Street, City, State, and ZIP Code)*	Telephone Number for Information
	Date Prepared
	Signature of Preparer *(optional)*

Section II — Hazardous Ingredients/Identity Information

Hazardous Components (Specific Chemical Identity; Common Name(s))	OSHA PEL	ACGIH TLV	Other Limits Recommended	% *(optional)*

Section III — Physical/Chemical Characteristics

Boiling Point		Specific Gravity (H_2O = 1)	
Vapor Pressure (mm Hg)		Melting Point	
Vapor Density (AIR = 1)		Evaporation Rate (Butyl Acetate = 1)	
Solubility in Water			
Appearance and Odor			

Section IV — Fire and Explosion Hazard Data

Flash Point (Method Used)		Flammable Limits	LEL	UEL
Extinguishing Media				
Special Fire Fighting Procedures				
Unusual Fire and Explosion Hazards				

(Reproduce locally)

OSHA 174, Sept. 1985

Figure 6.   Material Safety Data Sheet.

Section V — Reactivity Data

Stability	Unstable		Conditions to Avoid
	Stable		

Incompatibility (Materials to Avoid)

Hazardous Decomposition or Byproducts

Hazardous Polymerization	May Occur		Conditions to Avoid
	Will Not Occur		

Section VI — Health Hazard Data

Route(s) of Entry:	Inhalation?	Skin?	Ingestion?

Health Hazards (Acute and Chronic)

Carcinogenicity:	NTP?	IARC Monographs?	OSHA Regulated?

Signs and Symptoms of Exposure

Medical Conditions
Generally Aggravated by Exposure

Emergency and First Aid Procedures

Section VII — Precautions for Safe Handling and Use

Steps to Be Taken in Case Material Is Released or Spilled

Waste Disposal Method

Precautions to Be Taken in Handling and Storing

Other Precautions

Section VIII — Control Measures

Respiratory Protection (Specify Type)

Ventilation	Local Exhaust		Special
	Mechanical (General)		Other

Protective Gloves	Eye Protection

Other Protective Clothing or Equipment

Work/Hygienic Practices

Figure 6. Material Safety Data Sheet (continued).

Shell

MATERIAL SAFETY DATA SHEET

97367 (4-85)

MSDS NUMBER ▶   10-5

24 HOUR EMERGENCY ASSISTANCE	GENERAL MSDS ASSISTANCE	
SHELL: 713-473-9461 CHEMTREC: 800-424-9300	SHELL: 713-241-4819	**BE SAFE** READ OUR PRODUCT SAFETY INFORMATION ...AND PASS IT ON PRODUCT LIABILITY LAW REQUIRES IT!

ACUTE HEALTH	FIRE	REACTIVITY			
✚ 2	🔥 3	☢ 0	HAZARD RATING ▶	LEAST - 0 SLIGHT - 1 MODERATE - 2 HIGH - 3 EXTREME - 4	

=For acute and chronic health effects refer to the discussion in Section III

SECTION I   NAME

PRODUCT ▶ EPON(R) RESIN 872-X-75

CHEMICAL NAME ▶ REACTION PRODUCT OF BISPHENOL A/EPICHLOROHYDRIN RESIN AND DIMER

CHEMICAL FAMILY ▶ FATTY ACID; EPOXY RESIN

SHELL CODE ▶ 43322

SECTION II-A   PRODUCT/INGREDIENT

NO.	COMPOSITION	CAS NUMBER	PERCENT
P	EPON RESIN 872-X-75	MIXTURE	100
1	EPON RESIN 872	67989-50-0	75
2	XYLENE	1330-20-7	25

SECTION II-B   ACUTE TOXICITY DATA

NO.	ACUTE ORAL LD50	ACUTE DERMAL LD50	ACUTE INHALATION LC50
P	NOT AVAILABLE	NOT AVAILABLE	NOT AVAILABLE
1	NOT AVAILABLE	NOT AVAILABLE	NOT AVAILABLE
2	4.3 G/KG (RAT)		6700 PPM/4H (RAT)

SECTION III   HEALTH INFORMATION

THE HEALTH EFFECTS NOTED BELOW ARE CONSISTENT WITH REQUIREMENTS UNDER THE OSHA HAZARD COMMUNICATION STANDARD (29 CFR 1910.1200).

EYE CONTACT
BASED ON PRESENCE OF COMPONENT 1, PRODUCT IS PRESUMED TO BE IRRITATING TO THE EYES.

SKIN CONTACT
BASED ON PRESENCE OF COMPONENT 1 AND 2, PRODUCT IS PRESUMED TO BE MILDLY IRRITATING TO THE SKIN. PROLONGED OR REPEATED LIQUID CONTACT CAN RESULT IN DEFATTING AND DRYING OF THE SKIN WHICH MAY RESULT IN SKIN IRRITATION AND DERMATITIS.

INHALATION
BASED ON PRESENCE OF COMPONENT 2, PRODUCT IS PRESUMED TO BE MILDLY TOXIC AND MAY PRODUCE CNS DEPRESSION.

INGESTION
BASED ON PRESENCE OF COMPONENT 2, PRODUCT IS PRESUMED TO BE MODERATELY TOXIC AND MAY BE HARMFUL IF SWALLOWED; MAY PRODUCE CNS DEPRESSION.

Figure 6.   Material Safety Data Sheet (continued).

PRODUCT NAME: EPON(R) RESIN 872-X-75

SIGNS AND SYMPTOMS
IRRITATION AS NOTED ABOVE. EARLY TO MODERATE CNS (CENTRAL NERVOUS SYSTEM) DEPRESSION MAY BE EVIDENCED BY GIDDINESS, HEADACHE, DIZZINESS AND NAUSEA; IN EXTREME CASES, UNCONSCIOUSNESS AND DEATH MAY OCCUR.

AGGRAVATED MEDICAL CONDITIONS
PREEXISTING EYE AND SKIN DISORDERS MAY BE AGGRAVATED BY EXPOSURE TO THIS PRODUCT.

OTHER HEALTH EFFECTS
NONE IDENTIFIED.

SECTION IV — OCCUPATIONAL EXPOSURE LIMITS

NO.	OSHA PEL/TWA	OSHA PEL/CEILING	ACGIH TLV/TWA	ACGIH TLV/STEL	OTHER
P	NONE ESTABLISHED				
1	NONE ESTABLISHED				
2	100 PPM (SKIN)		100 PPM (SKIN)	150 PPM (SKIN)	

SECTION V — EMERGENCY AND FIRST AID PROCEDURES

EYE CONTACT
FLUSH EYES WITH PLENTY OF WATER FOR 15 MINUTES WHILE HOLDING EYELIDS OPEN. GET MEDICAL ATTENTION.

SKIN CONTACT
REMOVE CONTAMINATED CLOTHING/SHOES AND WIPE EXCESS FROM SKIN. FLUSH SKIN WITH WATER. FOLLOW BY WASHING WITH SOAP AND WATER. IF IRRITATION OCCURS, GET MEDICAL ATTENTION. DO NOT REUSE CLOTHING UNTIL CLEANED. CONTAMINATED LEATHER ARTICLES, INCLUDING SHOES, CANNOT BE DECONTAMINATED AND SHOULD BE DESTROYED.

INHALATION
REMOVE VICTIM TO FRESH AIR AND PROVIDE OXYGEN IF BREATHING IS DIFFICULT. GIVE ARTIFICIAL RESPIRATION IF NOT BREATHING. GET MEDICAL ATTENTION.

INGESTION
DO NOT INDUCE VOMITING. IF VOMITING OCCURS SPONTANEOUSLY, KEEP HEAD BELOW HIPS TO PREVENT ASPIRATION OF LIQUID INTO THE LUNGS. GET MEDICAL ATTENTION.°

NOTE TO PHYSICIAN
°IF MORE THAN 2.0 ML PER KG HAS BEEN INGESTED AND VOMITING HAS NOT OCCURRED, EMESIS SHOULD BE INDUCED WITH SUPERVISION. KEEP VICTIM'S HEAD BELOW HIPS TO PREVENT ASPIRATION. IF SYMPTOMS SUCH AS LOSS OF GAG REFLEX, CONVULSIONS OR UNCONSCIOUSNESS OCCUR BEFORE EMESIS, GASTRIC LAVAGE USING A CUFFED ENDOTRACHEAL TUBE SHOULD BE CONSIDERED.

SECTION VI — SUPPLEMENTAL HEALTH INFORMATION

COMPONENT #2: LABORATORY ANIMALS EXPOSED BY VARIOUS ROUTES TO HIGH DOSES OF XYLENE SHOWED EVIDENCE OF EFFECTS IN THE LIVER, KIDNEYS, LUNGS, SPLEEN, HEART AND ADRENALS. RATS EXPOSED TO XYLENE VAPOR DURING PREGNANCY SHOWED EMBRYO/FETOTOXIC EFFECTS. MICE EXPOSED ORALLY TO DOSES PRODUCING MATERNAL TOXICITY ALSO SHOWED EMBRYO/FETOTOXIC EFFECTS.

Figure 6. Material Safety Data Sheet (continued).

PRODUCT NAME: EPON(R) RESIN 872-X-75

SECTION VII   **PHYSICAL DATA**

BOILING POINT: NOT AVAILABLE   SPECIFIC GRAVITY: 1.02   VAPOR PRESSURE: 6
     (DEG F)   (H2O=1)   (MM HG)

MELTING POINT: NOT AVAILABLE   SOLUBILITY: NGELIGIBLE   VAPOR DENSITY: 3.7
     (DEG F)   (IN WATER)   (AIR=1)

EVAPORATION RATE (N-BUTYL ACETATE = 1): NOT AVAILABLE

APPEARANCE AND ODOR: LIGHT YELLOW LIQUID.

SECTION VIII   **FIRE AND EXPLOSION HAZARDS**

FLASH POINT AND METHOD: **FLAMMABLE LIMITS /% VOLUME IN AIR**
<100 DEG. F (SETAGLASH)   LOWER: 1.1   UPPER: 7.0

EXTINGUISHING MEDIA
USE WATER FOG, FOAM, DRY CHEMICAL OR CO2.

SPECIAL FIRE FIGHTING PROCEDURES AND PRECAUTIONS
WARNING. FLAMMABLE. CLEAR FIRE AREA OF UNPROTECTED PERSONNEL. DO NOT ENTER CONFINED FIRE SPACE
WITHOUT FULL BUNKER GEAR (HELMET WITH FACE SHIELD, BUNKER COATS, GLOVES AND RUBBER BOOTS),
INCLUDING A POSITIVE PRESSURE NIOSH APPROVED SELF-CONTAINED BREATHING APPARATUS. COOL FIRE EXPOSED
CONTAINERS WITH WATER.

UNUSUAL FIRE AND EXPLOSION HAZARDS
HANDLE AS FLAMMABLE LIQUID. CONTAINERS EXPOSED TO INTENSE HEAT FROM FIRES SHOULD BE COOLED WITH
WATER TO PREVENT VAPOR PRESSURE BUILDUP WHICH COULD RESULT IN CONTAINER RUPTURE. CONTAINER AREAS
EXPOSED TO DIRECT FLAME CONTACT SHOULD BE COOLED WITH LARGE QUANTITIES OF WATER AS NEEDED TO
PREVENT WEAKENING OF CONTAINER STRUCTURE.

SECTION IX   **REACTIVITY**

STABILITY: STABLE   HAZARDOUS POLYMERIZATION: WILL NOT OCCUR

CONDITIONS AND MATERIALS TO AVOID:
AVOID HEAT, SPARKS, FLAME AND CONTACT WITH STRONG OXIDIZING AGENTS AND STRONG LEWIS OR MINERAL
ACIDS.

HAZARDOUS DECOMPOSITION PRODUCTS
CARBON MONOXIDE, ALDEHYDES AND ACIDS MAY BE FORMED DURING COMBUSTION. REACTION WITH SOME CURING
AGENTS MAY PRODUCE CONSIDERABLE HEAT.

SECTION X   **EMPLOYEE PROTECTION**

RESPIRATORY PROTECTION
AVOID BREATHING VAPOR OR MISTS. IF EXPOSURE MAY OR DOES EXCEED OCCUPATIONAL EXPOSURE LIMITS (SEC.
IV) USE A NIOSH-APPROVED RESPIRATOR TO PREVENT OVEREXPOSURE. IN ACCORD WITH 29 CFR 1910.134 USE AN
ATMOSPHERE-SUPPLYING RESPIRATOR OR AIR-PURIFYING RESPIRATOR FOR ORGANIC VAPORS.

PROTECTIVE CLOTHING
AVOID CONTACT WITH EYES. WEAR CHEMICAL GOGGLES IF THERE IS LIKELIHOOD OF CONTACT WITH EYES. AVOID
CONTACT WITH SKIN AND CLOTHING. WEAR CHEMICAL-RESISTANT GLOVES AND PROTECTIVE CLOTHING.

ADDITIONAL PROTECTIVE MEASURES
USE EXPLOSION-PROOF VENTILATION AS REQUIRED TO CONTROL VAPOR CONCENTRATIONS. EYE WASH FOUNTAINS
AND SAFETY SHOWERS SHOULD BE AVAILABLE FOR EMERGENCY USE.

Figure 6.   Material Safety Data Sheet (continued).

PRODUCT NAME: EPON(R) RESIN 872-X-75

SECTION XI   ENVIRONMENTAL PROTECTION

SPILL OR LEAK PROCEDURES
WARNING. FLAMMABLE. ELIMINATE ALL IGNITION SOURCES. HANDLING EQUIPMENT MUST BE GROUNDED TO PREVENT SPARKING. *** LARGE SPILLS *** EVACUATE THE HAZARD AREA OF UNPROTECTED PERSONNEL. WEAR APPROPRIATE RESPIRATOR AND PROTECTIVE CLOTHING. SHUT OFF SOURCE OF LEAK ONLY IF SAFE TO DO SO. DIKE AND CONTAIN. "IF VAPOR CLOUD FORMS, WATER FOG MAY BE USED TO SUPPRESS; CONTAIN RUN-OFF. REMOVE WITH VACUUM TRUCKS OR PUMP TO STORAGE/SALVAGE VESSELS. SOAK UP RESIDUE WITH AN ABSORBENT SUCH AS CLAY, SAND OR OTHER SUITABLE MATERIAL; PLACE IN NON-LEAKING CONTAINERS FOR PROPER DISPOSAL. FLUSH AREA WITH WATER TO REMOVE TRACE RESIDUE; DISPOSE OF FLUSH SOLUTIONS AS ABOVE. *** SMALL SPILLS *** TAKE UP WITH AN ABSORBENT MATERIAL AND PLACE IN NON-LEAKING CONTAINERS; SEAL TIGHTLY FOR PROPER DISPOSAL.

WASTE DISPOSAL
UNDER EPA - RCRA (40 CFR 261.21). IF THIS PRODUCT BECOMES A WASTE MATERIAL, IT WOULD BE IGNITABLE HAZARDOUS WASTE. HAZARDOUS WASTE NUMBER D001. REFER TO LATEST EPA OR STATE REGULATIONS REGARDING PROPER DISPOSAL.

ENVIRONMENTAL HAZARDS
EPA - COMPREHENSIVE ENVIRONMENTAL RESPONSE, COMPENSATION AND LIABILITY ACT. UNDER EPA-CERCLA ("SUPERFUND") RELEASES TO AIR, LAND OR WATER WHICH EXCEED THE REPORTABLE QUANTITY MUST BE REPORTED TO THE NATIONAL RESPONSE CENTER, 800-424-8802.

THE REPORTABLE QUANTITY (RQ) FOR THIS PRODUCT IS 4000 LB (471 GAL) BASED ON COMPONENT #2 CONTENT.

SECTION XII   SPECIAL PRECAUTIONS

WARNING. FLAMMABLE LIQUID. KEEP LIQUID AND VAPOR AWAY FROM HEAT, SPARKS AND FLAME. SURFACES THAT ARE SUFFICIENTLY HOT MAY IGNITE EVEN LIQUID PRODUCT IN THE ABSENCE OF SPARKS OR FLAME. EXTINGUISH PILOT LIGHTS, CIGARETTES AND TURN OFF OTHER SOURCES OF IGNITION PRIOR TO USE AND UNTIL ALL VAPORS ARE GONE. VAPORS MAY ACCUMULATE AND TRAVEL TO IGNITION SOURCES DISTANT FROM THE HANDLING SITE; FLASH-FIRE CAN RESULT. KEEP CONTAINERS CLOSED WHEN NOT IN USE. USE (ONLY) WITH ADEQUATE VENTILATION. CONTAINERS, EVEN THOSE THAT HAVE BEEN EMPTIED, CAN CONTAIN HAZARDOUS PRODUCT RESIDUES.

CONTAINERS, EVEN THOSE THAT HAVE BEEN EMPTIED, CAN CONTAIN EXPLOSIVE VAPORS. DO NOT CUT, DRILL, GRIND, WELD OR PERFORM SIMILAR OPERATIONS ON OR NEAR CONTAINERS. STATIC ELECTRICITY MAY ACCUMULATE AND CREATE A FIRE HAZARD. GROUND FIXED EQUIPMENT. BOND AND GROUND TRANSFER CONTAINERS AND EQUIPMENT.

AVOID BODILY CONTACT WITH MATERIAL. WASH WITH SOAP AND WATER BEFORE EATING, DRINKING, SMOKING OR USING TOILET FACILITIES. LAUNDER CONTAMINATED CLOTHING BEFORE REUSE.

CONTAMINATED LEATHER ARTICLES, INCLUDING SHOES CANNOT BE DECONTAMINATED AND SHOULD BE DESTROYED.

SECTION XIII   TRANSPORTATION REQUIREMENTS

DEPARTMENT OF TRANSPORTATION CLASSIFICATION: FLAMMABLE LIQUID
D.O.T. PROPER SHIPPING NAME: RESIN SOLUTION

OTHER REQUIREMENTS:
DOT ID NO. UN1866; GUIDE NO. 27. TRANSPORTATION SPILLS WHICH CAN ENTER SURFACE WATERS ARE REPORTABLE IF THE "REPORTABLE QUANTITY" (RQ) IS RELEASED FROM ONE INDIVIDUAL PACKAGE OR INDIVIDUAL BULK TRANSPORT VEHICLE. SEE SECTION XI.

SECTION XIV   OTHER REGULATORY CONTROLS

THE COMPONENTS OF THIS PRODUCT ARE LISTED ON THE EPA/TSCA INVENTORY OF CHEMICAL SUBSTANCES.

Figure 6.   Material Safety Data Sheet (continued).

PRODUCT NAME: EPON(R) RESIN 872-X-78

THE INFORMATION CONTAINED HEREIN IS BASED ON THE DATA AVAILABLE TO US AND IS BELIEVED TO BE CORRECT.
HOWEVER, SHELL MAKES NO WARRANTY, EXPRESSED OR IMPLIED REGARDING THE ACCURACY OF THESE DATA OR THE
RESULTS TO BE OBTAINED FROM THE USE THEREOF.   SHELL ASSUMES NO RESPONSIBILITY FOR INJURY FROM THE
USE OF THE PRODUCT DESCRIBED HEREIN.

DATE PREPARED:NOVEMBER 20, 1985

                                                     JOHN P. SEPESI
    BE SAFE   ---------------------------------------

    READ OUR PRODUCT   SHELL OIL COMPANY
    SAFETY INFORMATION ...AND PASS IT ON   PRODUCT SAFETY AND COMPLIANCE
          (PRODUCT LIABILITY LAW   P. O. BOX 4320
             REQUIRES IT)   HOUSTON, TX 77210

Figure 6.   Material Safety Data Sheet (continued).

DOT Labeling

Each person who offers a hazardous material for transportation or transports a hazardous material must meet labeling and placarding requirements, unless the material is exempt. Examples of regulated hazardous shipments are oxidizers, explosives, corrosives, poisons, combustibles, flammables, and infectious substances. Examples of labels acceptable for DOT shipping are shown in Figures 7 and 8. The size and color of the labels must meet DOT requirements.

EMPLOYEE INFORMATION AND TRAINING

A widespread training requirement in the manufacturing industry is defined in OSHA's Hazard Communication Standard. This standard places requirements on chemical manufacturers, chemical distributors, and facilities that use chemicals. This standard is not applicable to workplaces that only use chemicals in the form of consumer products and where their use and frequency of exposure is the same as in normal consumer use.

The Hazard Communication Act requires employers to give workers adequate information about hazardous materials that they handle or use in the workplace. Specifically employers are required to provide employees with information and training on the hazards of chemicals used at the time of their initial agreement and as new hazards are introduced into the workplace.

This training includes information such as:

1. Methods and observations to detect the presence or release of a hazardous chemical in the workplace, such as monitoring conducted by the employer, continuous monitoring devices, and visual appearance or odor of hazardous chemicals when being released.

2. Physical and health hazard of the chemicals in the workplace.

3. Protective measures required when being exposed to the chemical such as use of protective equipment, specified work practices, and emergency procedures.

Figure 7.   Examples of acceptable DOT labels. Note: size and color must meet DOT requirements.   Appropriate division number and compatibility group must be added to an explosive label. From EMED Co., Buffalo, NY, with permission.

Figure 8. Examples of acceptable DOT labels. Note: size and color must meet DOT requirements. From EMED Co., Buffalo, NY, with permission.

4. Explanation of the labeling system, information and location of Material Safety Data Sheets (MSDS) or equivalent, and other information about the hazard communication program.

DOT has outlined training for those individuals involved in the transportation of hazardous materials. The training requirements include general awareness training, function specific training, and safety training. Certification of training is necessary, and refresher training is required at least once every two years. If an employee changes hazardous materials job functions, that employee must be trained in the new job function within 90 days. The employee must be supervised by a properly trained employee until the training is complete. A record of training, inclusive of the last two years, must be created and maintained for each employee as long as the employee works in a hazmat job and for 90 days thereafter.

Specialized training is required for workers who handle radioactive materials and waste and for workers who remove asbestos from equipment and buildings. Information about the contaminant, permissible exposure limits, personal protective equipment required, monitoring information, and other items are covered in the training.

Whatever training methods a company selects depends on several factors:

1. The type of operation; for example if the plant is a chemical company, what processes are used? Are the plant's facilities indoor or outdoor, batch or process, modern or older? If it is a multisite company, the chances are that all these conditions exist.

2. A factor that is often overlooked are plant sites that are divided into specific departments, work areas, or there is excessive job bumping or movement of employees from one area to another. This needs to be addressed by a dynamic training program.

3. Are employees well educated, highly technical or of average educational background? Are they young or old? How much job experience do they have? Most important, how effectively has management trained employees up to now? Are employees knowledgeable about hazardous substances where they work?

4. What kind of training personnel, facilities, and equipment are available? Is the site set up to train on shift, or is training scheduled on overtime?

5. How many hazardous substances exist on the plant site? Do hazardous substances exist throughout the site, or are they confined to specific work areas? What specific hazardous substances are present? Can the substances be generically grouped?

Management needs to logically examine company procedures and come up with the most efficient way to teach employees how to:

1. Know what hazardous substances are present in their work areas.

2. Know the hazards they present.

3. Know what precautions to take.

4. Know what to do if the precautions don't work.

TRADE SECRETS

The trade secret provisions of the standard are a study in contrast. They provide very broad protections for trade secrets but only limited protections for worker health. Manufacturers and employers can claim any chemical they choose as a trade secret, regardless of the chemical's hazards, withhold the specific chemical identity from the data sheet and the worker, if certain other requirements set forth in the standard are met. The trade secret protections for manufacturers and employers are so broad that they create a loophole within the standard.

Chemical manufacturers and employers must be able to support all trade secret claims. For chemicals alleged to be trade secrets, general information on the properties and effects of the chemicals must be disclosed, and the MSDS must indicate that specific chemical identity is being withheld on trade secret grounds.

Workers and union representatives have no right of access to specific chemical identities claimed as trade secrets. Limited access is provided only to health care professionals. The Hazard Communication Act appears to be in direct conflict with the OSHA Access to Medical and Exposure Record rule which provides for workers and unions access to specific chemical identities claimed trade secret by the employer if the worker or union signs a confidentiality agreement.

Under standard health care, professionals have limited access to secret chemical identities in emergency and nonemergency situations. In emergency situations, treating physicians or nurses may request and obtain trade secret identities needed for diagnosis or treatment. The manufacturer must provide the information but may require a written statement of need and confidentiality agreement after the fact.

The procedures of nonemergency access to trade secret identities are complicated, burdensome, and unworkable. Access is limited to health professionals, including physicians, industrial hygienists, toxicologists, and epidemiologists. Health professionals must request the trade secret information in writing and state in reasonable detail why the information if needed for one of the occupational health purposes set forth in the standard. The request must detail why the specific chemical identity is needed and why the specific types of information are inadequate.

The health professionals must sign a written confidentiality agreement stating that the information won't be used for other purposes and agree not to release the information to anyone, including the exposed or affected worker, unless such release is authorized in the agreement.

For these confidentiality agreements, manufacturers or employers may restrict specific legal remedies if the information is disclosed, including the manufacturer's or employer's estimate of the damages.

Health care professionals who decide that the trade secret information should be disclosed to OSHA must inform the chemical manufacturer or employer of the action.

Chemical manufacturers and employers may deny requests for trade secret identities. The manufacturer or employer must respond in writing within 30 days of the request. The denial must state why the request is being denied and why other alternative information may satisfy the occupational health needs.

OSHA is supposed to determine whether there is a legitimate health reason for withholding the information. Citations against the manufacturer or employer are to be issued for noncompliance.

However, the manufacturer may still contest the citation and withhold the information until the case is decided by the Occupational Safety and Health Review Commission. This procedure invites denials of trade secret request, contestation of OSHA findings, and will result in years of delay.

There are no provisions in the standard for workers or union representatives to challenge overly broad trade secret claims or to request a chemical identities claimed trade secret. All workers and union representatives must work through a health professional. However, few local unions have access to a health professional they trust. The practical effect of the standards trade secret provisions will be that manufacturers or employers can claim anything they choose as a trade secret and withhold the chemical identity from workers and their representatives.

2 THE OCCUPATIONAL SAFETY AND HEALTH ACT: HAZARDOUS WASTE OPERATIONS AND EMERGENCY RESPONSE

INTRODUCTION

Over the years, industry, government, and the general public have become more aware of the need to respond to hazardous waste problems which seem to be increasing each year. Over the past two decades Congress passed key environmental legislation strictly targeted at the hazardous waste industry. The Resource Conservation and Recovery Act (RCRA) was passed to better track the generation, transportation, treatment, storage, and disposal of hazardous waste from "cradle-to-grave." The Comprehensive Environmental Response, Compensation, and Liability Act (CERCLA), also known as Superfund, was passed in 1980 to provide liability, compensation, cleanup, and emergency response for hazardous substances released into the environment and the cleanup of inactive hazardous waste sites. The Superfund Amendments and Reauthorization Act (SARA) addressed several issues including protection for workers engaged in hazardous waste operations.

Hazardous waste sites pose a number of health and safety concerns which could easily lead to serious injury or even death. Types of hazards may include chemical exposure, fire and explosion, oxygen deficient environments, radiation hazards, and a number of other health and safety related concerns. Improper control of hazardous substances can result in a severe threat to site workers and to the general public. In addition to potential exposures, workers are also subject to hazards associated with working in personal protective clothing.

Due to the seriousness of the safety and health hazards related to hazardous waste operations, the Occupational Safety and Health Administration (OSHA) issued a standard specifically developed to

49

protect workers from chemical and physical hazards and to help them handle hazardous wastes safely and effectively. The standard, known as 29 Code of Federal Regulations Part 1910.120, provides employers and employees with the information and training necessary to improve workplace safety and health in an attempt to reduce the number of injuries and illnesses resulting from exposure to hazardous materials.

Regulatory History and Types of Industries Targeted

The Superfund Amendments and Reauthorization Act (SARA) gave the Secretary of Labor 60 days to issue interim final regulations which would provide no less protection for workers employed by contractors and emergency response workers than the protection contained in the Environmental Protection Agency Manual "Health and Safety Requirements for Employees Engaged in Field Activities" and existing standards under the Occupational Safety and Health Act found in Subpart C of Part 1926 of the Code of Federal Regulations. Those interim final regulations were to take effect upon issuance and would apply until final regulations became effective. SARA also instructed the Secretary of Labor to promulgate, within one year after the date of the enactment of Section 126 of SARA and pursuant to Section 6 of the Occupational Safety and Health Act, standards of the health and safety protection of employees engaged in hazardous waste operations. Informal public hearings in the subject of the rule making were scheduled and held to afford interested parties the opportunity to comment on OSHA's proposals. In addition to the public hearings and the testimony received in response to those hearings, OSHA received over 125 written comments on its proposed language for a final rule.

On March 6, 1989, OSHA issued the final Hazardous Waste Operations Standard (HAZWOPER) under the authority of Section 126 of SARA. The final rule became effective one year later on March 6, 1990. EPA and OSHA promulgated identical health and safety standards to protect workers engaged in hazardous waste operations and emergency response. The types of industries affected by the standard include:

- Cleanup operations at uncontrolled hazardous waste disposal sites that have been identified for cleanup by a governmental health or environmental agency.

- Routine operations at hazardous waste treatment, storage, and disposal facilities or the portions of any facility regulated by 40 CFR Parts 264 and 265.
- Emergency response operations at sites where hazardous substances have been or may be released.
- Corrective actions at Resource Conservation and Recovery Act (RCRA) sites.
- Voluntary cleanups at any uncontrolled hazardous waste sites recognized by a government body.

The firms affected by this standard are as follows: contractors that perform hazardous waste site cleanups; engineering or technical services firms that perform hazardous waste preliminary assessments or site investigations and remedial investigations or feasibility studies for hazardous waste site cleanups; RCRA-regulated commercial treatment, storage, and disposal facilities that are operated by a hazardous waste generator; state and local police departments; fire departments; private hazardous materials (HAZMAT) response teams; and manufacturers that use in-house personnel to respond to emergency spills of hazardous materials with the facility.

SAFETY AND HEALTH PROGRAMS

An effective and comprehensive safety and health program is essential in reducing work-related injuries and illnesses and in maintaining a safe work environment. According to 1910.120, employers are required to develop and implement a written safety and health program for their employees involved in hazardous waste operations. The program shall be designated to identify, evaluate, and control safety and health hazards, and provide for emergency response operations.

This program must include specific information with respect to the following topics:

- An organizational workplan.
- Site evaluation and control.
- A site-specific health and safety plan.
- Information and training program.
- Personal protective equipment.

- Monitoring.
- Medical surveillance program.
- Decontamination procedures.
- Emergency response program.

Site-Specific Health and Safety Plan

The purpose of a site-specific Health and Safety Plan (HASP) is to provide guidelines and procedures required to ensure the health and safety of persons working at waste sites is part of the overall Health and Safety Program. Before on-site activities are initiated, a Health and Safety Office (typically a Certified Industrial Hygienist (CIH)), prepares a site-specific HASP. The site safety and health plan, which must be kept on site, shall address the safety and health hazards of each phase of site operation and include the requirements and procedures for employee protection.

The HASP must include all of the basic requirements of the overall health and safety program, but with attention to those characteristics unique to the particular site. Pre-entry briefings must be conducted prior to site entry and at other times as necessary to ensure that employees are aware of site safety and health plan and its implementation. The employer must also ensure that periodic safety and health inspections are made of the site and that all known deficiencies are corrected prior to work at the site.

Site Evaluation and Control

Site evaluation, both initially and during the course of field activities, is imperative to the safety and health of workers. Site evaluation provides employers with the information needed to identify site hazards so that appropriate protection methods can be selected.

It is a requirement of the standard that a trained person conduct a preliminary evaluation of an uncontrolled hazardous waste site before entering the site. The evaluation must include all suspected conditions that are immediately dangerous to life or health (IDLH) or that may cause serious harm to employees. Examples of these hazards include confined space entry, explosive or flammable situations, visible vapor clouds, or areas where biological indicators such as dead animals or stressed vegetation are observed.

The evaluation must include the location and size of the site, hazardous substances and health hazards, a description of the work to be performed, the duration or time needed to complete the task, site topography and accessibility, pathways for hazardous substance dispersion, and the status and capabilities of response teams.

Controlling the activities of workers and the movement of equipment is an important aspect of the overall safety and health program. Effective control of the site will minimize potential contamination of workers, and protect the public from hazards. The following information is useful in implementing the site control program: a site map, site work zones, site communication, safe work practices, and the name, location, and phone number of the nearest medical assistance. Refer to Figure 1.

Establishment of Work Zones

To reduce the accidental spread of hazardous substances by workers from the contaminated area to the clean area, zones should be delineated on the site where different types of operations will occur, and the flow of personnel among the zones should be controlled. The three most common zones include:

The Exclusion Zone--is the portion of the sites where the contamination is the highest and the greatest potential for exposure exists. The outer boundary is known at the Hotline and all personnel who enter the Exclusion Zone must wear the appropriate level of personal protective equipment (PPE) for the degree and types of hazards at the site.

Contaminant Reduction Zone--is the transition area between the contaminated area and the clean area. Decontamination procedures occur in the CRZ, and the degree of contamination should decrease the further one moves from the Hotline to the Support Zone. The boundary between the CRZ and the Support Zone is the Contamination Control Line.

Support Zone--is the area known as the uncontaminated zone where workers should not be exposed to site-related hazards. The Support Zone is usually located upwind and is the appropriate location for the command post, medical station, equipment, supplies, and other administrative support functions.

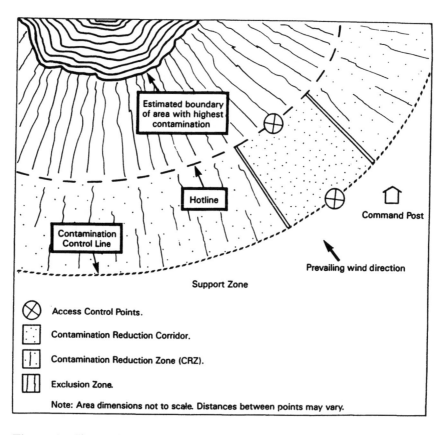

Figure 1. Site work zones.

EMPLOYEE TRAINING AND MEDICAL SURVEILLANCE PROGRAM

Anyone who enters a hazardous waste site must recognize and understand the potential hazards to health and safety associated with the cleanup of a particular site. Personnel actively involved in cleanup must be thoroughly familiar with programs and procedures contained in the Site Safety Plan and must be trained to work safely in contaminated areas. Visitors to a site must receive adequate training on hazard recognition and on the site's Standard Operating Procedures to enable them to conduct their visit safely.

The objectives of training programs for employees involved in hazardous waste site activities are:

- To make workers aware of the potential hazards they may encounter.
- To provide the knowledge and skills necessary to perform the work with minimal risk to worker health and safety.
- To make workers aware of the purpose and limitations of safety equipment.
- To ensure that workers can safely avoid or escape from emergencies.

The level of training provided should be consistent with the worker's job function and responsibilities. All workers are required to complete refresher training, at least annually, to reemphasize the initial training and to update workers on any new policies or procedures.

Types of Hazardous Waste Training

29 CFR 1910.120 defines four different levels of training depending upon the job function or types of hazards one may encounter, and the extent of responsibility. The training need not be repeated if the employee goes to work at a new site; however, the employee must receive additional training to work safely at a new site which may require specialized training for confined space entry or radioactively contaminated sites. The various levels of training are as follows:

Initial Training--General site workers engaged in hazardous substance removal or other activities which expose or potentially expose workers to hazardous substances and health hazards must receive a minimum of 40 hours of instruction off site, and a minimum of three days actual field experience.

Occasional or Routine Site Employees--Workers on site only occasionally for a specific limited task and who are likely to be exposed over permissible exposure limits (PEL) must receive 24 hours of instruction and one day of actual field experience. Routine employees regularly on site who work in areas which have been monitored and fully characterized indicating that exposures are under the PEL where respirators are not required, and where there are no health hazards, must meet the same training requirements as an occasional employee.

Supervisor Training--On-site management and supervisors directly responsible for, or who supervise employees engaged in hazardous operations, must fulfill the initial training requirements in addition to 8 hours of specialized training.

Emergency Response--Employees who respond to hazardous emergency situations at hazardous waste cleanup sites that may expose them to hazardous substances must be trained in how to respond to such expected emergencies.

Within the emergency response category, five levels of training exist which include: (1) First Responder Awareness Level would most likely witness a release and notify the proper authority; (2) First Responder Operations Level would respond by trying to contain the release from a safe distance; (3) Hazardous Materials Technician would attempt to plug, patch, or otherwise stop the release of a hazardous substance; (4) Hazardous Materials Specialist would respond by providing support to hazardous materials technicians and would act as the liaison with federal, state, or local government authority; and (5) On-Scene Incident Commander would assume control of the incident scene beyond the first responder awareness level.

Medical Surveillance Program

Workers handling hazardous wastes may be exposed to toxic chemicals, biological hazards, radiation, or physical dangers. They may develop heat stress while wearing protective equipment or working under temperature extremes, or face life-threatening emergencies such as explosions and fires. Therefore, a medical program is essential to assess and monitor workers' health and fitness both prior to employment and during the course of work, to provide emergency and other treatment as needed, and to keep accurate records for future reference.

The OSHA standard states than an employer must establish a medical surveillance program for all employees exposed or potentially exposed to hazardous substances or health hazards above the PEL for more than 30 days per year; workers exposed above the published exposure levels for 30 days or more per year; workers who wear approved respirators for 30 days or more per year; workers who are exposed to unexpected or emergency releases of hazardous wastes above exposure limits; and members of hazardous materials (HAZMAT) teams.

All examinations must be performed under the supervision of a licensed physician, without cost to the employee, without loss of pay, and at a reasonable time and place. These examinations must be given as follows:

- Prior to job assignment and then annually (can be every 2 years if determined by a physician).
- At the termination of employment or reassignment if an examination has not been given within the last 6 months.
- If the physician believes that a periodic follow-up is medically necessary.
- As soon as possible for employees injured or becoming ill from exposure to hazardous substances during an emergency, or who develop signs or symptoms of overexposure from hazardous substances.

Proper record keeping is essential at hazardous waste sites because of the nature of the work and risks. Records allow medial care providers to be informed about workers' previous and current exposures. The employer must maintain and preserve medical records on exposed workers for 30 years after they leave employment and must make available results of medical records at the employee's request.

RESPIRATORY PROTECTION AND PERSONAL PROTECTIVE EQUIPMENT

Anyone entering a hazardous waste site must be protected against potential hazards. The purpose of personal protective clothing and equipment (PPE) is to shield or isolate individuals from the chemical, physical, and biological hazards that may be encountered at a hazardous waste site. Careful selection and use of adequate PPE should protect the respiratory system, skin, eyes, face, hands, feet, head, body, and hearing. No single combination of protective equipment and clothing is capable of protecting against all hazards.

According to 29 CFR 1910.120(c)(5)(i), "Based upon the results of the preliminary site evaluation, an ensemble of PPE shall be selected and used during initial site entry which will provide protection to a level of exposure below permissible exposure levels for known or suspected hazardous substances and health hazards, and which will provide protection against other known and suspected hazards identified during the preliminary site evaluation . . ."

Selection of Respiratory Protective Equipment

Respiratory protection is of primary importance in an industrial facility handling toxic materials and on a hazardous waste site because inhalation is one of the major routes of exposure to chemicals. Respiratory protective devices (respirators) consist of a face piece connected to either an external air source or an air-purifying device. Respirators with an external air source are known as atmosphere-supplying, or air-supplied respirators. Two types include: (1) self-contained breathing apparatus (SCBA), which supplies air from a bottle carried by the user, and (2) a supplied-air respirator (SAR), which supplies air from a source located some distance away, connected to the user by an air-line hose.

Air-purifying respirators (APRs) do not have a separate air source, instead, ambient air is purified through the use of disposable filters.

APRs selectively remove specific contaminants such as particulates, gases, vapors, or fumes from ambient air by filtration, absorption, or chemical reactions. APRs have limited use and cannot be used in oxygen deficient environments (below 19.5% oxygen), in confined spaces, and when IDLH conditions exist, in environments where unknown contaminants exist, when concentrations of contaminants exceed designated maximum use concentrations, or when the relative humidity is elevated.

There are a total of four levels of personal protective equipment based on the level of protection afforded. OSHA defines all four levels ranging from "Level A" (the most protective) to "Level D" (the least protective) in Appendix B of the standard. Table 1 provides a description of each level of respiratory and protective clothing, and the rationale for choosing the proper level of protection.

Two basic objectives of any PPE program should be to protect the wearer from safety and health hazards, and to prevent injury to the wearer from incorrect use and/or malfunction of the PPE. To accomplish these goals, personnel should develop a comprehensive PPE program, including hazard identification, medical monitoring, environmental surveillance, training and selection, use, and decontamination of PPE.

In addition, the use of PPE can itself create significant hazards such as heat stress, physical stress, psychological stress, and impaired vision, mobility, and communication. In general, the greater the level of PPE used, the greater the associated risks. For any given situation, PPE should be selected to provide maximum protection using the minimum amount of equipment.

DECONTAMINATION PROCEDURES

Decontamination protects workers from hazardous substances that may contaminate and possibly permeate clothing, respiratory equipment, tools, vehicles, and other equipment used on site. Proper decontamination procedures protects all site personnel by minimizing the transfer of harmful materials into clean areas, protects the local community by preventing the release of contaminants outside the work area, and protects workers' families by minimizing the possibility of subjecting family members to contaminants on personal clothing.

TABLE 1

LEVELS OF PERSONAL PROTECTIVE EQUIPMENT (OSHA 1910.120 Appendix B)

Level of Protection	Respiratory Protection	Protective Clothing	Rationale for PPE Selection
Level A	Positive pressure, full face-piece self-contained breathing apparatus (SCBA), or positive pressure supplied-air respirator with escape SCBA	Totally encapsulating chemical-resistant suit (with attached outer gloves and boots which have a steel toe and shank), and hard hat (to be worn under suit)	-- When the highest level of respiratory, skin, and eye protection is required -- For unknown hazardous substances -- When high concentrations of hazardous substances are present in the atmosphere -- For substances which present a high degree of hazard to the skin -- When working in confined spaces or poorly ventilated areas
Level B	Positive pressure, full face-piece self-contained breathing apparatus (SCBA), or positive pressure supplied-air respirator with escape SCBA.	Hooded chemical-resistant clothing, inner and outer chemical-resistant gloves, boots with a steel toe and shank, disposable outer boot covers, and hard hat	-- When the highest level of respiratory protection is necessary but a lesser level of skin protection is needed -- When high concentrations of hazardous substances are present in the atmosphere -- When the atmosphere contains less than 19.5% oxygen -- When low concentrations of vapors have been detected by direct reading instruments

Level C	Full or half-face air-purifying respirator with approved disposable cartridges	Hooded chemical-resistant clothing, inner and outer chemical-resistant gloves, boots with a steel toe and shank, disposable outer boot covers, and hard hat	– When the concentrations and types of airborne contaminants are known – When exposure through direct contact will not adversely affect or be absorbed through any exposed skin – When an air purifying respirator is capable of removing contaminants – When all criteria for the use of APRs are met (i.e., minimum 19.5% oxygen, low humidity, non IDLH conditions)
Level D	None	Coveralls, gloves, boots with a steel toe and shank, hard hat, and safety glasses	– When the atmosphere contains no known hazard – When work functions do not include splashes or the potential for unexpected inhalation or contact with hazardous levels of any contaminants

The OSHA standard states that a decontamination procedure must be developed, communicated to employees, and implemented before any workers or equipment may enter areas on the site where potential exists for the exposure to hazardous substances. The site safety and health officer must require and monitor decontamination of the employee and decontamination and disposal of the employee's clothing. In addition, any solvents, water, materials, or contaminated equipment used during the decontamination process must also be properly disposed.

Decontamination involves physically removing contaminants and/or chemically converting them into innocuous substances. The extent of decontamination depends on a number of factors, the most important being the type of contaminants involved. The more harmful the contaminant, the more extensive and thorough decontamination must be. Other factors may include: the levels of protection selected, topography, availability of equipment and supplies, and weather conditions.

The initial decontamination plan is based on the assumption that all personnel and equipment leaving the Exclusion Zone are grossly contaminated. A system is then set up for personnel decontamination to wash and rinse, at least once, all the protective equipment worn. Typically several stations are set up in a line whereby a worker begins at one end and removes portions of PPE after being properly decontaminated. Contamination should decrease as a person moves from one station to another further along in the line.

An area of the Contamination Reduction Zone (CRZ) is generally designated for decontamination procedures, and is known as the Contamination Reduction Corridor (CRC). The size of the corridor depends on the number of decontamination stations and the amount of space available at the site.

AIR MONITORING AND INSTRUMENTATION

Airborne contaminants can present a significant threat to worker safety and health, thereby making air monitoring an important component of an effective safety and health program. The employer must conduct monitoring before site entry at uncontrolled hazardous waste sites to identify IDLH conditions such as oxygen deficient environments and areas where chemical and quantification of airborne contaminants is useful for:

- Selecting personal protective equipment.
- Delineating areas where protection and controls are needed.
- Assessing the potential health effects of exposure.
- Determining the need for specific medical monitoring.

According to 29 CFR 1910.120(h)(3), periodic monitoring must be conducted when the possibility of an IDLH condition or flammable atmosphere has developed or when there is indication that exposures may have risen over the PELs. These situations may occur when work begins on a different portion of the site, when contaminants other than those previously identified are being handled, when a different type of operation is initiated, and when workers are handling leaking drums or containers or working in areas with obvious liquid contamination.

After a hazardous waste cleanup operation begins, the employer must monitor workers who are likely to have higher exposures to determine if they have been exposed to hazardous substances in excess of permissible exposure limits.

Direct Reading Instruments

Several types of portable direct reading instruments are available for air monitoring. Some display an immediate readout when each sample is taken, others contain a built-in alarm to signal a potential hazard. The types of direct reading instruments that may be used during a preliminary evaluation of a site may include:

- **Combustible Gas Indicator (CGI)** is one of the most useful instruments. CGI readings are normally taken concurrently with O_2 level readings. The indicator measures the concentration of a flammable vapor or gas in air, registering the results as a percentage of the lower explosive limit (LEL) of the calibration gas.

- **Photoionization Detector (PID)** is used to detect many organic and some inorganic gases and vapors by using ionization as the detection method caused by UV light. The most common PID is the HNU which responds to aromatics, unsaturated chlorinated hydrocarbons, unsaturated hydrocarbons, paraffinic hydrocarbons, and other compounds.

- **Flame Ionization Detector (FID)** is used to detect organic vapors and gases. The FID also uses ionization as the detection method, except that ionization is caused by a hydrogen flame. The most common FID is the Foxboro Organic Vapor Analyzer (OVA) which can operate in a survey mode or a gas chromatography mode.

- **Oxygen Meters** are used to detect the amount of oxygen in a work environment. They are especially useful in confined space situations which may be oxygen deficient. The normal ambient oxygen concentration is 20.9%.

- **Colorimetric Indicator Tubes** are used to quickly measure a specific vapor or gas. The contaminant reacts with the indicator chemical in the type producing a stain whose length is proportional to the contaminants concentration. Detector tubes are normally chemical specific to detect such compounds as benzene, toluene, or xylene; however, tubes are also available for groups of gases such as aromatic by hydrocarbons or alcohols.

EMERGENCY RESPONSE OPERATIONS

The nature of work at hazardous waste sites make emergencies a continual possibility, no matter how infrequently they may occur. Emergencies happen quickly and unexpectedly and require immediate response. Three of the most severe nationwide hazards involve (1) hazardous material highway incidents, (2) hazardous material railroad incidents, and (3) stationary hazardous material incidents at storage facilities. Some statistics report that approximately 50% of hazardous material incidents occur on land, 25% in air or water, and 25% as other.

Proper emergency planning and response are important elements of the safety and health program that help minimize employee exposure and injury. The standard requires that the employer develop and implement a written emergency response plan to handle possible emergencies before performing hazardous waste operations. For uncontrolled hazardous waste sites and TSD facilities, the plan must include the following:

- Personnel roles, lines of authority, and communication procedures.
- Pre-emergency planning.
- Emergency recognition and prevention.
- Emergency medical treatment and first aid.
- Procedures for alerting on-site employees.
- Safe distances and places of refuge.
- Site security and control.
- Decontamination procedures.
- Critique of response and follow-up.
- PPE and emergency equipment.
- Evacuation routes and procedures.

The plan requirement also must be rehearsed regularly, reviewed periodically, and amended, to keep them current with new or changing site conditions or information.

The senior emergency response official responding to an emergency shall become the individual in charge of a site-specific Incident Command System (ICS). All emergency responders and their communications shall be coordinated and controlled through the individual in charge of the ICS. As more senior officers arrive, the position is passed up the line of authority.

ENFORCEMENT AND INTERACTION WITH OTHER REGULATIONS

Enforcement of OSHA and Modes of Defense

If OSHA is to have an effective enforcement program, it must increase the probability that dangerous workplaces will be inspected and, if violations are found, the stiff penalties will be assessed. The willingness of employers to comply with OSHA regulations is a function of the size of the fines that violators are willing to pay. There are two kinds of violations OSHA may find during an inspection. There may be violations of specific OSHA standards, or there may be violations under the "general duty clause" which states that employers must keep the workplace free from recognized hazards that are causing or likely to cause death or serious physical harm.

Some violations may not have any direct or immediate effect on job safety and health such as posting or record keeping requirements. In those cases, there are no citations or penalties, but they must be corrected. Other violations defined by OSHA which do have fines and penalties include:

- **Other than serious**--OSHA can propose a fine up to $7000 per violation for health and safety hazards which probably would not be likely to cause death or serious physical harm if an accident occurred. If the violation has a greater probability of resulting in an injury or illness, then a base penalty of $1000 is imposed.

- **Serious**--OSHA must propose a fine of up to $7000 for violations which are likely to cause death or serious physical harm if an accident occurs which the employer knew, or could have known about.

- **Willful or repeated**--OSHA may impose fines up to $70,000 for each violation if any employer willfully or repeatedly violates the OSHA regulations. If an employer is convicted of a willful violation which caused a worker's death, the maximum punishment is a fine of $70,000, imprisonment of up to 6 months, or both. A second conviction doubles these maximum penalties. In order to ensure that the most flagrant violators are in fact fined at an effective level, a minimum penalty of $5000 has recently been adopted.

- **Failure to abate**--Failure to correct prior violation within the prescribed abatement period could result in a penalty for each day the violation continues beyond the abatement date. The daily penalty is equal to the amount of the initial penalty (up to $7000). The employer may contest the penalty amount as well as the citation within a statutory 15-day contest period. The penalty may be heard by an independent Occupational Safety and Health Review Commission, or OSHA may negotiate with the employer to settle for a reduced penalty.

One mode of defense an employer may request is a "good faith effort" reduction by demonstrating that they are making every effort to provide a safe work environment for their employees. For example, they may inform the OSHA inspector that previous problems (unrelated to the noncompliances currently noted) within a facility have been corrected, or that the facility maintains open lines of communication between workers and management to identify potential health and safety issues, or that the same problem has been repaired or addressed in other parts of the facility.

Other modes of defense an employer may emphasize is an excellent history of regulatory compliance with respect to previous violations or lost employee work days. Finally, an employer can also argue that they should receive extra consideration based on the size of their business, if appropriate.

Interaction Between OSHA and Other Federal Regulations

The Occupational Safety and Health Administration issued the standard known as 29 CFR 1910.120 to protect workers engaged in hazardous waste operations and emergency response.

The types of sites which would be most affected would include cleanup operations at uncontrolled hazardous waste disposal sites, routine operations at hazardous waste treatment, storage, and disposal facilities, corrective actions at RCRA sites, and emergency response operations at sites where hazardous substances may be released. Therefore, the federal regulations associated with each of these sites that would interact with the 1910.120 would include CERCLA, RCRA, SARA, and the Hazardous Material Transportation Act (HMTA).

The Comprehensive Environmental Response, Compensation, and Liability Act, also known as Superfund, required OSHA to issue a standard to provide workers involved in hazardous waste operations with protection at least equivalent to that provided by EPA. The standard was mandated to cover various areas of worker protection, site analysis, training, medical surveillance, and protective equipment. The final rule protects workers who work with toxic wastes, including those who respond to emergency spills, such as firefighters, police officers, and ambulance and hazardous materials personnel.

The Resource Conservation and Recovery Act was passed to better track the generation, transportation, treatment, storage, and disposal of hazardous waste from the point of generation to ultimate disposal. Workers involved in any stage of the RCRA process are subject to the requirements of OSHA 1910.120.

The Superfund Amendments and Reauthorization Act, Title III, established specific requirements for federal, state, local governments, and industry to plan, notify, and report on hazardous and toxic chemicals storage, emissions, and spills. Under SARA, facilities must comply with emergency planning requirements, community right-to-know reporting requirements, and training requirements for workers.

The Hazardous Material Transportation Act applies to the transportation of hazardous materials by rail car, aircraft, vessel, interstate and foreign carriers by motor vehicle, and intrastate carriers by motor vehicle as it relates to hazardous wastes and hazardous substances. The HMTA specifically addresses shipping papers, marking, labeling, placarding, training, and emergency response information for hazardous materials shipments. Each HAZMAT employee must receive OSHA training to comply with the hazard communication programs required under 1910.120.

In the last 20 years, environmental legislation has turned its focus more toward the hazardous waste portion of the environmental industry with the passage of RCRA, CERCLA, and SARA. The Occupational Safety and Health Administration has responded with the issuance of the Hazardous Waste Operations Standard (HAZWOPER) to protect workers in these portions of the hazardous waste industry.

SUGGESTED READING

1. Federal Register/Vol. 54 No. 42/Monday March 6, 1989/Rules and Regulations, p. 9295.
2. Federal Register/Vol. 59 No. 161/Monday, August 22, 1994/Rules and Regulations, p. 43268.
3. Federal Register/Vol. 54 No. 42/Monday, March 6, 1989/ Rules and Regulations, p. 9311.
4. 29 CFR 1910.120 Ch. XVII (7-1-93 Edition), Occupational Safety and Health Administration, Labor p. 389.
5. 29 CFR 1910.120 Ch. XVII (7-1-93 Edition), Occupational Safety and Health Administration, Labor p. 392.
6. NIOSH/OSHA/USCG/EPA Occupational Safety and Health Guidance Manual for Hazardous Waste Site Activities, October 1985, p. 9-1.

7. NIOSH/OSHA/USCG/EPA Occupational Safety and Health Guidance Manual for Hazardous Waste Site Activities, October 1985, p. 4-1.

8. 29 CFR 1910.120 Ch. XVII (7-1-93 Edition), Occupational Safety and Health Administration, Labor pp. 392, 406-408.

9. Hazardous Waste Handbook, Martin, Lippitt, and Prothero, Butterworth Publishers, Ann Arbor, Mich., 1987, p. 237.

10. 29 CFR 1910.120 Ch. XVII (7-1-93 Edition), Occupational Safety and Health Administration, Labor pp. 393-395.

11. NIOSH/OSHA/USCG/EPA Occupational Safety and Health Guidance Manual for Hazardous Waste Site Activities, October 1985, p. 8-1.

12. Health and Safety Training for Hazardous Waste Site Workers, ACEC Research & Management Foundation, October 1987, p. 168.

13. NIOSH/OSHA/USCG/EPA Occupational Safety and Health Guidance Manual for Hazardous Waste Site Activities, October 1985, p. 7-1.

14. Worker Protection During Hazardous Waste Remediation, The Center For Labor Education and Research, Van Nostrand Reinhold, 1990, p. 291.

15. 29 CFR 1910.120 Ch. XVII (7-1-93 Edition), Occupational Safety and Health Administration, Labor p. 405.

16. New OSHA Civil Penalties Policy, U.S. Department of Labor Program Highlights, Fact Sheet No. OSHA 92-36, 1992.

17. Occupational Safety & Health, Options for Improving Safety and Health in the Workplace, U.S. General Accounting Office, August 1990, pp. 67-68.

18. 49 CFR Ch. 1 (10-1-92 Edition) Research and Special Programs Administration, DOT p. 67.

3 THE CLEAN AIR ACT

INTRODUCTION

The Clean Air Act was and still is a fundamental promulgation regarding the emissions of hazardous and toxic waste in the atmosphere. Since its enactment, the abatement of air pollution, though slow-paced, has been markedly decreased with respect to industry growth. This chapter briefly describes the major points of the Clean Air Act and its amendments related to emissions and standards for hazardous and toxic wastes.

HISTORICAL PERSPECTIVE

Air pollution concern goes back at least as far as the fourteenth century when restrictions were placed on burning certain types of coal in London. From that time until the late 1960s, practically all air pollution control was directed to the reduction of the visible smoke from stationary sources. For the most part, the early abatement of emissions was secured through judicial proceedings which designated the smoke as a nuisance under common law.

The first promulgated regulations regarding air pollution began with the passage of smoke control ordinances by the cities of Chicago and Cincinnati in 1881. This regulation sparked many cities in the United States to pass similar ordinances and by 1912, twenty-three other American cities had done so. The focus of these local control efforts was to reduce the amount of the dense smoke resulting from the combustion of coal. Commonly called "smoke chasing," the primary function of local air pollution officers was to control stack emissions from coal burning plants. Unfortunately, these ordinances and accompanying air quality control efforts had limited success in controlling smoke problems. By the late 1930s and early 1940s air pollution became more severe and cities such as Pittsburgh and St.Louis enacted tougher smoke ordinances providing more effective enforcement, thereby abating smoke problems significantly in those cities. Until the 1950s, however, when the Clean Air legislation was passed, there was no federal involvement in air pollution control. Air pollution control was for the most part a local responsibility.

Federal air pollution legislation was enacted in 1955, authorizing the Public Health Service in the Department of Health, Education, and Welfare to perform air pollution research and training programs, and to provide technical assistance to state and local governments. This legislation affirmed that state and local governments had the primary responsibility for air pollution control. Amendments to this law in 1960 and 1962 called for special studies relating to health effects from motor vehicle emissions.

The nation's first Clean Air Act (CAA), modest in scope and limited in effect, was enacted in 1963 due to the further deterioration of the nation's air quality. In the initial statute, Congress provided very little federal authority by assigning the federal government a mostly advisory role in air protection efforts such as: preparing air quality reports, identifying the health effects of pollutants, and issuing guidance emission control technologies. Specifically, it provided for:

1. Awarding grants for program development and improvement of state and local air pollution control agencies.

2. Accelerated research, training, and technical assistance.

3. Federal enforcement authority to abate interstate air pollution problems.

4. Federal research responsibility for automobile and sulfur oxides pollution.

5. Air quality criteria development for the protection of public health and welfare.

In this first Clean Air Act, there was no provision for hazardous materials emission and standards, and primary responsibility for pollution control still remained with state and local governments.

Even though considerable progress was made with the enactment of the Clean Air Act of 1963, air pollution problems continued to escalate and enforcement levels continued to be ineffectual. In 1967, Congress passed the Comprehensive Air Quality Act, which further strengthened the country's air pollution control effort. Many of the pollution control concepts developed in 1967 still offer the basic framework for our current control efforts. Perhaps the most significant of these concepts was the development and implementation of air quality standards.

The act required the Secretary of Health, Education, and Welfare to designate Air Quality Control Regions and issue air quality criteria and control technique information. Moreover, individual states were required to delineate state air quality standards and plan for their implementation on a fixed time schedule. The federal government was given even more authority, but still, it didn't compare to the primary responsibility of enforcement that lied in the states' hands.

The basic regulatory structure of today's Clean Air Act was formed when Congress passed the CAA of 1970, which federalized air pollution control due to the public's dissatisfaction with the progress of the Air Quality Act. It made health protection the basis for much of that regulation, clarified the emission standards for automotive and other industries and created a pervasive regulatory system. Specific provisions of the Amendments of 1970 included:

1. National Ambient Air Quality Standards.

2. Immediate designation of Air Quality Control Regions.

3. State Implementation Plans (SIPs) to achieve National Ambient Air Quality Standards.

4. Stringent new automobile emission standards and standards for fuel additives.

5. Emission standards for new or modified sources and for hazardous pollutants.

6. Right to citizen's suits.

7. Federal enforcement authority in air pollution emergencies and interstate and intrastate air pollution violations.

However, this approach dealt with one pollutant at a time and involved the use of extensive risk assessment. The risk assessment became very controversial because of the early stages of development of the Environmental Protection Agency's (EPA) risk assessment capabilities and the debate about the appropriate role of such estimates in the regulatory decision process. The result was a very low output for the program, with only seven toxic air pollutants regulated in a twenty year period. In fact, in 1971, the list only included three hazardous air pollutants: asbestos, beryllium, and mercury.

Congress has since amended and strengthened the CAA significantly, first in 1977 and again in 1990. The 1977 amendments did not provide for any significant air pollution control initiative. Specifically it postponed compliance deadlines for national primary air quality standards, modified federal automobile emission standards, included the concept of Prevention of Significant Deterioration into the language of the act, and provided a mechanism by which the EPA could have flexibility in allowing some growth in the "dirty air" regions. The 1990 amendments added several new titles to the CAA, bolstered its existing regulatory programs, and led to the use of the market-based incentives and other innovative approaches.

The Clean Air Act of 1963 expanded the research and technical assistance programs begun in 1955, provided for development of air quality criteria by the Department of Health, Education, and Welfare (HEW), and created federal investigative and abatement authority similar to that already in effect for water pollution. These provisions were more bark than bite, however, criteria, which were to reflect the most recent scientific knowledge on air pollution effects, were simply advisory guidelines to be used or ignored by state and local government.

With the passage of the CAA, congressional attention focused more closely on automotive emissions. With an increasing number of automobiles and the population rising, vehicle emissions became more and more important.

The CAA currently represents a comprehensive programmatic and regulatory system on the subject of air pollution. To begin, the act, as amended through 1990, is subdivided into eleven titles.

Title I: Provision for Attainment and Maintenance of National Ambient Air Quality Standards
Title II: Provisions Relating to Mobile Sources
Title III: Hazardous Air Pollutants
Title IV: Acid Deposition Control
Title V: Permits
Title VI: Stratospheric Ozone Protection
Title VII: Provisions Relating to Enforcement
Title VIII: Miscellaneous Provisions
Title IX: Clean Air Research
Title X: Disadvantaged Business Concerns
Title XI: Clean Air Employment Transition Assistance

Viewed somewhat differently, the act can be subdivided on a fundamental basis: it has a programmatic element, and a regulatory element. Quite different from the original intentions, the Clean Air Act's primary purpose is to protect human health.

In addition to emission limitations on new sources, the Clean Air Act Amendments of 1970 established a program of emission limitations on sources which emit hazardous pollutants. Some pollutants by definition were deemed to pose a significant threat to public health and were not regulated under ambient air quality standards. Hazardous pollutants are considered to pose a more immediate health threat than other regulated pollutants. Central to this idea is that permissible levels cannot be established because the hazard is so great. Consequently, emissions must be limited to the lowest possible level.

In theory, the National Environmental Standards for Hazardous Air Pollution (NESHAP) provisions can be applied more expeditiously to a health-threatening air pollution problem than other provisions such as air quality standards. In actual practice, the designation and subsequent regulation of hazardous pollutants has been very slow. Initially, three hazardous pollutants (asbestos, mercury, and beryllium) were designated and regulated in the early 1970s. After a considerable pause, vinyl chloride (1976) and then benzene (1984) were regulated as hazardous pollutants. Arsenic and radioactive isotopes were both listed as hazardous pollutants, with EPA deciding not to regulate them. Currently there are 190 listed hazardous pollutants. Air pollution by hazardous pollutants is usually not a community-wide problem. With the exception of asbestos, hazardous pollutants are usually very localized problems associated with relatively few point sources. The hazardous natures of designated pollutants are usually identified from diseases in those occupationally exposed.

NATIONAL AMBIENT AIR QUALITY STANDARDS

As part of the 1990 Amendments to the Clean Air Act, National Ambient Quality Standards were assigned. According to Volume 40 of the Code of Federal Regulations, Section 50 (40 CFR 50), ambient air is defined as:

> ". . . that portion of the atmosphere external to buildings, to which the general public has access."

In other words, it is the air that we are surrounded by. The following is a list, taken from the CFR for Air Quality Standards for various compounds and elements.

- **Sulfur oxides**--80 micrograms per cubic meter annual arithmetic mean concentration per year. 365 micrograms per cubic meter maximum 24 hour concentration, not to be exceeded more than once per year.
- **Particulate matter**--50 micrograms per cubic meter annual arithmetic mean concentration per year. 150 micrograms per cubic meter, 24 hour average concentration, not be exceed more than once per year.
- **Carbon monoxide**--10 milligrams per cubic meter maximum 8-hour average, not to be exceeded more than once per year. 40 milligrams per cubic meter 1-hour average concentration, not to be exceeded more than once per year.
- **Nitrogen dioxide**--100 micrograms per cubic meter annual arithmetic mean concentration per year.

- **Lead**--1.5 micrograms per cubic meter quarterly arithmetic mean concentration, not to be exceeded more than once per year.

TITLE III - HAZARDOUS AIR POLLUTANTS

Title III of the CAA deals primarily with Hazardous Air Pollutants and the means to control them. To begin, Section 112 of the act describes sources and types of hazardous pollutants. A major source, according to the act, is described as any source emitting 10 tons per year or more of any one hazardous air pollutant or 25 tons per year or more of any combination of hazardous air pollutants. It also lists 189 hazardous pollutants, which the EPA is required to periodically review and amend when necessary. A pollutant found to pose an adverse human or environmental threat, through inhalation or other routes of exposure may be added to the list. According to the "Clean Air Act Amendments of 1990, A Detailed summary of Titles," an adverse health effect is as follows:

> "Substances causing 'adverse human health effects' include, but are not limited to those that are known to be or that may reasonably be anticipated to be carcinogens, mutagenic, teratogenic, neurotoxic, or that cause reproductive dysfunction or that are acutely or chronically toxic."

In order to add or delete a hazardous substance from the list, any person may petition the EPA to modify the list. The EPA, in turn, must grant or deny the petition within 18 months of receipt.

Emission Standards

Due to the rate of technology growth, the EPA must also review and revise as necessary the emission standards at least every 8 months, under the act. Emission standards must achieve the maximum degree of emissions reduction deemed achievable by the EPA for new or existing sources in the applicable category or subcategory considering cost of achieving the emissions reduction, and non air quality health and environmental impacts and energy requirements, through application of measure, processes, methods, systems, or techniques. If a health threshold can be established for a pollutant, the EPA may consider such a level with an ample margin of safety when establishing the emissions standards.

STATE HAZARDOUS POLLUTANT CONTROL PROGRAMS

Each state may submit to the EPA for approval, a program for partial or complete delegation of authority for implementation and enforcement of the hazardous pollutant control requirements of the accidental release provisions. However, states do not have the authority to set standards that are less stringent than the federal requirements. Upon receipt of a state implementation program, the EPA must approve or disapprove state programs within 180 days. The program must be disapproved if the EPA determines that (1) the state program authorities are not adequate to assure compliance by all sources, (2) the program lacks adequate authority or adequate resources to implement the program, (3) the schedule for implementation and compliance is not sufficiently expeditious, and (4) the program is otherwise not in compliance with EPA guidance. Upon notification that the state program is disapproved, the state may revise and submit the program for approval.

Prevention of Accidental Releases

The EPA promulgated an initial list of 100 substances that, in the event of an accidental release, are known to cause or may reasonably be anticipated to cause death, injury, or serious adverse human health or environmental effect. The EPA's original list incorporated the list of extremely hazardous substances established under the Superfund Amendment Reauthorization Act (SARA). The initial list includes:

chlorine	anhydrous ammonia
methyl chloride	ethylene oxide
vinyl chloride	methyl isocyanate
hydrogen cyanide	ammonia
hydrogen sulfide	toluene disocyanate
phosgene	bromine
anhydrous hydrogen chloride	hydrogen fluoride
anhydrous sulfur dioxide	sulfur trioxide

The list may be revised when necessary, but must be revised every 5 years, and it may not include any pollutant for which a national ambient air quality standard has been established or any substance practice of process or activity regulated under Title IV - Acid Deposition Control. The EPA must establish procedures for addition and deletion of substances from the list consistent with those for listing substances as a hazardous air pollutant. The EPA must establish, by rule, a threshold quantity at the time a substance is listed, considering toxicity, reactivity, volatility, dispersability, combustibility, or flammability and the amount known or reasonably anticipated to cause, as a result of accidental release, death or injury of serious adverse human health effects.

SOLID WASTE COMBUSTION

The EPA is required to establish performance standards and other requirements for each category of solid waste incineration units, including emissions limits and other requirements for new units and guidelines and other requirements for existing units. The emission standards set in the operating permit are required to reflect the maximum degree of emission reduction technologically feasible for new or existing units. For new units, the emission reduction deemed achievable must not be less stringent than the control achieved in probative by the best controlled similar unit. For existing units, the emission standards may be less stringent than the average emissions limit achieved by the best performing 12% of units in the category. For solid waste incineration units, the standards must be based on removal or destruction technologies before, during, or after combustion and must include siting requirements for new units to minimize risks to human health or the environment, on a site-specific basis. Numerical emissions limits must be established for particulate matter, opacity, sulfur dioxide, hydrogen chloride, NO_x, CO, lead, cadmium, mercury, dioxins, and dibenzofurans. The EPA may promulgate numerical emissions limits for other pollutants as well. The EPA must review and revise the performance standards or other requirements within 5 years after promulgation and every 5 years thereafter. States are required to submit plans to implement and enforce the guidelines providing for compliance.

As part of the performance standards, the EPA promulgated regulations requiring owners or operators of each solid waste incineration unit to monitor emissions and report findings. The regulations include requirements on the frequency of monitoring, and it is illegal for any person having control over processes affecting emissions to operate the solid waste incineration units and high capacity fossil fuel fired plants, unless the person has adequately completed an approved training program. For solid waste incineration units that combust municipal waste, the EPA must review the availability of acid gas scrubbers as a control technology for small new units or existing units before promulgating performance standards. The management, handling, storage, treatment, transportation, reuses, recycling, and disposal of ash from solid waste incineration of municipal waste is not subject to Subtitle C of the Solid Waste Disposal Act prior to the date two years after enactment.

INCINERATORS

An incinerator burning hazardous waste must be designed, constructed, and maintained so that, when operated according to the requirements specified in 40 CFR 264, it will achieve a Destruction Removal Efficiency (DRE) of 99.99% for each principal organic hazardous constituent (POHC). If an incinerator is burning severely hazardous wastes listed in the Code of Federal Regulations, it must achieve a DRE of 99.9999% for each POHC designated in the permit. Also, the owner or operator must notify the Regional Administrator of its intent to burn these wastes.

Moreover, incinerators burning hazardous waste and producing 4 pounds per hour (lb/hr) of hydrochloric acid (HCl), the emissions must be less then 4 lb/hr or less than 1% of HCl in the stack gas prior to entering pollution control equipment. Furthermore, particulate matter, must be less than 180 milligrams per cubic meter.

Operating Requirements

Permits for incinerators and types of waste allowable for burning are determined on a case by case basis. However, all permits specify the allowable composition of waste feed and allowable operating limits for each incinerator including: carbon monoxide levels in stack exhaust gases, waste feed rate, combustion temperature, appropriate indicator of combustion gas velocity, and allowable variations in incinerator system design or operating procedures. Furthermore, fugitive emissions controls are compulsory. They are controlled by sustaining a fully sealed compression zone, combustion zone pressure that of less than atmospheric pressure, and providing an alternate means of control should a failure of equipment occur. Moreover, a fail-safe waste feed must cut off when operating conditions exceed limits designated in permit.

The owner or operator must also monitor assorted aspects of the incineration process including: waste feed, combustion temperature, indicator of gas velocity, carbon monoxide at a point in the incinerator downstream of the combustion zone, and prior to release to the atmosphere. All incinerators are subject to periodic inspections, and must submit reports of stack emission to the EPA. In the event of a closure of an incinerator, the owner or operator must remove all hazardous waste and hazardous waste residue from the incinerator site.

ENFORCEMENT

Under the Clean Air Act, the Environmental Protection Agency (EPA) is given a modest role in enforcement, while substantial authority remains in the hands of the states. There are three types of enforcement powers that outline the laws for enforcement: miscellaneous powers of the EPA, the underrated role of the EPA's civil administrative orders, and the state enforcement powers.

Miscellaneous Powers

As with most administrative schemes in the environmental domain, the EPA Administrator has authority to require recordkeeping and gain entry pursuant to his enforcement and other powers. Therefore, in Section 114, the Administrator may require the owner or operator of any emission source, to establish or maintain records, make reports, install, use or maintain monitoring equipment of methods, sample emissions in accordance with methods and at places and intervals as he may reasonably require. Authorized employees also are given a right of entry upon presentation of credentials, into a premises where an emission source or records required to be maintained are located. At reasonable times, he may have access to and copy records, inspect monitoring equipment or methods, and sample emissions the owner is required to control. These are broad powers to make unannounced inspection and secure information potentially of evidential value in an enforcement proceeding. Moreover, the information is made public except to the extent the Administrator is persuaded it is entitled to trade secret protection.

The posers are confined by constitutional considerations but are not seriously so. The question of the reasonableness of warrantless searches turns upon whether frequent, unannounced inspections are essential to effective enforcement of the pollution control scheme. The answer ordinarily is in the affirmative, given the possibilities of rapid-fire curtailments, manipulation of monitoring equipment, and so on. Where a conspicuous plume brings the inspectors, this too would establish probable cause without the normal warrant procedure. The normal difficulties of compulsory self incrimination raised by the mandatory recordkeeping and sampling requirements are overcome largely by the fact that the privilege extends only to private citizens and no corporations. Even as to individuals, it is stated authoritatively, that routine record keeping requirements of the kind imposed in the provisions from the federal air and water pollution control acts are not likely to lead to cognizable Fifth Amendment claims.

Each state may develop and submit to the Administrator a procedure for carrying out the inspection, monitoring, and entry provisions of Section 114, and with limited expectations, and the Administrator is authorized to delegate these powers to the state without relegating his own authority to enforce the section.

One popular, yet not outstanding, enforcement technique is the opacity standard. Opacity, the degree which pollution reduce the transmission of light, it established at 20 percent for particulate matter discharges from coal fired steam generators. Many owners have argued that trained inspectors often vary widely in their estimates of the opacity of known plumes. The courts were determined to assure that the EPA standards be objective so that the owner could be sure his own tests could be duplicated by the agency.

Somewhere between fines or jail sentence lies the essence of an administrative effort, which the EPA has some deficiencies of under the Clean Air Act. Most prominent among these is the power to impose fines administratively, which offers wide flexibility and judgement in lieu of established fines.

Administrative Orders

The Administrator has a great amount of power in evaluating the extent of the law. For example, under Section 113, upon 30 days notice, the Administrator may issue an order requiring compliance with any applicable implementation plan; and a similar order may be issued-- without 30 days notice--to correct violations or new source, hazardous emission and energy-related standards. The power to issue orders is controlled by other procedural requirements. While the Administrator has discretion, due to the subjectivity, his jurisdiction is many times argued in court.

State Enforcement

The primary responsibilities of the states with regard to air pollution is a settlement with origins in the Air Pollution Control Act of 1955. The supreme court has accorded substance to the state role by its holding in the Train vs. Natural Resources Defense Council, Inc., approving the grant of minor variances by state authorities that do not jeopardize the national standards, and in Union Electric Co. vs. Environmental Protection Agency, permitting state consideration of claims of economic and technological infeasibility that are not open to question when the Administrator approves an implementation plan. The EPA agrees substantial honor to state enforcement actions, although conflict between jurisdictions is inevitable. In fact with the significant expectation of the EPA's issuance of administrative orders, the instatement of most enforcement action rests in the hands of state and local authorities.

An understanding of enforcement, even more so than other administrative practices, requires a look beneath the surface of formally prescribed powers and reported cases. Many analyses of state air pollution laws have been undertaken, and these usually emphasize enforcement. The reported cases while misleading to the extent they represent the tip of the iceberg, are fully in mainstream of the close scrutiny doctrine of judicial review. Predictable, the case law often deals with delineating the basic powers of administrative agencies to move against one or another type of air pollution, such as emissions from automobiles, fuel additives, or burning. A common objection is that the legislators or administrators engaged in unreasonable classification decisions are moving against some sources of air pollution while ignoring others.

CASE STUDIES

Union Electric Company vs. Environmental Protection Agency (1975)

Due to financial reasons, Union Electric declared that it couldn't comply with control standards contained in the Missouri State Implementation Plan (SIP) approved by the Administrator of the EPA. Furthermore, Union Electric alleged that the most technologically advanced removal equipment has not been determined to operate successfully or satisfactorily. The cost for the inadequate equipment would be $500 million, virtually impossible to raise that much money. In the alternative, it contends that the cost is so high in relation to the benefits that public interest would be adversely affected. Union Electric seeks relief from compliance with these standards pursuant to Section 307 of the Clean Air Act. The SIP restricts the emissions of sulfur dioxide into the ambient air. Union Electric Co. operated three coal-burning electric power generating plants in the metropolitan St. Louis area which are subject to the sulfur dioxide restriction contained in the SIP. Union Electric Co. alleges that it cannot comply with these restrictions without shutting down the plants, resulting in a power loss across the area. Union Electric petitioned the Administrator of the Environmental Protection Agency for a variance 30 days after promulgating. However, on May 31, 1974, before the petition was reviewed and granted, Union Electric was notified that the coal burning plants were in violation of the sulfur dioxide regulations.

The Administrator argues that he cannot consider economic and technological feasibility in approving or disapproving implementation plans; Congress could not have intended to allow these questions to be raised in a petition for review. Essentially, the court left it up to the state to grant a variance to Union Electric showing even more evidence that the federal government does not have absolute power.

Natural Resources Defense Council, Inc. vs. EPA

In 1974, Georgia petitioned the EPA with its State Implementation Plan for meeting the national ambient air quality standards for particulate matter and sulfur dioxide. The plan, as submitted and approved by the Administrator, made the allowable amounts of particulates and sulfur dioxide emissions dependent on the heights of smoke stacks at sources, i.e., the higher the smoke stacks, the more a source was permitted to emit. This tall stack approach represents a form of dispersion enhancement technique. Dispersion enhancement techniques are methods to reduce concentrations of pollutants by altering the conditions under which substances are emitted in order to enhance their dispersion throughout the atmosphere, instead of reducing the actual amount of pollutants into the air. According to the Natural Resources Defense Council, Georgia's choice of such a strategy is in conflict with 42 USCA's 1857-c5(a)(2)(B) which says:

> The Administrator shall approve such [state implementation] plan, or any portion thereof, if he determines that. . . (B) it includes emission limitations, schedules, and timetables for compliance with such limitations, and such other measure as may be necessary to insure attainment and maintenance of such primary or secondary standard, including but limited to, land uses and transportation controls.

The use of dispersion techniques is at odds with the nondegradation policy. Dispersion enhancement techniques operate by keeping pollutants out of areas of high pollutant concentration, and dispersing them to lower concentration areas. Inevitably, however, the pollutants emitted into the atmosphere must end up somewhere and the atmosphere at their destination, wherever that may be, will be degraded, in violation of congressional policy. The only techniques fully capable of guaranteeing nondegradation are emission limitation techniques. Refer to Table 1 for listed hazardous wastes.

TABLE 1

190 LISTED HAZARDOUS WASTES

Acetylhyde	Acetamide	Acetonitrile
Acetophenene	2-Acetylaminofluorene	Acrolein
Acrylamide	Acrylonitrile	Allyl chloride
4-Aminobiphenyl	Aniline	o-Anisidine
Asbestos	Benzene	Benzidene
Benzotrichloride	Benzyl chloride	Biphenyl
Bis(2-ethylhexyl)phthalate	Bis(chloromethyl)ether	Bromoform
1,2-Butadiene	Calcium Ctanamide	Caprofactam
Captan	Carbaryl	Carbon disulfide
Carbon Tetrachloride	Carbonyl sulfide	Catechol
Chloramben	Chlordane	Chlorine
Chloracetic acid	2-Chloroacetophenome	Chlorobenzene
Chlorobenzilate	Chloroform	Chloromethyl methyl ether
Chloroprene	Cresols/cresylic acid	o-Cresol
m-Cresol	p-Cresol	Cumene
2,4-D, salts and esters	DDE	Diazomethane
Dibenzofurans	1,2-Dibenzofurans	1,2-Dibromœ3-chloropropane
Dibutriphthalate	1,4-Dichlorobenzene(p)	3,3-Dichlorobenzidene
Dichloroethyl ether	1,2 Dichloropropene	Dichlorvos
Diethanolamine	N,N-Diethyl aniline	Diethyl sulfate
3,3-Dimethhyoxybenzidine	Dimethyl aminoazobenzene	3,3-Dimethyl benzidine
Dimethyl carbomoyl chloride	Dimethyl formamide	1,1-Dimethyl hydrazine
Dimethyl sulfate	4,6-Dinitro--cresol and salts	2,4 Dinitrophenol
2,4-Dinitrotoluene	1,4-Dioxane	1,2-Diphenylhydrazine
Epichlorohydrin	1,2-Epoxybutane	Ethyl acrylate
Ethyl benzene	Ethyl carbamate	Ethyl chloride
Ethylene dibromide	Ethylene dichloride	Ethylene glycol
Ethylene imine	Ethylene oxide	Ethylene thiorea
Ethylidene dichlored	Formaldehyde	Heptachlor
Hexachlorobenzene	Hexachlorobutandiene	Hexachlorocyclopentadiene
Hexachloroethane	Hexamethylene-1,60 disocyanate	Hexamethylphosphoramide
Hexane	Hyzadrine	Hyydrochloric acid
Hydrogen fluoride	Hydrogen sulfide	Hydroquinine
Isophorone	Lindane	Maleic anhydride
Methanol	Methoxychlor	Methyl bromide
Methyl chloride	Methyl chloroform	Methyl ethyl ketone
Methyl hydrazine	Methyl iodide	Methyl isobutyl
Methyl isocyanate	Methyl methylacrylate	Methyl tert-butyl ether
4,4-Methylene bid	Methylene chloride	Methylene diphenyl
4,4-Methylenodianiline	Naphthalene	Nitrobenzene
4-Nitrobiphenyl	4-Nithrophenol	2-Nitropropane
N-Nitroso-N-methylurea	N-Nitrosodimethylamine	N-Nitrosomorpholine
Parathion	Pentachloronitrobenzene	Pentacholophenol
Phenol	o-Phenlyenediamine	Phosgene
Phosphite	Phosphorus	Phthalic anhihydride
Polychlorinated phenyls	1,2-Propane sultone	b-Propiolactone

TABLE 1

190 LISTED HAZARDOUS WASTES
(Continued)

Propionaldehyde	Propozur	Propylene dichloride
Propylene oxide	1,2-Propylenimine	Quinoline
Quinone	Styrene	Styrene oxide
2,3,7,8-Tetracholodibenzo-p-dioxin	1,1,2,2-Tetrachoroethane	Tetrachloroethylene
	Titanium tetrachloride	Toluene
4-Toluene diamine	2,4-Toluene diisocyanate	o-Toluidine
Toxaphene	1,2,4-Trichlorobenzene	1,1,2-Trichloroethane
Trichloroethylene	2,4,5-Trichlorophenol	2,4,6-Trichlorophenol
Triethylamine	Trifluralin	2,2,4-Trimethylpentane
Vinyl acetate	Vinyl bromide	Vinyl chloride
Vinylidene chloride	Xylene	o-Xylenes
m-Xylenes	p-Xylenes	Antimony compounds
Arsenic compounds	Cadmium compounds	Berrylium compounds
Chromium compounds	Cobalt compounds	Coke oven compounds
Cyanide compounds	Fine mineral fibers	Glycol ethers
Lead compounds	Manganese compounds	Mercury compounds
Nickel compounds	Polycyclic organic matter	Radionuclides
Selenium compounds		

SUGGESTED READING

1. Clean Air Act of 1963 and Amendments; United States Congress.
2. Code of Federal Regulations; Office of Federal Registrar, National Archives and Records Administration; July 1, 1993.
3. Godash, Thad, *Air Quality*, Lewis Publishers, Inc., Boca Raton, Florida, 1985.
4. IEEE Press, *Clearing the Air: The Impact of the Clean Air Act on Technology*, The Institute of Electrical and Electronics Engineers, Inc., 1971.
5. Krier and Ursin, *Pollution and Policy*, University of California Press, New York, 1977.
6. Shaw, Bill, *Environmental Law: People, Pollution, and Land Use*, West Publishing Co., New York, 1976.

4 THE CLEAN WATER ACT

INTRODUCTION

The Clean Water Act (CWA) is a product of many years of historical development. It can also be said that its information is a result of a lengthy period of trial and error. CWA functions as a complex and reasonably efficient mixture of all the regulatory and enforcement tools that have proven efficient in the environmental contexts.

The federal law controlling discharges of waste to surface waters was first enacted around 1899, namely the Refuse Act. The law was enforced primarily for the protection of navigation. People were not concerned over the quality of water in streams and lakes until the late 1940s, when the diseases due to the bacterial contamination of drinking water affected public health. It led the federal government to assist states in the construction and operation of wastewater treatment plants. In 1965, the federal government implemented a Water Quality Act, and helped states and interstate compact organizations to establish and enforce water quality standards for surface waters in their jurisdictions. The approach sounded good, but it did not work so well in most states, the major reasons being:

- Inability to determine precisely when a discharge violated applicable standards.
- Inapplicability of federal-state water quality standards to intrastate waters.
- Lack of state initiative in making load allocations required to set enforceable discharge standards.
- Cumbersome enforcement mechanisms and the requirement of state consent for federal enforcement.

Because of these, prior to 1970, the federal state programs to protect surface waters were largely ineffective, and there was no workable program to directly enforce environmentally oriented limits on the discharge of pollutants to water. It took such events as the Santa Barbara oil spills of 1969 and Earth Day 1970 to make Congress interested in getting into the business of protecting the environment. The Water Quality Improvement Act was passed in 1970. It dealt with oil spills and treatment of sewage from toilets and recreational boats. Also, in 1970 the Nixon administration declared that all industrial discharge had to obtain Refuse Act Permits and meet appropriate treatment standards in order to continue operations. Because the Refuse Act was not drafted as a comprehensive water pollution control statute, it encountered severe problems. Among these were:

- The act provided no standards for the grant or denial of permits, nor were any regulations promulgated to provide such standards.
- As a result of a court decision, Kalur vs. Resor, environmental impact statements had to be prepared for every permit decision, further taxing the inadequate staff in charge of processing the applications.
- Penalties under the act were thought to be inadequate.
- The relationship of the act to other federal and state water pollution control efforts was unclear and created considerable confusion.

The government was finally convinced of the need for a comprehensive modern pollution statute. In 1972, a landmark legislation was passed in which it empowered the EPA with the responsibility for setting nationwide effluent standards on an industry-by-industry basis. The EPA was charged to set the standards on the basis of the capabilities of both pollution control technologies and costs to the regulated industry as a whole. Overall, the 1972 act--the Federal Water Pollution Control Act, concerning the national effluent limitations, water quality standards, the permit program, special provisions for oil spills and toxic substances, and a publicly owned treatment works construction program--was proved to be reasonably sound and remains in place with some modifications, even today.

The EPA did not implement the Federal Water Pollution Control Act correctly as it was intended. Wrong pollutants were regulated. It concentrated on oxygen demanding materials and failed to emphasize the toxics that were perceived to threaten human health. The agency was taken to court by an organized group of environmentalists. The case was settled by a 1976-consent decree, called the Flannery Decree, which refocused all of the CWA enforcement tools on toxics control and spelled out a detailed toxics strategy. *This is the heart of the CWA program as it exists today.* The Flannery Decree is still in effect, and its mandates to the EPA about how to administer the CWA remain an important factor in its implementation. In 1987, the act was amended to tighten the focus on toxic dischargers, force action on toxics oriented water quality standards, attempt to resolve long-standing problems with the effort to regulate storm runoff, and strengthen the enforcement mechanisms. Discharge standards today are quite stringent to assure that water quality standards for toxic pollutants are met.

The Clean Water Act consisted of six main titles as described below:

Title I   -   Research and Related Programs
Title II   -   Grants for Construction of Treatment Works
Title III -   Standards and Enforcement
Title IV   -   Permits and Licenses
Title V   -   General Provisions
Title VI   -   State Water Pollution Control Revolving Funds

The following is a brief description and overall view of the above-mentioned titles.

Title I. This title mainly deals with the goals, objectives, policies, and programs for water pollution control. It also calls for the establishment of research programs on the effects of pollutants. It establishes grants for R&D projects, pollution control demonstrations, and reclamation demonstration projects.

Title II. As new law was enforced in the early 1970s, a federal assistance program was a major necessity if most municipalities were able to construct the facilities necessary to comply with the new effluent limitations. Title II revolves around the grants and loan guarantees for the construction of treatment works, area-wide waste treatment management, design for cost-effective waste treatment program, and minimization of the cost of industrial participation in a loan or grant funded Publicly Owned Treatment Works (POTWs). The user charge regulation, also a part of the title, defines that a treatment works operator is obliged to develop a user charge mechanism, which requires that each user is supposed to pay its proportionate share of the cost of waste treatment. In general, POTWs are not insulated from the substantial costs of complying with CWA requirements. Cost effectiveness is a major factor in wastewater treatment. The topics will be discussed in more detail.

Title III. Through this title, many effluent limitations have been established nationwide, for treatment of existing direct and indirect discharge sources in every industrial category. Also, Title III defines standards for waste water quality and enforcement laws to control them. It discusses effluent limitations, toxic and pretreatment effluent standards, water quality standards, EPA's interest in end-of-pipe treatment, and required level of treatment, based on available technology. It also includes water quality related effluent limitations and toxic water quality standards. It sets National Standards of Performance and standards for toxic pretreatment effluents. How the act is to be enforced to be effective is specified in this title.

Title IV. Probably the core program of the CWA is this title which specifies the permits and licenses requirements for discharges. It is through this program that the intent of the law can be achieved. The basic intention in this is to regulate *all* pollutants discharged from all facilities and most contaminated areas into virtually all waters in the United States. The National Pollution Discharge Elimination System (NPDES) permit program defines individual permissible level of release into the waters of the United States. It also discusses criteria for ocean discharge, disposal or use of sewage sludge, discharge to ground water, and many companies that are exempt from CWA by not discharging in navigable water are subject to requirements governing waste water discharges pursuant to the Safe Drinking Water Act. Permits and licenses will be discussed in Chapter 5 of this book.

Title V. General provisions such as the administration of the act, employee protection under the law, issues on state authority to adopt or enforce the effluent or other limitations, labor standards, citizen suits, federal procurement, effluent standards, etc., are included in this title. Title V also discusses about body and authorities of water pollution control advisory board, and assistance to Indian tribes to develop waste treatment management plans, and to construct treatment programs under this act by Indian tribes can be maximized.

Title VI. In this title, topic areas such as grants to states for establishment of revolving funds, allotment of funds, corrective action, audit reports and fiscal controls, and intended use plan are covered. It describes how the EPA Administrator should distribute capitalization grants to each state for the purpose of establishing a water pollution control revolving fund for providing assistance, for POTWs, for implementing a waste treatment management program. It includes corrective action such as a notification of noncompliance if the state had not complied with its agreement or any requirement of this title. There is provision for conditionally withholding of the payments to a state. It also covers allotment of withheld payment to a state under certain conditions.

THE CAUSES OF WATER POLLUTION

As defined by the Bureau of National Affairs, Inc., water pollution refers to the alteration of aquatic ecosystems to the extent that the aquatic life may be impaired or destroyed, human health may be threatened, or water may become so foul that recreation and use no longer are possible. This alteration may occur when substances such as chemicals, sediments, and excess nutrients are introduced into surface waters.

In 1988, the EPA conducted a survey of state water quality problems. It reported that the leading causes of impaired water uses are excess nutrients and sediments. These and other types of pollution are contributed by a variety of other sources. The following is a list of sources of water pollution as categorized by the EPA:

- **Industrial**--including pulp and paper mills, chemical manu-facturers, and food processing plants.
- **Municipal**--including POTWs that may receive indirect dis-charges from small factories or businesses.
- **Combined sewers**--including storm and sanitary sewers that when combined, may discharge untreated wastes during storms.
- **Agricultural**--including crop production, pastures, rangeland, or feedlots.

- **Silvicultural**--including forest management, harvesting, and road construction.
- **Construction**--including highways building and land development.
- **Resource extraction**--including mining, petroleum drilling, and runoff from mini tailing sites.
- **Land disposal**--including leachate or discharge from septic tanks, landfills, and hazardous waste disposal sites.
- **Hydrologic**--including channelization, dredging, dam construction, and steambank modification.

Water may be polluted from the above sources and by the following pollutants:

- **Nutrients**--such as the nitrates found in fertilizers and the phosphates found in detergents, over-stimulate growth of aquatic plants and algae. These organisms begin to choke up the water, use up dissolved oxygen, and cut off light in deeper waters.
- **Sediments**--When it rains, silt and other suspended solids wash over plowed fields, construction and logging sites, urban areas, and strip-mined land. As these solids enter rivers, lakes, and coastal waters, the accessibility of oxygen to fish is reduced, plant productivity is diminished, and aquatic habitats are smothered.
- **Bacteria/Viruses/Protozoans**--Certain waterborne bacteria, viruses, and protozoans can cause human illness that range from typhoid and dysentery to minor respiratory and skin diseases. These organisms can enter surface water via a number of routes, including sewers, stormwater drains, septic systems, and runoff from livestock pens. Since it is impossible to test water for every type of organism, fecal coliform bacteria are measured as indicators that the water may have been contaminated with untreated sewage and that other, more dangerous organisms are present.
- **Organic enrichment**--Organic material may enter the water in many forms, including sewage, leaves and grass clippings, or runoff from livestock feedlots and pastures. As the natural bacteria in the water breaks down this organic material, they begin to use up the oxygen dissolved in the water. When levels

of dissolved oxygen drop, many types of fish and bottom-dwelling animals cannot survive.

- **Toxic chemicals/Heavy metals**--Metals, such as mercury, lead, selenium, and cadmium and toxic organic chemicals such as PCBs and Dioxin, may originate in industrial discharges, runoff from city streets, mining activities, leachate from landfills, and a variety of other sources. These contaminants can cause death or reproductive failure in fish and wildlife, as well as pose human health risks.
- **Pesticides/Herbicides**--Rainfall and irrigation can wash pesticides and herbicides used on farmland, lawns, and in termite control into groundwater and surface waters. These contaminants generally are very persistent in the environment and may accumulate in fish, shell-fish, and wildlife.

REGULATORY INTENT

As written in the beginning of the act, the following is the declaration of goals and policy:

a. The objective of this act is to restore and maintain the chemical, physical, and biological integrity of the nation's waters. In order to achieve this objective, it is hereby declared that, consistent with the provisions of this act--

1. It is the national goal that the discharge of pollutants into navigable waters be eliminated by 1985.
2. It is the national goal that whatever attainable, an interim goal of water quality which provides for the protection and propagation of fish, shellfish, and wildlife and provides for recreation in and on the water be achieved by July 1, 1983.
3. It is the national policy that the discharge of toxic pollutants in toxic amounts be prohibited.
4. It is the national policy that federal financial assistance be provided to construct publicly owned waste treatment works.
5. It is the national policy that area-wide waste treatment management planning processes be developed and

implemented to assure adequate control of sources of pollutants in each state.

6. It is the national policy that a major research and demonstration effort be made to develop technology necessary to eliminate the discharge of pollutants into the navigable waters, waters of contiguous zones, and the oceans.

7. It is the national policy that programs for the control of nonpoint sources of pollution be developed and implemented in an expeditious manner so as to enable the goals of this act to be met through the control of both point and nonpoint sources of pollution.

In summary, **the Federal Water Pollution Control Act (commonly referred to as the Clean Water Act) was intended to protect the surface water of the United States**. The means to accomplish this intention is the requirement that the discharge of pollutants to the U.S. waters be controlled and prevented.

Although the Federal Water Pollution Control Law has routes extended back in time as far back as 1899 with the Refuse Act, the law and its many revisions was unworkable and ineffectual to directly enforce environmentally oriented limits of the discharge of pollutants. Then in 1972, Congress passed an important legislation, Public Law 92-500 to clearly provide a basic framework for an effective national environmental act, which are: national effluent limitations, water quality standards, the permitting program, special provisions for oil spills and toxic substances, and a publicly-owned treatment works construction grant program. The CWA in its current form, which has been hailed as the most tested, most innovative, and most enforceable of all the federal environmental statutes, was amended in 1987 to tighten the focus on toxics dischargers, force action on toxics-oriented water quality standards, attempt to resolve long-standing problems with the effort to regulate storm water runoff, and to strengthen the enforcement mechanisms.

The CWA is to "restore and maintain the chemical, physical and biological integrity of the nation's water." It is to eliminate the discharge of pollutants into surface waters with a national policy that the discharge of toxic pollutants in toxic amounts be prohibited.

Perhaps what is important here is the meaning of the term toxics as it is applied to the regulated and hazardous waste substances. Let us first define some of these related terminologies.

- **Pollutant**--means all waste material, whether or not that material has value at the time it is discharged.
- **Toxicity**--the degree of danger posed by a substance to animal or plant life. Toxicity is the fourth category for identifying a waste as hazardous.
- **Toxic substances**--chemicals or mixtures that may present a unreasonable risk of injury to health or the environment. Many are soluble in water and are often present in household, municipal and industrial wastes, and stormwater. Among the most common are organic compounds (pesticides and herbicides) and heavy metals such as lead, mercury, and cadmium.

How is the intent of the CWA to be accomplished? It is to be accomplished through a regulatory system which involves, at its core: (1) the effluent limitations and standards and (2) the permitting program.

To this end, it is *the objective of the CWA to prohibit virtually all discharges without permit*. It is to regulate all pollutants discharged from all facilities and most contaminated areas into virtually all waters in the United States. This is a rather broad, ambitious, and aggressive agenda, but its success is a must if this environmental medium is to be protected and preserved for our quality of life.

The regulatory intent of the CWA is the basis of the standards and enforcement program which is Title III of the act. The title provides the framework for the development and the implementation of the effluent guidelines. Nationally consistent discharge limits for individual pollutants in industrial discharges and sewage treatment plants were developed. Since the passage of the Federal Water Pollution Control Act in 1972 and the Clean Water Act in 1977, wastewater discharge requirements or standards in the U.S. have evolved significantly. The resulting regulations have developed a system of wastewater discharge standards, permits, and enforcement mechanisms. Wastewater discharges to waters of the U.S. require substantial levels of treatment in order to satisfy the CWA intents and objectives.

From these effluent guidelines, which were developed for each specific industry, the intent of the CWA is achieved through the permitting program. The permitting program is designed for two (2) general types of discharge. If the discharge is direct, it is governed by the National Pollutant Discharge Elimination System (NPDES) in accordance with 40 CFR Part 122-125. If the discharge is indirect, that is through POTWs, it is governed by federal and local standards via sewer use permit. In either case, the intent of the CWA is satisfied in that no discharge is allowed without control and/or permit.

The EPA has defined 10 major categories which are classified as the contributors of water pollution. These are industrial, municipal, combined sewers, storm sewers/runoff, agricultural, silvicultural, construction, resource extraction, land disposal, and hydrologic. Also, there are 7 major types of water pollutants: nutrients, sediments, bacteria/viruses/protozoans, organic enrichment, toxic chemicals/heavy metals, pesticides/herbicides, and habitat modifications.

INDUSTRIES AFFECTED BY THE ACT

The following is a list of NPDES Primary Industry. These discharges shall include effluent limitations and a compliance schedule to meet the requirements of section 301(b)(2)(A), (C), (D), (E), and (F) of CWA if their permits issued after June 30, 1981.

Adhesives and sealants	Organic chemicals, plastics, and
Aluminum forming	and synthetic fibers
Auto and other laundries	Paint and ink formulation
Battery manufacturing	Pesticides
Coal mining	Petroleum refining
Coil coating	Pharmaceutical preparations
Copper forming	Photographic equipment and
Electrical and electronic component	supplies
Electroplating	Plastics processing
Explosives manufacturing	Porcelain enameling
Foundries	Printing and publishing
Gum and wood chemicals	Pulp and paper mills
Inorganic chemicals manufacturing	Rubber processing
Iron and steel manufacturing	Soap and detergent manufacturing
Leather tanning and finishing	Steam electric power plants

Mechanical products manufacturing
Nonferrous metals manufacturing
Ore mining processing

Textile mills
Timber products

The industries subjected to the CWA stormwater discharge regulations are:

Metal mining
Coal mining
Oil and gas extraction
Mining and/or quarrying nonmetallic minerals
Food and kindred products
Tobacco products
Textile mill products
Apparel and other finished products made from fabrics and similar materials
Lumber and wood product (except furniture)
Wood kitchen cabinets
Furniture and fixtures
Paper and allied products
Paperboard containers and boxes
Converted paper and/or paperboard products
Printing, publishing, and allied industries
Chemical and allied products
Drugs
Paints, varnishes, lacquers, enamels etc.
Petroleum refining and related industries
Local/suburban transit and major interurban highway passenger transportation
Motor freight transportation/warehousing
Farm product warehousing and/or storage

Refrigerated warehousing and/or storage
General warehousing and/or storage
Rubber and miscellaneous plastic products
Leather and leather products
Leather tanning and finishing
Stone, clay, glass, concrete products
Glass products made of purchased glass
Primary metal industries
Fabricated metal products, machinery, and transportation equipment
Fabricated structural metal
Industrial and commercial machinery and computer equipment
Electronic and other electrical equipment components
Transportation equipment
Ship and boat building and repairing
Measuring, analyzing, controlling instruments; photographic, medical, optical goods; watches and clocks
Miscellaneous manufacturing
Railroad transportation
U.S. Postal Service
Water transportation
Transportation by air
Wholesale trade, durable goods
Used motor vehicle parts
Scrap and waste metals
Wholesale trade, nondurable goods
Petroleum bulk stations and terminals

THE PERMIT AND ENFORCEMENT PROGRAM

The "NPDES permit program" is coordinated with the prohibition to translate the comprehensive prohibition of all discharges into a set of sixty thousand or so specific conditional authorizations to discharge. It defines for each individual discharger his permissible level of release into waters of the United States.

NPDES stands for "National Pollutant Discharge Elimination System." It is a system of requirements to obtain permits for the continuation of any discharge of pollutants to surface water. Since the discharge of pollutants without a permit is prohibited, we have to notify the authorities about the nature and circumstances of any anticipated discharge by filing a permit application. Those conditions generally include requirements to reduce the discharge through monitoring and reporting obligations. The permit program has proven to be an almost indispensable element of effective environmental regulation. It gives the issuing authority correct information concerning the discharger's activities and permits precise advice to the discharger as to what is permissible. Thus, the requirement to file applications and report compliance or noncompliance based on accurate and current data is jealously maintained and strenuously enforced. The NPDES program is described in two categories: (1) the typical permit conditions and (2) the procedures by which the permits are issued.

Permit Condition

The NPDES permit performs two basic roles in the Clean Water Act regulatory process. It provided specific levels of performance the discharger must maintain, and it requires the discharger to report failures to meet those levels to the appropriate regulatory agency.

Monitoring and Reporting

The monitoring requirements in an NPDES are significantly important, especially the biomonitoring. The effectiveness of the permit program will depend on the effectiveness of monitoring and data maintenance requirements. The owner or operator of any point source is required to establish and maintain specific records, make specified reports, install, use, and maintain information which the EPA may reasonably require. As with the permit program in general, the states have the chance to administer their own monitoring program; the state becomes the monitoring authority for all point sources within its own jurisdiction. The enforcing authority will have the right to enter the premises of the discharger at any reasonable time, inspect the records required to be maintained, take test samples, and so forth.

The NPDES regulations specify the manner in which effluent limitations are to be included in permits and thus imposed on permitees. The monitoring requirements in various sections of Part 122 are intended to assure compliance imposed, and monitoring is to take place at the point of discharge except in limited situations where monitoring at point of discharge is infeasible. A permittee is required to monitor waste streams, as specified in his permit, to determine (1) compliance with the limitations on amounts, concentrations, or other pollutant measures specified in the permit, (2) the total volume of effluent discharged from each discharge point, and (3) otherwise as required by permit. Monitoring records, including charts from continuous monitoring devices and calibration and maintenance records, must be maintained for a minimum period of three years.

The result of monitoring must be reported periodically to the permit issuing authority on forms provided by the authority. Frequency of reporting is governed by the terms of each individual permit and must be at least annual. In addition to the periodic reporting requirement, certain toxic discharges must be reported within 24 hours. Failure to properly monitor and to report is a violation of the permit, and any person who knowingly makes any false statements in monitoring records, monitoring reports, or compliance or noncompliance notifications is subject upon conviction to substantial fines and criminal penalties.

Schedules of Compliance

Although the act itself establishes firm deadlines for the achievement of the required levels of treatments, the issuing authority has considerable latitude to require compliance or interim steps towards compliance at earlier dates. The act also provides mechanisms to extend compliance deadlines in limited situations.

Effluent Limitation

The determination of the precise effluent limitations to be included in the permit are to be based on "professional judgement," which is obviously more flexible than published rules where a permit is issued prior to the publication of effluent limitations for a particular pollutant or applicable industrial category or subcategory. This situation is most likely to arise pending promulgation of new limitations for toxic pollutants, or in connection with facilities for which no specific set limitations is wholly applicable.

Even after promulgation of limitations, the applicant may in certain cases seek modification of limits in the permit, and there is also considerable opportunity for the permitting authority to impose discharge limitations more stringent than the "base-level" quality related effluent limitations. Thus, there is considerable room for discussion regarding limits to be imposed in permits and a careful engineering analysis of proposed permit requirements.

Additional Effluent Limitation

Until recently, NPDES permits normally specified four or five pollutants as being subject to effluent limitations; a far greater number are now included in permits as a result of the EPA's toxic strategy. The EPA's NPDES permit application and related regulations (Section 122.21) required extensive waste streams analysis in order to file permit applications, extensive cataloging in the application of virtually all chemicals in the waste stream, and imposition of controls on the discharges of those chemicals. Implementation of these requirements complicates the permit process and requires more extensive monitoring than was true in the past.

Duration and Revocation

Permits may be valid for terms up to five years and may be subject to revocation and modification based on a very minimal showing of "cause." A discharge's interest in connection with the permit process will generally be best served by obtaining a permit with the maximum duration and with as much specificity as is obtainable in regard to the possible grounds of revocation or modification. On the permit's expiration, the permittee, in order to obtain reissuance, must demonstrate compliance with any more stringent criteria which have been promulgated during the term of the original permit.

Permitting Procedures

Under the Clean Water Act, the EPA is the issuing authority for all NPDES permits in a state. The state can be the program's administration and obtains EPA approval of its program. Where the state is the issuing authority, permission procedures are generally comparable to the EPA procedures, with certain exceptions. Where the state is the issuing authority, procedures for judicial review of permit issuance are those provided under the state's Administrative Procedure Act rather than under the Clean Water Act and the federal Administrative Procedure Act.

The EPA has the authority to withdraw its approval of a state program and take over the entire program administration if it finds that the state is not carrying out the program in accordance with the act's requirements.

Procedures for permit issuance are generally as follows: a permit application must be submitted to the EPA regional administrator (or the state) at least 180 days in advance of the data on which a proposed discharge is to commence or the expiration of the present permit, as the case may be. Where the EPA is the issuing authority, it will require for new dischargers the submission of a new source questionnaire before it will process the permit application.

After the application is filed, the district engineer of the Corps of Engineers must be given an opportunity to review the application to evaluate the impact of permit issuance upon anchorage and navigation. Where the EPA is the issuing authority, the state in which the discharge will occur must be provided with an opportunity to review the application. Based on that review, the state is asked to certify that the permitted discharge will comply with the act.

In processing the application, the issuing authority makes tentative determinations as to whether a permit should be issued, and, if so, as to the required effluent limitations, schedules of compliance, monitoring requirements, and so forth. These tentative determinations are organized into a draft permit, and the discharger is normally given an opportunity to review and comment on this draft. The public is given notice of the permit application proceeding and the issuing authority's preliminary determinations with respect there to.

Contested provisions of the permit become effective, and a final permit is issued, upon completion of these review proceedings. The issuance of a permit under the Clean Water Act will be account to fulfill the permit requirements of the Refuse Act of 1899, as well as those under the act itself.

Once the permit is issued, careful attention must be paid to the business of monitoring and demonstrating compliance with the permit as well as with making certain that the facility continues to operate in the manner portrayed in the permit application.

Determining Discharge Limits

Three basic criteria are commonly used to determine the level of control required at the point of discharge. We can require that the effluent be treated to the limits of available or practicable technology; we may require that the discharge be limited so as to maintain a specified level of water quality in a stream which is suitable for maintenance of a balanced population of fish, shellfish, and wildlife and for water contact recreation; or we may limit the discharge as may be necessary to minimize health or other risks such as toxicity.

The Clean Water Act's approach uses all three methods. Technology-based effluent limitations--"best available" or "best conventional" technology--establish the baseline level of treatment to be met for all discharges. More stringent treatment--beyond the technological baseline--may be required if necessary to achieve water quality standards or avoid effluent toxicity problems. Similar technology, toxicity, or water quality-based "pretreatment" requirements are applicable to facilities that send their waste to public owned treatment plants rather than discharging directly to surface water. The technology-based effluent limitations are established by the EPA in notice and comment rulemaking proceedings based in evidence as to the level of treatment achieved by exemplary operations. Until applicable limits are formally established, the limits put in permits are determined on the basis of "best professional judgement."

Where adopted limits are applicable, there is limited ability to vary the application to individual facilities even where those facilities may differ fundamentally from those considered in establishing the effluent limits.

Water-quality-based effluent limits are harder to establish because of the need to model or assess the fate of pollutants in specific receiving waters in order to determine the limits to put in the permits. This is particularly true for permits which must address numerous pollutant parameters in the permitted discharge.

Limits based on effluent toxicity are also difficult due to the absence of agreement as to what is meant by the "toxic pollutants in toxic amounts." Biological monitoring methods to determine "whole effluent toxicity" are touted some as an approach to resolution of this difficulty.

The CWA mandates a two part approach to establishing effluent limitations for industrial discharge: (1) nationwide base-level treatment to be established through an assessment of what is technologically and economically achievable for particular industry and (2) more stringent treatment requirements for specific plants where necessary to achieve water quality objectives for the particular body of water into which the plant discharges.

Toxic Pollutant

Although the CWA broadly defined pollutants subject to regulation and permitting, it furnished little guidance before 1977 with regard to toxic pollutants. For that reason, and because the 1972 act imposed unrealistic deadlines on EPA's limited staff, the EPA focused almost entirely on high-volume "conventional" pollutants such as biochemical oxygen demand (BOD), suspended solids (SS) and the acidity and alkalinity (pH) when it developed the effluent limitations required by the act. The effluent limits and permit requirements failed to address the dangers posed by more toxic pollutants such as chlorinated organic chemicals, heavy metals, pesticides and so forth, and at the same time may have overemphasized removal of solids and oxygen-demanding materials contained in conventional wastes.

EPA's failure to develop an effective toxics control strategy under the 1972 act led the Natural Resources Defense Council (NRDC), an environmental organization, to sue the Environmental Protection Agency. The lawsuit was settled, and in the process of settlement, the EPA and NRDC developed a policy which focused all of the regulatory mechanisms provided by the 1972 act upon effective regulation of toxic and priority pollutant discharges. In developing this policy, the parties identified that (1) the pollutants which would be the primary subject of regulation, (2) the industries which would be the primary concern in applying the regulations, and (3) the methods of regulating toxic discharges with the act's existing legal mechanism. The agreements reached in these negotiations were embodied in a settlement decree, and were adopted by Congress as a blueprint for a toxics control strategy in the 1977 amendments, and to a certain degree in the 1987 amendments.

The decree mandated full use of all the available regulatory tools under the act with a specific focus on the identified "priority pollutants." Pursuant to the decree, the EPA was to develop a program to regulate the discharge of 65 categories of "priority pollutants." The current 65 toxic pollutants group and their toxic categories are listed in Tables 1 and 2.

The consent decree required adoption of the best available technology effluent limitations for each priority pollutant in each industrial category by June 30, 1983. These limitations had to be applicable to at least 95 percent of the point sources in each identified industry category or subcategory. Similar technology-based requirements had to be adopted for new sources and sources discharging into publicly owned treatment works. The basis for excluding a category of point sources from the toxic-focused system of technology-based effluent limitation is quite limited.

In addition to these stringent industry-by-industry toxic effluent limits, the consent decree made specific provision for full implementation for the waterway segment-by-segment approach, discussed below.

The NRDC consent decree provided a judicial mandate for full use of the Clean Water Act's enforcement mechanisms in a carefully tailored effort to reduce discharges of toxic pollutants.

The 1977 amendments largely adopted by the technology-based aspects of this mandate and enacted them into federal statutory law. The amendments:

- Adopted the consent decree list of priority pollutants as the list of toxic substances to be given primary emphasis in the implementation of the Clean Water Act.
- Required adoption of best available technology (BAT) effluent limitations for each listed substance.

TABLE 1

LIST OF 65 TOXIC POLLUTANT GROUPS

Acenaphthene
Acrolein
Acrylonitrile
Aldrin/Dieldrin
Antimony and compounds
Arsenic and compounds
Asbestos
Benzene
Benzidine
Beryllium and compounds
Cadmium and compounds
Carbon tetrachloride
Chlordane (technical mixture and metabolites)
Chlorinated benzenes (other than dichlorobenzenes)
Chlorinated ethanes (including, 1,2-dichloroethane, 1,1,1-trichloroethane, and hexachloroethane)
Chloroalkyl ethers (chloroethyl, and mixed ethers)
Chlorinated naphthalene
Chlorinated phenols (other than those listed elsewhere; includes trichlorophenols and chlorinated cresols)
Chloroform
2-chrolophenol
Chromium and compounds
Copper and compounds
Cyanides
DDT and metabolites
Dichlorobenzenes (1,2-,1,3-, and 1,4-dichlorobenzenes)
Dichlorobenzidine
Dichloroethylenes (1,1-, and 1,2-dichloroethylene)

2,4-dimethylphenol
Dinitrotoluene
Diphynylhydrazine
Endosulfan and metabolites
Endrin and metabolites
Ethylbenzene
Fluoranthene
Haloethers (other than those listed elsewhere; includes chlorophynylphynyl ethers, bromophynylphynyl ether, bis-(dischloroisopropyl) ether, bis(dischloroethoxy)methane and polychlorinated diphynyl ethers)
Halomethanes (other than those listed elsewhere; includes methylene chloride, methyl-chloride, methylbromide, bromoform, dichlorobromo-methane)
Heptachlor and metabolites
Hexachlorobutadiene
Hexachlorocyclohexane (all isomers)
Hexachlorocyclopentadiene
Isophorone
Lead and compounds
Mercury and compounds
Naphthalene
Nickel and compounds
Nitrobenzene
Nitrophenols (including 2,4-dinitrophenol, dinitrocresol)
Nitrosamines
Pentachlorophenol
Phenol
Phthalate esters

TABLE 1 (Continued)
LIST OF 65 TOXIC POLLUTANT GROUPS

Polychlorinated biphenyls (PCBs)	2,3,7,8-Tetrachlorodibenzo-p-dioxin (TCDD)
Polynuclear aromatic hydrocarbons (including benzanthracenes, benzopyrenes, benzofluoranthene, chrysenes, dibenzanthracenes, and indenopyrenes)	Tetrachloroethylene
	Thallium and compounds
	Toluene
Selenium and compounds	Toxaphene
Silver and compounds	Trichloroethylene
	Vinyl chloride
	Zinc and compounds

TABLE 2
TEN CATEGORIES OF PRIORITY TOXIC POLLUTANTS

Pollutant	Characteristics	Sources
Pesticides Generally clorinated hydrocarbon	Readily assimilated by aquatic animals, fat soluble, concentrated through the food chain (biomagnified), persistent in soil and sediments	Direct application to farm and forestlands, runoff from lawns and gardens, urban runoff, discharge in industrial wastewater
Polychlorinated biphenyls (PCBs) Used in electrical capacitors and transformers, paints, plastics, insecticides, other industrial products	Readily assimilated by aquatic animals, fat soluble subject to biomagnification, persistent, chemically similar to the chlorinated hydrocarbons	Municipal and industrial waste discharges disposed of in dumps and landfills

	TABLE 2 (Continued)	
	TEN CATEGORIES OF PRIORITY TOXIC POLLUTANTS	
Pollutant	**Characteristics**	**Sources**
Metals Antimony, arsenic, beryllium, cadmium, chromium, copper, lead, mercury, nickel, selenium, silver, thallium, and zinc	Not biodegradable, persistent in sediments, toxic in solution, subject to biomagnification	Industrial discharges, mining activity, urban runoff, erosion of metal-rich soil, certain agricultural uses (e.g., mercury as a fungicide)
Other inorganics Asbestos and cyanide	*Asbestos* May cause cancer when inhaled, aquatic toxicity not well understood	*Asbestos* Manufacture and use as a retardant, roofing material, brake lining, etc.; runoff from mining
	Cyanide Variable persistent, inhibits oxygen metabolism	*Cyanide* Wide variety of industrial uses
Halogenated aliphatics Used in fire extinguishers, refrigerants, propellants, pesticides, solvents for oils, and greases and in dry cleaning	Largest single class of "priority toxics," can cause damage to central nervous system and liver, not very persistent	Produced by chlorination of water, vaporization during use

TABLE 2 (Continued)		
TEN CATEGORIES OF PRIORITY TOXIC POLLUTANTS		
Pollutant	**Characteristics**	**Sources**
Phthalate esters Used chiefly in production of polyvinyl chloride and thermoplastics as plasticizers	Common aquatic pollutant moderately toxic but tertogenic and multagenic properties in low concentrations; aquatic invertebrates are particularly sensitive to toxic effects; persistent; and can be biomagnified	Waste disposal vaporization during use (in nonplastics)
Monocyclic aromatics (excluding phenols, cresols and phthalates) Used in the manufacture of other chemicals, explosives, dyes and pigments, and in solvents, fungicides, and herbicides	Central nervous system depressant; can damage liver and kidneys	Enter environment during production and byproduct production states by direct volatization, wastewater
Ethers Used mainly as solvents for polymer plastics	Potent carcinogen, aquatic toxicity and fate not well understood	Escape during production and use

TABLE 2 (Continued)		
TEN CATEGORIES OF PRIORITY TOXIC POLLUTANTS		
Pollutant	**Characteristics**	**Sources**
Phenols Large volume industrial compounds used chiefly as chemical intermediates in the production of synthetic polymers, dyestuffs, pigments, pesticides, and herbicides	Toxicity increases with degree of chlorination of the phenolic molecule; very low concentrations can taint fish flesh and impart objectionable odor and taste to drinking water, difficult to remove from water by conventional treatment: carcinogenic in mice	Occur naturally in fossil fuels, wastewater from coking ovens, oil refineries, tar distillation plants, herbicide manufacturing, and plastic manufacturing; can all contain phenolic compounds
Polycyclic aromatic hydrocarbons Used as dyestuffs, chemical intermediates, pesticides, herbicides, motor fuels, and oils	Carcinogenic in animals and indirectly linked to cancer in humans; most work done on air pollution; more is needed on the quatic toxicity of these compounds; not persistent and are biodegradable though bioaccumulation can occur	Fossil fuels (use, spills, and production, incomplete combustion of hydrocarbons)

TABLE 2 (Continued)		
TEN CATEGORIES OF PRIORITY TOXIC POLLUTANTS		
Pollutant	**Characteristics**	**Sources**
Nitrosamines Used in the production of organic chemicals and rubber: patents exist on processes using these compounds	Tests on laboratory animals have shown the nitrosamines to be some of the most potent carcinogens	Production and use can occur spontaneously in food cooking operations

- Permitted the EPA to add to or remove items from the list of "toxic" substances.
- Required compliance with BAT effluent limitations for toxic pollutants.
- Provided a new system for upgrading and enforcing pretreatment regulations based on both the effluent limitations on the discharge from publicly owned treatment works and the intended use of the sludge from the facilities.
- Authorized the EPA to adopt the regulations establishing best management practices to control the discharge of toxic pollutants in the form of runoff or other uncontrolled discharges from industrial plant sites, parking lots, and so forth.

The consent decree as confirmed by the 1977 and 1987 amendments and modified by the court has transformed the entire Clean Water Act Program and focused the EPA and industry attention on the most dangerous pollutants. The industry-by-industry technology-based effluent limitations have been transformed from limited requirements focused on three or four conventional pollutants to a very specific system of limitations potentially applicable to 126 or more different pollutants as well as whole effluent toxicity for each especially now that the technology-based standards have been promulgated, and deadlines set to impose more stringent discharge limitations on the basis of water quality standards.

Industrial Effluent Limits: Required
Level of Treatment-Technology-Based Limits
for "Existing" Discharges

Section 301(b) of the 1972 act provided for the establishment of nationally applicable technology-based effluent limitations on an industry-by-industry basis. These effluent limitations were to establish a nationwide base-level of treatment for existing direct discharge sources in every significant industrial category. This level of treatment was to be achieved in two phases. For "existing" industrial discharges, Section 301 directs the achievement.

By July 1, 1977, of effluent limitation which will require application of the best practicable control technology currently available, and by July 1, 1983, of effluent limitations which will require application of the best available technology economically achievable.

As the time for achievement of best practicable technology (BPT) is long past, its primary relevance now is as a basic for setting subsequent standards or, as discussed below, in regulating discharges of conventional pollutants. The EPA defined BPT as the "average of the best existing performance by well-operated plants within each industrial category or subcategory." In developing the BPT limitations, the EPA was required to make what amounted to a cost-benefit balancing test that took into account a broad range of specific engineering factors relating to the ability of plants within a category or subcategory to achieve limits.

The EPA defined best available technology (BAT) as the "very best control and treatment measures that have been or are capable of being achieved." The BAT effluent limitations now focus primarily on the priority pollutants listed in the NRDC consent decree and on additional toxic pollutants identified pursuant to Section 307(a) of the act. For pollutants not listed in the consent decree but identified as toxic pollutants under Section 307(a)(1) of the act, compliance with BAT effluent limits is required not later than three years after the date on which the limitations are established.

The best conventional pollutant control technology (BCT) effluent limitations were, like the BBT and BAT limitations, to be adopted on an industry-by-industry basis but were to apply for each affected industry only to pollutants which were identified as "conventional." The 1977 amendments specifically included within the definition of "conventional pollutants" biological oxygen demand (BOD), suspended solids (SS), fecal coliform bacteria, and pH. The EPA is authorized to include additional pollutants within the definition of conventional pollutants. The act specifically excludes heat from the conventional pollutant definition as there are special statutory provisions for thermal discharges.

Congress anticipated that the EPA, in developing the best conventional technology limits, would review the old BAT limits for conventional pollutants and reduce the stringency of such limits to the extent indicated by the economic justification and cost comparability with secondary treatment requirements.

The last of the three categories of technology-based effluent limits for existing industry direct charges provided in the 1977 amendments in the system of effluent limitations to be adopted for "nonconventional nontoxic" pollutants. This is essentially an "everything else" category which applies to all pollutants other than those identified as priority pollutants, toxic pollutants, or conventional pollutants under the preceding sections of Section 301.

Required Level of Treatment-Technology-Based Limits for "New Source" Direct Discharges

The establishment of effluent limitations for "new source" (defined as any facility or major modification, the construction of which is commenced "after the publication of proposed regulations" prescribing an applicable standard of performance) is separately dealt with in Section 306 of the act. Although the general approach for establishment of new source performance standards under Section 306 is similar to the approach for the establishment of Section 301 effluent limitations (discussed in the previous section), there are significant differences both as to the level of treatment required and the manner of applying the limitations established. These differences remain important because of the EPA's delay in promulgating many BPT and BAT limitations, and because the new source standards will govern the addition or replacement of certain equipment at an existing discharger.

The primary difference between these criteria and the Section 301 criteria is the requirement in Section 306 that the EPA consider not only pollution control techniques, but also various alternative production processes, operating methods, in-plant control procedures, and so forth.

A second major difference regarding criteria for development of a new source performance standards is the absence of the kind of requirements for detailed consideration of economic and technological factors which are established by Section 301 for existing source effluent limitations procedure.

Effluent Guidelines for Additional Source Categories

The act requires the EPA to make periodic revisions of effluent guidelines for industry categories as technology improves and economy changes. The initial set of industry categories was specified in part by Section 306(b)(1)(A) and has been expanded with time by the EPA. Section 304(m) was added by the 1987 amendments. This provision requires the EPA to establish and to publish a schedule for the annual review and revisions of existing effluent guidelines.

Required Level of Treatment-Technology-Based Limits for Indirect Dischargers (Pretreatment)

Industrial facilities that discharge into publicly owned treatment works (POTWs) are regulated not by the requirements governing direct dischargers, but rather by comparable treatment requirements-- pretreatment standards--adopted pursuant to Section 307(b) of the act. Pretreatment standards are calculated to achieve two basic objectives: (1) to protect the operation of POTWs and (2) to prevent the discharge of pollutants which pass through POTWs without receiving adequate treatment. The dual objectives of the treatment program result in a two-part system of controls under the applicable EPA regulations.

The first part of the general pretreatment regulation focuses primarily on preventing the discharge into POTWs of pollutants which will interfere with the proper operation of the receiving treatment works. This "protection" standard prohibits the introduction into any POTWs of:

1. Pollutants which create a fire or explosion hazard in the POTW, including but not limited to, waste streams which meet the RCRA test for characteristic inflammable waste.

2. Discharges with a pH lower than 5.0 unless the works is specifically designed to accommodate such discharges.

3. Solid or viscous pollutants in amounts which obstruct the flow in a sewer system.

4. Discharges, including discharges of conventional pollutants, of such volume and concentration that they upset the treatment process and cause a permit violation (for example, unusually high concentration of oxygen demanding pollutants such as BOD).

5. Heat in amounts which will inhibit biological activity in the POTWs resulting in interference, but in no case heat is such quantities that the temperature influent at the treatment works exceeds 40°C unless the works are designed to accommodate such heat.

6. Petroleum oil, nonbiodegradable cutting oil, or products of mineral oil origin in amounts that will cause interference or pass through.

7. Pollutants which result in the presence of toxic gases, vapors of fumes within the POTW in quantity that may cause acute worker health and safety problem.

8. Any trucked or hauled pollutants, except at discharge points designated by the POTW.

The second major objective of the pretreatment regulations preventing the discharge into POTWs of pollutants which pass through those treatment works without receiving adequate treatment is to be achieved by "categorical" pretreatment regulations. These categorical regulations are applicable only to "incompatible" pollutants that is pollutants other than BOD, SS, pH, and fecal coliform bacteria, which are not adequately treated in the POTW treatment process. These categorical pretreatment regulations, like the BAT regulations and new source performance standards, focus primarily on the 34 industries and 65 toxic pollutant categories specified in the NRDC consent decree. For each discharger into a POTW, these categorical standards are intended to result in the same level of treatment prior to discharge from the POTW as that which would have been required had the industrial facility discharged those pollutants directly to the receiving waters. The stringency of these categorical standards can theoretically be reduced through the mechanism of removal credits, which takes into account the removal of these pollutants consistently achieved by the POTW in question. Removal credits, however, will not be available under the statute until the EPA completes promulgation of its sewage sludge regulations under Section 405 of the act.

Accordingly, the industrial facility discharging into a POTW will be required to achieve, in meeting the applicable pretreatment limits, a level of treatment performance equivalent to the applicable BAT effluent limitations or new source performance standards unless the receiving POTW has an approved pretreatment limit. This removal credit is to be based on the POTW's demonstrated capability to consistently remove the pollutant in its treatment process. In order to qualify for revision, the POTW must provide consistent removal of each pollutant for which a discharge limit revision is sought, and its sludge use or disposal practices must, at the time of the application and thereafter, remain in compliance with all applicable criteria, guidelines, and regulations for sludge use and disposal. The EPA modified its regulations in an effort to account for process variations by POTWs, but that effort was reversed by the Court of Appeals.

Pretreatment requirements are directly enforceable by the EPA and states with NPDES permit issuance authority, but the EPA regulations contemplate eventually delegation of primary enforcement responsibility to individual POTWs, with the EPA and the states receding into the background.

Under the regulations, any POTW (or combination of POTWs operated by the same authority) having a total design flow greater than five million gallons per day must have developed and implemented a pretreatment program by July 1, 1983, if it receives incompatible industrial waste. A POTW must have an approved pretreatment program in order to grant removal credits, although it may grant conditional removal credits while the EPA is considering approval of the POTW's pretreatment program. A POTW program must meet funding, personnel, legal, and procedural criteria sufficient to ensure that the POTW's enforcement responsibilities can be carried out. Once the program is developed and approved, the POTW will be responsible for enforcement of the national pretreatment standards. A POTW may exercise enforcement authority through a number of methods including contracts, joint power agreements, ordinances, or otherwise.

Finally, pretreatment regulations establish extensive reporting requirements for both industrial users and POTWs in order to monitor and demonstrate compliance with categorical pretreatment standards.

The pretreatment regulations significantly affect industries subject to categorical pretreatment standards, as well as other industrial users of POTWs which will have to comply with general pretreatment requirements.

Best Professional Judgement

As in the case with many regulatory efforts, the comprehensive, detailed and formal process outlined above is, as a practical matter, inapplicable in a very large number of situations where discharge limits are established since, very often if not always, a significant part of the process leading to the discharge under consideration will be in a category for which applicable effluent limits or performance standards are yet to be promulgated. In such cases, the regulation authorize the permit issuer to establish discharge limits based on the exercise of professional or engineering judgement.

Variance

The statutory mechanisms to authorize variances from technology-based standards are exceedingly limited. The most broadly applicable of these variances is the fundamentally different factors (FDF) variance. There are also variance mechanisms to recognize use of innovative control technology and in certain limited circumstances from BAT limitations. Other than the FDF variance, there are no variances allowed from toxic pollutant standards and none at all from discharge limitations set to meet water quality standards.

Technology-Based Treatment Standards-Publicly Owned Treatment Works

For discharges from publicly owned treatment works (POTWs), Section 301 directed that by July 1, 1977, they achieve effluent limitations based on secondary treatment, as defined by the EPA, and any more stringent limitations necessary to comply with water quality standards or treatment standards imposed by state law.

The effluent levels prescribed by regulations are as follows:

	% Removal	Concentration	
	30-Day average	Monthly average	Weekly average
BOD (5 day)	85	30	45
Suspended solid (SS)	85	30	45
Coliform	200/100 ml	400/100 ml	
pH	6.0-9.0		

The regulation makes special provisions for upward revision of the "secondary treatment" effluent limits (1) where necessary to take into account storm water infiltration into combined sewers during wet weather periods and (2) where necessary to take into account the fact that the Section 301 and Section 306 effluent limitations applicable to major industrial dischargers into the treatment works would permit an industrial user to directly discharge greater concentrations than those set forth in the table. In the latter case, the permitted discharge from the POTW which is attributable to the industrial waste would be permitted under the applicable effluent limitations if the industrial facility were discharging directly into a waterway.

Water Quality-Related Effluent Limitations

The 1972 act made technology-based effluent limitations the nationwide minimum or base level of treatment. The act provides several mechanisms by which these discharge limitations are to be tightened in order to protect or maintain adequate water quality in specific bodies of water. These more stringent water quality-based limitations are most often an issue along bodies of water where there is heavy concentration of industrial dischargers, where receiving water needs to be maintained at a very high quality for recreational or other purposes, or where hydrologic modifications for navigation have reduced the capacity of receiving waters to assimilate pollutants.

Toxic Water Quality Standards

The NRDC consent decree contained several provisions addressing water quality standards as they related to priority pollutants. These provisions required the EPA to issue water quality guidance for states to use in setting additional discharge standards for toxic or priority pollutants. In addition, the decree required identification of so-called toxic "hot spots," where the application of technology-based standards would not achieve water quality standards. For a variety of reasons, including the EPA's slow pace in issuing the technology-based standards and the inherent scientific complexity of establishing water quality standards, these provisions of the consent decree had little effect at the level of the individual discharger.

Water Quality Standards Affecting All Pollutants

Section 301(b)(1)(C) required industrial dischargers and POTWs to achieve no later than July, 1977, any effluent limitations more stringent than the minimum technology-based standards which may be necessary to meet applicable federal/state water quality standards. This requirement is incorporated into permits issued by the EPA through the state certification requirement under Section 401 of the act. Section 302 of the act authorizes the EPA directly to establish effluent criteria more stringent that the applicable BAT limits where necessary for the attainment or maintenance in a specific water body of water quality which "shall assure protection and propagation of a balanced population of shellfish, fish and wildlife, and allow recreational activities in and on the water."

The water quality standards and water-quality related effluent limitations imposed by these two mechanisms can require levels of treatment considerably higher than those required by the technology-related effluent limits, particularly for water bodies with heavy concentrations of dischargers, or exceptionally poor water quality and correspondingly stringent water quality standards, or water bodies with limited assimilative capacity because of hydrologic factors such as dams or navigation and water supply.

Preventing, Giving Notice of, Responding to Spills

Although the permit mechanism is highly effective for dischargers that can be anticipated and controlled, it is of limited use in dealing with events like accidental spills. To meet this need, the CWA imposes pollution prevention planning requirements to minimize the potential for a spill, mandatory spill notification requirements, and provisions assessing responsivity for responding to spills or paying the cost of response.

Spill Prevention

Pollution prevention mandates under the Clean Water Act include requirements to maintain and implement up-to-date Spill Prevention, Control, and Countermeasure (SPCC) plans for facilities with total oil storage capacity in excess of 1320 gallons above ground or 40,000 gallons below ground. These plans, which included physical requirements such as secondary containment capacity, are intended to minimize the consequences of a spill if one does occur.

The SPCC plan must be a plan to respond, to the maximum extent practicable, to a worst case discharge, and to a substantial threat of such discharge, of oil or hazardous substances. The plan must:

1. Be consistent with the NCP and area contingency plans.

2. Identify the person in charge of the facility, who must have authority to implement the plan.

3. Immediate communication between the person in charge and appropriate federal officials and response action contractor.

4. Ensure by contract (or other means allowed under regulation) that adequate private personnel and equipment be available to remove the worst case discharge to the maximum extent practicable.

5. Require training, periodic unannounced drills, and equipment testing necessary to assure effective response actions.

6. Be updated periodically.

7. Be submitted for approval with each significant change.

The SPCC plans are to be submitted for federal approval, and the language of the statute suggests all such plans must be submitted and approved in order for facilities to continue operations. The legislative history suggests that only a fraction of these will have to be submitted and reviewed; otherwise the Coast Guard or EPA, as the case may be, would be overwhelmed with SPCC plans for small facilities.

Spill Notification

Section 311 of the CWA requires the owner or person in charge of any vehicle, vessel, or facility from which there is a discharge or threatened discharge of oil or a reportable quantity of a hazardous pollutant to notify the National Response Center.

Hazardous Substance Spills

Section 311 governs the discharge of hazardous substances. The amendments of Section 311 were directed toward the two most significant problems identified by the court: the elements necessary to establish what was to be considered a harmful quantity and the relationship of Section 311 to the NPDES program.

Pursuant to these amendments, the EPA has designated approximately 300 substances as hazardous and thus subject to the Section 311 program. The second principal feature of the EPA's hazardous substance regulations is the exclusion of discharges made in compliance with an NPDES permit. A facility owner or operator who spills a harmful quantity of a hazardous substance must report the spill; failure to do so will subject him to criminal penalties.

Controlling Non-Industrial Process Waste Discharges

Although the system of effluent limits imposed through the NPDES permit program is an effective means of regulating waste discharges which result from normal industrial or municipal processes and which are amendable to treatment prior to discharge, this system is an inappropriate means of regulation and controlling accidental and unanticipated discharges or discharges which, by their nature, are not subject to confinement and treatment. For this latter class of discharges, the focus of regulation must be on preventing the discharge or on minimizing the volume of pollutants carried. Since accidental spills and "non-point source" discharges are responsible for a large percentage of the total pollutants introduced into the nation's waterways, the act provides a number of mechanisms, supplemental to the NPDES permit program, to control discharges which are unrelated to industrial process wastes. This system of supplemental regulatory controls is the subject of this section and the next one.

Non-Point Source Pollution

The 1987 amendments of 208 plans made non-point sources of toxic pollutants an important aspect of water quality planning under Sections 303 and 304(1). As part of identifying water bodies which fail to attain and maintain water quality standards after compliance with technology-based standards, the state must also identify those waters which fail to meet standards for toxic pollutants as a result of non-point sources of such pollutants.

Storm Water Discharges and Best Management Practices

The regulation of municipal and industrial storm water discharges from industrial areas and municipalities above a certain size. The 1987 amendments adopted but substantially modified this approach. The amended act now requires that five categories of municipal or industrial storm water discharges be regulated as NPDES discharges:

1. Discharges which have NPDES permits issued as of February 1987.

2. Discharges "associated with industrial activity."

3. Discharges "from municipal separate storm sewer system serving a population of 250,000 or more."

4. Discharges "from municipal separate storm sewer system serving a population of 100,000 or more but less than 250,000."

5. Other discharges designated by the EPA administrator or the state if such discharge "contributes to a violation of a water quality standard or is a significant contributor of pollutants to waters of the United States."

For discharges associated with industrial activities, NPDES permits must meet the applicable effluent limitations imposed upon the industry in question. Storm water discharges "associated with industrial activity" which do not currently have permits are subject to different procedures. The permittee must comply with the permit "as expeditiously as practicable but in no event later than three years after the date of issuance of such permit."

Large municipal discharges in discharges from storm water sewer systems serving more than 250,000 people must meet the same schedule for permit application regulations, application filing, and deadlines for permit issuance and compliance. Smaller municipal storm sewer systems are subject to these same substantive permit requirements, the timetable for applications, permit decisions, and compliance is longer than for large municipal systems.

The EPA published its final regulations concerning NPDES permits for storm water discharges in November 1990. These rules address application requirements for storm water discharges associated with industrial activity and for discharges from municipal separate storm water rules, the final rules require direct permit coverage for all storm water discharge associated with industrial activity, including those that discharge through municipal separate storm sewers.

The storm water regulations do not apply to all discharges of storm water by industries but only to storm water discharges associated with industrial activity. The regulations define "storm water discharge associated with industrial activity include, but are not limited to, storm water discharges from:

1. Industrial plant yards.

2. Immediate access roads and rail lines used or traveled by carriers of raw materials, manufactured products, waste materials, or by-products used or created by the facility.

3. Material handling sites.

4. Refuse sites.

5. Sites used for the application or disposal of process waste waters.

6. Sites used for the storage and maintenance of material handling equipment.

7. Sites used for residual treatment, storage, or disposal.

8. Shipping and receiving areas.

9. Manufacturing buildings.

10. Storage areas for raw material, and intermediate and finished products.

11. Areas where industrial activity has taken place in the past and significant materials remain and are exposed to storm water."

However, for a limited number of categories of industries, storm water discharges associated with industrial activity include only storm water discharges from all the areas listed in the previous sentences where material handling equipment or activities, raw materials, intermediate products, final products, waste materials, by-products, or industrial machinery are exposed to storm water.

A discharger of storm water associated with industrial activity may apply for an NPDES permit through a notice of intent to be covered by a general permit. Facilities not eligible for coverage under a general permit are required to file either a group or individual permit application. If a facility cannot seek coverage under a general permit and cannot obtain a group permit, then it must bear the considerable burden of applying for an individual permit.

This permit system is the primary mechanism for regulating plan site runoff where toxic and hazardous pollutants are not involved. In addition, Section 304(e) authorizes the EPA to require permittees to adopt "Best Management Practices" to control toxic pollutants resulting from ancillary industrial activities. The EPA also is authorized to prescribe regulations to control plant site runoff, spillage or leaks, and sludge or waste disposal. The permittee must submit BMP as part of the permit application and which will be subject to all permit issuance procedures.

Enforcement

The Federal Clean Water Act imposes complex and difficult requirements on both direct dischargers and indirect dischargers to POTWs. If dischargers do not comply with the enforceable requirements of the act, including compliance with all discharge permit limitations under the National Pollutant Discharge Elimination System permit program, enforcement actions by the appropriate state or federal agency, or in some cases, by private individuals will be taken. Enforcement mechanisms include civil suits brought against a discharger by private individuals, civil enforcement procedures generated by the permitting authority, criminal sanctions, and a variety of administrative actions and penalties (p. 911:81). The NPDES compliance and enforcement program is managed by the EPA regional offices and state permitting authorities.

The important tools of enforcement programs are listed as follows:

1. Discharge Monitoring Report (DMR)--Each permittee is expected to complete a DMR monthly based on its self-monitoring. The permitting authority relies largely on the honesty and integrity of dischargers to report their permit violations.

2. Enforcement Management System

3. Noncompliance Reporting
      Category I Noncompliance
      Category II Noncompliance

4. Significant Noncompliance

5. List of Violating Facilities

The CWA provides a number of enforcement options to the EPA and the states, as well as a heavily-used citizen suit provision. As companies' potential exposure under these enforcement and penalty provisions can be staggering, even for infractions causing little actual harm, it is important for regulated entities to understand what their potential exposure is under the CWA's criminal, civil, and administrative penalty provisions, as well as for citizens suits.

Criminal Penalties

Criminal penalties range up to $50,000 per day of violation and/or imprisonment of three years. The following are the types of violations that may subject to criminal penalties:

1. Negligently violating are potentially criminal, negligently violating Sections 301, 302, 306, 308, 318, or 405 of the Clean Water Act. The scope of potential criminal violations under the amended CWA is extremely broad, and reason for diligent attention to compliance.

2. Negligently introducing into a sewer system or POTWs any pollutant or hazardous substance that may cause personal injury or property damage, or cause the POTW to violate its permit. For these violations, up to $25,000 per day of violation or 1 year of imprisonment or both.

3. Knowingly violating the same sections of the act. It may include many situations where a discharger is aware of a violation but continues to operate while seeking to abate the violation.

4. Knowingly violating any permit condition or limit.

5. Knowingly violating any requirement imposed by a pretreatment system. Up to $100,000 per day violating or 6 years of imprisonment or both (after the first conviction).

6. Knowingly endangerment is a class offense where a person knowingly violates a permit or other requirement and who knows at that time that he thereby places another person in imminent danger of death or serious bodily harm. Up to $250,000 or 15 years or both. For an organization, up to a $1,000,000 fine could be imposed.

7. Knowingly making a false statement.

Civil Enforcement Options

Under the Clean Water Act as amended in 1987, the EPA acting through the Department of Justice, has a number of civil enforcement options to address violations of the act, the implementing regulations, and NPDES and other permits. The fine for violating is $25,000 per day.

The EPA may seek injunctive relief under either Section 504 or Section 309(b). Section 504 applies to discharges which present an imminent and substantial endangerment to the health, welfare, or livelihood of persons. Section 309(b) is used when the EPA seeks injunctive relief. This section empowers the district courts to enter preliminary and permanent injunctions to restrain and abate violations of the statute, regulations, and permits, including state NPDES permits. In settlement of several recent cases, the EPA has obtained consent decree requiring the discharger to undertake substantial remedial action to address the environmental problems caused by the discharge.

Administrative Orders and Penalties

Section 309(a) authorizes the EPA to issue administrative compliance orders to persons in violation of their permits or other Clean Water Act obligations. The order may require compliance with an interim compliance schedule or an operation and maintenance requirement in not more than 30 days; permanent compliance is to be required in a time that the EPA determines is reasonable under the circumstance. The state is to be sent copies of any such order.

The issuance of an EPA compliance order is a serious matter for a discharger, since failure to comply or at least to make good faith orders to do so may be the basis to initiate a criminal prosecution for "knowing" violations, or to initiate a civil penalty proceeding where the claim is made that the discharger is recalcitrant or acting in bad faith.

Compliance Strategy

The broad enforcement discretion enjoyed by the EPA and the states, coupled with the requirement of publicly filing monitoring reports, including those showing violations, requires potential violators to take the initiative in dealing with compliance problems. The discharger should attempt to accomplish the following three goals:

1. Convince the relevant enforcement agencies of its good faith.

2. The discharger must attempt to deal with any major health or environmental problems caused by a potential violation in a manner that is as serious as the manner the EPA or the state enforcement agency would adopt.

3. The discharger must realistically appraise not only its chances of ultimately prevailing in any enforcement litigation, but the

bureaucratic constraints such as the civil penalty policy on the EPA and the Department of Justice in settlement negotiations.

Citizen Suits

Section 505 of the act provides an additional impetus to vigorous enforcement of the act's provisions. It authorizes any person "having an interest which is or may be adversely affected" to commence civil actions either against a discharger, for violation of any effluent standard or limitation under the act, or against the EPA for failure to proceed expeditiously to enforce the act's provisions.

THE CWA AND CURRENT WASTEWATER TREATMENT TECHNOLOGIES

Wastewater treatment issues involve the following: wastewater characteristics, discharge requirements, source controls, preliminary treatment methods, and a range of end-of-pipe treatment technologies. These include biological treatment and physical/chemical processes addressing organics or metals. Biological treatment processes include fixed film and suspended growth aerobic and anaerobic systems. Physical/chemical methods are adsorption, stripping, chemical oxidation, metals precipitation, and ion exchange as well as membrane processes including micro/ultrafiltration and reverse osmosis. Residual management are stabilization, dewatering, and disposal.

The following is an overview of the current wastewater treatment technologies as they are used by industries to comply with the Clean Water Act.

Stripping Technology

Gas stripping may be accomplished using one of two stripping media: air or steam. Volatile organics may be stripped readily by air stripping, while steam stripping may be necessary for semi-volatiles or to provide a condensate for reuse or disposal as opposed to treating an air stream as with air stripping. Highly soluble substances like alcohols and ketones are not readily stripped.

The following is an actual Upjohn Company's steam stripping system and the P&ID.

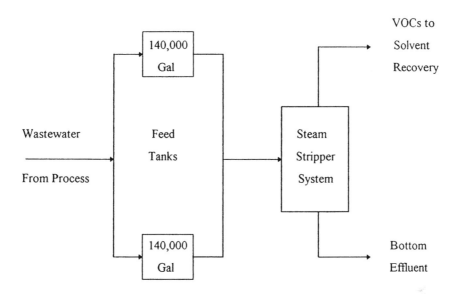

Upjohn Company reported the following resulting benefits:

- Reduction of 2,000,000 lbs/yr of VOC emissions.
- Recovery of ~200,000 gallons of MeCl$_2$.
- Tool for comparing actual removal data with EPA proposed pretreatment standards.

The above system design was from one of the Upjohn Company's project. Its design basis were:

- 50 gpm wastewater feedrate.
- 5 wt% miscellaneous solvents in feed (2 wt% MeCl$_2$).
- 2 wt% suspended solids (tar like).
- 3 wt% chlorides.

- 2-12 pH range.
- Wide variations in feed composition.

Steam stripping technology is one of the technology that the EPA is going to emphasize particularly for industries such as the pharmaceutical manufacturing industry. Its technical challenges and solutions are the following:

- Variability of feed stream composition.
  -- two 140,000 gallon agitated feed tanks
- Chloride corrosion of process equipment.
  -- lined feed tanks
  -- teflon lined pipe
  -- Hastelloy C stripper, pumps, and heat exchangers
- Fouling from suspended solids.
  -- on-line/off-line heat exchangers
  -- automated HCl cleaning off-line heat exchangers
  -- sieve trays in stripper
  -- steam contact feed preheater
  -- bottoms flash tank
  -- overhead condenser/feed preheater
- Energy conservation.
  -- bottoms flash tank
  -- overhead condenser/feed preheater

Biological Treatment Technology

Aerobic biological treatment is the most common major unit operation of pharmaceutical manufacturing wastewater, whether employed with an on-site pretreatment facility or off-site at a POTW treatment plant. Aerobic biological treatment involves the conversion of organics to carbon dioxide, water, and new bacterial cell growth by employing a heterogeneous mixture of microorganisms under aerobic conditions. The major types of aerobic biological treatment include:

- Aerated lagoon.
- Facultative lagoon.
- Oxidation ditch.
- Activated sludge.
- Extended aeration.
- Rotating biological contractors (RBCs).
- Trickling filters.
- Squenching batch reactors (SBRs).

Anaerobic biological treatment is often well suited for treatment of pharmaceutical manufacturing process wastewater. Anaerobic systems offer several advantages over aerobic systems for high-strength wastewater: (1) energy requirements and costs are significantly lower, (2) sludge generated and nutrient requirements are lower, and (3) space requirements are lower for high-rate anaerobic systems. Disadvantages of anaerobic systems may include relatively long start-up times, efficiencies of less than 90 percent, and the need to treat biogas for sulfur compounds. Other potential disadvantages include a greater sensitivity to inhibition or upset than aerobic systems for some substances such as methylene chloride, and the need to control sulfide toxicity for effluents with elevated sulfate levels. Several types of anaerobic systems may be employed including:

- Anaerobic lagoons.
- Anaerobic contact systems.
- Upflow anaerobic sludge blanket systems.
- Anaerobic filters (both downflow and upflow).
- Fluidized bed systems and hybrid systems.

Advanced biological treatment, as defined by the EPA, encompasses nitrification of ammonia, possible dentrification of nitrate and nitrite to nitrogen gas, and possibly biological phosphorus removal. There are both generic and proprietary approaches to advanced biological nutrient removal, but each has the common characteristic of employing aerobic biological treatment.

Review of Upflow Biological Reactors

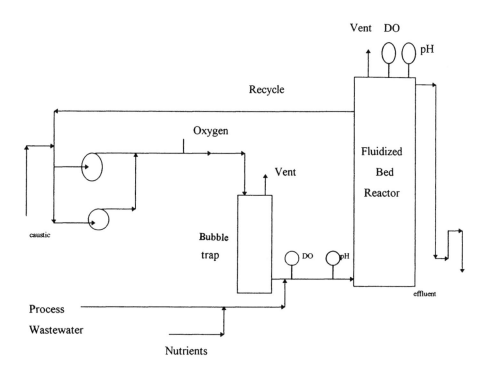

Upflow biological reactor is a fixed-cell system designed to degrade aromatics and polycyclic aromatic hydrocarbons. The Celgene process uses an upflow reactor that maximizes microbial degradation and minimize volatilization. DO is used in the reactor to facilitate rapid degradation and minimize VOC loss. This is particularly beneficial for industrial plants in major source emissions categories located in non-attaintment areas. A key to the process is immobilization of proprietary biocatalysts onto a support medium of granular activated carbon. Celgenes's upflow reactor design and use of proprietary, immobilized biocatalysts targeted to remove specific organics have advantages over traditional biological processes including:

- A closed-loop design that minimizes vapor stripping and, in some cases, enables mass balance determinations.

- Durable operation and a wide range of treatment options.
- Potential for water reuse.
- Rapid startup and recovery from upsets.
- Ability to achieve low contaminant levels.

Physical/Chemical Treatment Technology

This category of technology includes the following:

- Coagulation/flocculation.
- Dissolved air flotation.
- Gravity clarification.
- Filtration.
- Evaporation.
- Adsorption.
- Membrane processes.
- Chemical precipitation.
- Ion exchange.

Coagulation/flocculation might be applied to a waste stream containing calcium or other cations including heavy metals. **Dissolved air flotation, clarification, and filtration** are appropriate for solids-laden waste stream. **Evaporation** may be used for a high solids, low volume waste stream that is difficult to treat by more conventional means. Several **adsorption** processes are available for organics treatment, including activated carbon and polymeric resins. **Activated carbon** is best suited to removal of nonpolar, high molecular weight compounds. **Membrane processes** such as microfiltration, ultra-filtration, and reverse osmosis may be utilized to remove colloidal particles, or molecules and ions in the case of RO. RO, while highly effective in effluent polishing, requires a substantially pretreated effluent and is costly to install and operate. **Chemical precipitation and ion exchange** are applicable to metals.

Membrane Processes

Well-established membrane processes are finding application in wastewater treatment. The processes operate by separating contaminants on the basis of their molecular weight and size. They appeal to industry because the apparatus is relatively easy to maintain, uses few moving parts, and does not involve the use of hazardous substances other than some cleaning agents. In addition, the separated water often is of sufficiently high quality to be reused in the process, and in certain processes, some of the separated materials also can be reused. Membrane processes are a range of related technologies that share a fundamental design approach, but operate differently and are appropriate for different parts of the waste stream. The most widely used membrane processes can be described in three broad categories: microfiltration, ultrafiltration, and reverse osmosis. In each of the categories, wastewater is pumped at a fixed pressure through porous membranes that efficiently separate contaminants within a certain size range from the wastewater stream.

Reverse Osmosis

Osmosis is defined as the spontaneous transport of a solvent from a dilute solution to a concentrated solution across an ideal semipermeable membrane which impedes passage of solute but allows solvent flow. Solvent flow can be reduced by exerting pressure of the solution side of membrane. At certain pressure, the osmotic pressure, equilibrium is reached and the amount of solvent which passes in each direction is equal. Reverse osmosis, is somewhat similar to filtration, according to Weber--both involve removing a liquid from a mixture by passing it through a device which retains the other components. The process has also been termed hyperfiltration. However there are at least three important differences. First, the osmosis pressure plays a very important role in the process. Second, filter cakes with low moisture are unfeasible in RO. Third, filters separate mixtures primarily on the basis of size, whereas the semipermeability of desalination RO membranes depends significantly upon other factors as well.

Advanced Oxidation System

For facilities seeking ways to reduce toxic organics compounds in their wastewater advanced oxidation technologies offer an effective, self-contained, on-site approach. The prime advantage of these technologies is that their end products pose few disposal problems, enabling the wastewater to be treated by conventional technologies or to be discharged to a POTW.

Innovative/NICH Technologies

Many potentially wastewater treatment technologies are at various stages of development. They are:

- Deep shaft aeration.
- Electrolytic oxidation.
- Polyethylene strip media trickling filters.
- Solvent extraction.
- Supercritical water oxidation.
- Thermophilic activated sludge.
- UV/chemical oxidation.
- Wet air oxidation.

Residuals Management

Residuals management is a critical aspect of wastewater treatment effectiveness and economics. Residuals may include both solids (as a slurry, cake, or ash) and gases. The following descriptions will focus on major solids handling components. Residual solids generally require thickening, stabilization, dewatering, and final disposal.

Thickening may be accomplished by gravity, flotation, or mechanically (belt thickening). Aerobic digester may serve as a thickener. It is generally cost-effective to employ thickening prior to separate digestion or dewatering.

Stabilization methods most commonly employed are lime stabilization, aerobic digestion, and anaerobic digestion. Stabilization serves to limit the potential for putrefaction and odors, while digestion also provides mass reduction. Composting is another stabilization method which could potentially be employed.

Sludge dewatering reduces sludge volume and enhances sludge handling given the relative ease of handling a semi-solid versus a largely liquid sludge. Sludge dewatering may be accomplished by sludge drying beds, lagoons, centrifuges, belt filters, filter presses, and vacuum filters. Drying beds and lagoons are land-intensive and climate-dependent.

In summary, aerobic biological treatment technology remains the predominant treatment technique, at least in the pharmaceutical manufacturing industry. The SBR process is an emerging aerobic treatment technique which warrants a close look when considering biological treatment. Anaerobic treatment for high-strength wastewater treatment offers several advantages, which can lead to cost advantages compared to aerobic treatment. Anaerobic lagoons, upflow anaerobic sludge blanket systems, and anaerobic fluidized beds have been used for pharmaceutical manufacturing wastewater treatment.

Technology Option for Direct Dischargers BPT

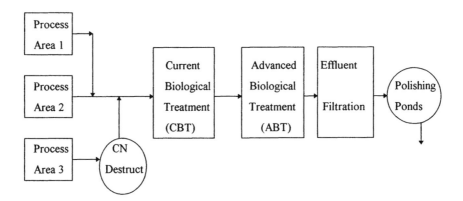

Technology Options for Direct Dischargers BAT, NSPS, PSES, PSNS

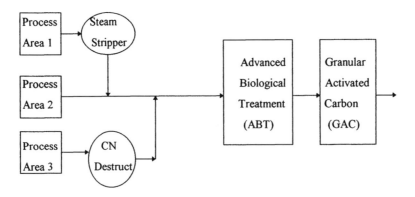

THE CWA AND ITS RELATIONS TO OTHER STATUTES

RCRA and CWA

Section 1006(a) of RCRA its "Administrator to integrate all provisions of this act for purposes of administration and enforcement and shall avoid duplicate, to the maximum extent practicable, with the appropriate provisions of the CAA, CWA, FIFRA, SDWA, Marine Protection Research and Sanctuaries."

Public Owned Wastewater Treatment Works (POTWs) can be affected by hazardous waste regulations under RCRA in a number if ways. According to the BNA, these areas include the following:

1. **Pretreatment**--Under CWA pretreatment regulations, industrial users of POTWs may be considered hazardous waste generators under RCRA, as well. In evaluating RCRA wastes, POTWs must determine whether dischargers must pretreat their wastes prior to discharging to the POTWs.

2. **Acceptance of hazardous waste for disposal**--POTWs that accept hazardous wastes by truck, rail, or separate pipe are considered to be hazardous waste treatment, storage, and disposal facilities (TSDFs) and are subject to some requirement of RCRA.

3. **Permit-by-rule**

4. **Hazardous waste/sludge generation**

5. **Non-hazardous sludge disposal**

The following flowchart depicts the decision-making process in determining whether the POTWs can be accepted into the wastestream:

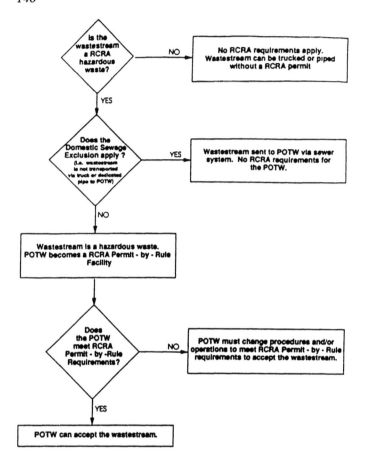

CERCLA and CWA

In CERCLA, "federally permitted release" is defined as discharges in compliances with a permit under Sections 402 and 404 of the Clean Water Act. Also, in CERCLA the National Contingency Plan (NCP) and the Reportable Quantity (RQ) are referred to that under Section 311 (c) and (b)(4) of the Clean Water Act, respectively. As a matter of fact, because CERCLA is very much concerned with the potential contamination of the water supplies from the nation's abandoned contaminated sites, its HRS (Hazardous Ranking System) seriously takes into account one of the four principle pathways as groundwater and surface water. CERCLA site employing waste water treatment technologies are required to meet BCT/BAT requirements as applicable and relevant and appropriate requirements (ARARs), because the EPA has established no effluent guidelines for CERCLA sites, technology-based effluent limitations are determined using best professional judgement. CWA and CERCLA authorizes the U.S. Coast Guard (USCG) and the EPA to respond to releases of pollutants into the nation's water. The CWA established the NCP to govern these responses, such as the massive response effort for the recent oil spill from the Exxon Valdez in Alaska. The CWA provided the EPA with its first response authority and was a precursor to CERCLA.

SDWA and CWA

The Clean Water Act has increased pressure to dispose of waste materials on or below land and the consequential increased threat of groundwater contamination. The act, in Section 402(b)(1)(D) requires states, as a precondition to approval of their NPDES programs, to "control the discharge of pollutants into wells." The EPA relies on the authority of RCRA and the Safe Drinking Water Act (SDWA) to regulate such discharges and is encouraging the states to develop underground injection control programs pursuant to 40 CFR Part 146. Section 208 of CWA is to protect the sole source aquifer. And because between 20 to 50% of the drinking water in the U.S. is derived from underground aquifer, SDWA and CWA are highly interrelated.

FIFRA and CWA

Section 202(c) of FIFRA requires procedures to be established to monitor activities which includes, among other media, water. This is especially true when rinsing of pesticide containers and its residue. Also agrichemical usage has potential effects on water discharge and runoff.

CAA and CWA

Sections 103(c)(2) and (e)(4) of the Clean Air Act specifies "Evaluation of the effects of air pollution on water quality, including assessments of the short-term and long-term ecological effects of acid deposition and other atmospherically derived pollutants on surface water and ground water." (Ref 42 USC 7403).

Air pollution may cause acid deposition and this deposition potentially could find its way to the water supplies, either surface water or ground water. This is where CAA and CWA are interrelated in the federal statutes.

EPCRA and CWA

Section 305 of the EPCRA mandates public warning regarding release of harmful substances to the atmospheres including surface and groundwater.

TSCA and CWA

Section 309 of TSCA mandates investigation of radon contamination of the water supplies, both surface and aquifer.

PENDING REAUTHORIZATION OF THE CWA

The Clean Water Act was well on schedule for a congressional reauthorization in 1994 until it suffered a sudden death near the end of the legislative year in October. At the time of its fatal defeat, the legislation S.1114 has been through subcommittee and full committee markup. H.R.3048 was still in subcommittee, and there was no plan for subcommittee hearings. Of the two bills, the Senate version was further along and appeared to be the more stringent.

The reauthorization is expected to come up in 1996 again when Congress convenes for a new term. The following is an overview of what could be expected as a result of the reauthorization.

1. It is expected to strengthen efforts to reduce toxic pollution.

2. It is expected to implement tougher enforcement provisions.

3. It is expected to increase authority of the EPA to influence product contents, production, processes, and quantities produced.

4. It is expected to provide an even more watchful eye on the industries.

The Clean Water Act Reauthorization consists of the following:

Title I--Water Program Funding

- It reauthorizes state revolving funds. A total of $2.5 billion/year through the year 2000. It establishes a new allotment formula.
- It expands fees collected from industrial dischargers. Both indirect and direct dischargers are required to pay annual fees. Fees are used by the states to cover at least 60% of the cost for water quality programs.
- EPA is required to develop fee program if states do not. The federal program will recover 100% of cost.
- Maximum fee will be $125,000.

Summary of impact: Significant increase in the cost of permitting because of the range of the programs to be funded.

Title II--Toxic Pollution Prevention and Control Effluent Guidelines and NSPS

- The reauthorization will increase regulation of conventional and non-conventional pollutants as well as toxics. Conventional pollutants are required to meet: Best Practicable Control Technology Available ("Currently" is deleted). As far as non-conventional pollutants are concerned, they are required to meet: BAT (Best Available Technology Economically Achievable).
- The reauthorization gives the EPA authority to:
  -- include in guidelines source reduction practices including changes in production processes.

-- prohibit cross-media transfer of pollutants.
-- require elimination of discharge.

Summary of impact: Standards will be increasingly tighter; zero discharge is the target; manufacturing will have even more regulatory interventions.

Toxic Pollutant Phaseout

- The reauthorization strengthens EPA authority where pollutants are toxic or bioaccumulative.
- The reauthorization requires EPA strategy to reduce loadings by 85% in seven years.
- It gives citizens the right to petition for effluent standards or prohibition of toxic pollutants.

Summary of impact: The number of toxic pollutant standards will increase; challenging of guidelines will be made more difficult.

Pretreatment

The reauthorization:

- Gives highest priority to source reduction.
- Eliminates discharge to POTW except where treatment will meet industrial direct discharge guidelines.
- Limits removal credits to biodegradable materials that do not harm sludge.

Summary of impact: Pretreatment standards will be as stringent as direct discharge guidelines. There will be a limit on removal credits.

Pollution Prevention Planning

The reauthorization:

- Adds broad multi-media pollution planning requirement. As a result, the facilities are subject to emergency planning. And the facilities are to report releases or transfers of 200,000 pounds each year.
- Requires progress report submission each year with the toxic chemical release form.
- Makes progress reports available to the public.

Summary of impact: Cost and time associated with planning and reporting.

Antidegradation Policy

- States must designate uses of waters within 3 years.
- States must develop antidegradation policy and implementation plan.
- Citizens are given the right to petition for classification of water as outstanding national resource.

Title III--Watershed Planning and Non-Point Source Control

- It provides the EPA with broader authority for comprehensive management of water quality.
- It specifies authority to require dischargers to monitor receiving waters.
- It requires the EPA to publish guidelines for non-point source programs including management measures and implementation criteria.

Title IV--Municipal Pollution Control

- It addresses key sources of urban water pollution such as combined sewer overflows and municipal storm water discharges.

Summary of impact: It may lead to more restrictive regulation of runoff from industrial facilities.

Title V--Permitting and Enforcement

- State is required to collect permit fees.
- Permit prior to construction is required.
- Permit applicants are required to characterize discharge.

Summary of impact: Cost of permitting process is increased; allows citizens actions to be brought for past violation within 5 years; judges may order fines used for environmental projects; minimum penalty is set to be equal to benefit derived by the violator; raised cap on administrative penalty to $200,000.

Title VI--Program Management

- It expands employee protection for "whistle blower."
- It authorizes programs to demonstrate new treatment technologies, practices, and processes.
- It requires the EPA to establish information programs.
- It extends point source definition to include landfill leachate collection systems.

Title VII--Wetlands

The reauthorization:

- Strengthens wetlands conservation programs.
- Sets a national policy of achieving no net loss of wetlands.

Title VIII--Other Major Provision

- Stormwater permits are required for both commercial and light industries.
- Assessment of phosphates in detergents.

KEY CWA TERMINOLOGIES AND CONCEPTS

Like any other federal statutes and public laws, the Clean Water Act is unique that it possesses many terminologies and concepts of its own. It is therefore fundamental to the understanding of the CWA to understand the CWA language. The following is a selected few of these terms.

Absorption--the surface penetration of a substance into or through another, such as takes place when nutrients and water enter plants.

Activated carbon--a highly absorbent form of carbon used to remove odors and toxic substances from water and wastewater.

Activated sludge loading--the pounds of biochemical oxygen demand (BOD) in the applied liquid per unit volume of aeration capacity or per pound of activated sludge per day.

Adsorption--the adhesion of a substance to the surface of a solid or liquid; often used to extract pollutants by causing them to be attached to such absorbents as silica gel.

Advanced waste treatment--any process of water renovation that upgrades water quality to meet specific reuse requirements; may include general cleanup of water or removal of specific parts of wastes insufficiently removed by conventional treatment processes.

Aeration--the circulation of oxygen through a substance by spraying a liquid in the air, bubbling air through a liquid or agitating a liquid to promote surface absorption of air. In wastewater treatment, it aids in biological and chemical purification.

Aerobic--life or processes that depend on the presence of oxygen.

Aerobic decomposition--decomposition and decay or organic material in the presence of "free" or dissolved oxygen.

Aerobic digestion--as it pertains to sewage sludge, it is the biochemical decomposition of organic matter is sewerage sludge to convert the waste into carbon dioxide and water by microorganisms in the presence of air.

Allowable headworks loading--the maximum pollutant that may be received at the headworks of a specific treatment works calculated to ensure the prevention of interference or pass through from that pollutant.

Annual pollutant loading rate (APLR)--the maximum amount of a pollutant that can be applied to unit area of land during a 365-day period. The rate is calculated by dividing the cumulative pollutant loading rate for an inorganic pollutant by 20 years.

Anti-backsliding--an EPA policy that prohibits the renewal or reissuance of NPDES permits containing interim effluent limitations less stringent than those imposed in the previous permit.

Applicable standards and limitations--all state, interstate, and federal standards and limitations to which a discharge or related activity is subject under the Clean Water Act, effluent limitations, water quality standards, standards of performance, toxic effluent standards or prohibitions, best management practices, and treatment standards under Sections 301, 302, 303, 304, 306, 307, 308, 403, and 405 of the Water Act.

Ash transport water--water used in the hydraulic transport of either fly ash or bottom ash.

Assimilation--the process by which food is converted to cell protoplasm.

Autotrophic--an organism that produces food from inorganic substances.

Bailing--compacting solid wastes into block to reduce volume.

Ballistic separator--a machine that sorts organic matter from inorganic matter for composting.

BAT or BATEA (Best Available Technology Economically Achievable)--Subject to economic and engineering feasibility limitations, BAT should incorporate the top-of-line current technology, with a capacity up to and including no discharge of pollutants. Considerations include the age of the equipment and facilities involved; the process used; the engineering aspects of applying various types of control techniques; process changes; the cost of achieving the effluent reduction from applying the technology; and non-water quality environmental aspects such as energy use.

BCT (Best Conventional Pollutant Control Technology)--measures and practices for point sources of conventional pollutant, determined with consideration of the reasonableness of attainment costs versus effluent reduction benefits, the age of equipment and facilities involved, and energy impacts.

BMPs (Best Management Practice)--schedules of activities, prohibitions of practices, maintenance procedures, and other management to prevent or reduce water pollution. The term also includes treatment requirements, operating procedures, and practices to control plant site runoff, spillage or leaks, sludge or water disposal, or drainage from raw material storage.

BPT or BPTTA (Best Practicable Technology)--technology based on the average of the best existing performance levels achieved by exemplary plants of various sizes, ages, and unit processes with an industry.

Bioaccumulation--the process by which a compound is taken up by an aquatic organism, both from water and through food.

Bioassay--estimating the toxicity of an effluent by testing its effects on living organisms.

Bioflocculation--a condition whereby organic material tends to be transferred from the dispersed form in wastewater to settleable material by mechanical entrapment and assimilation.

Biological oxidation--the way that bacteria and microorganisms decompose complex organic materials; used in self-purification of waterbodies and activated sludge wastewater treatment.

Biological wastewater treatment--form of wastewater treatment in which bacterial or biochemical action is intensified to stabilize, oxidize, and nitrify the unstable organic matter present. Intermittent sand filters, contact beds, trickling filters, and activated sludge processes are examples.

Blackwater--wastewater containing human excreta.

BOD (Biochemical Oxygen Demand)--the dissolved oxygen required to decompose organic matter in water. It is a measure of pollution because heavy waste loads have a high demand for oxygen.

BOD$_5$--the amount of dissolved oxygen consumed in 5 days by biological processes breaking down organic matter in an effluent.

Capitalization--the actual federal funds received by a state environmental protection agency for deposit in its state revolving fund.

Carbonaceous oxidation--biochemical process by which heterotrophic microorganisms derive energy from organic wastes. This process renders more stable organics or organics as end products.

Categorical industrial user--an industrial facility subject to regulation by a national categorical pretreatment standard established by the EPA.

Categorical standards--pollutant discharge standards that apply to users in specific industrial categories determined to be the most significant sources of toxic pollutants discharged to the nation's treatment works. These standards are based on the best technology available to treat the pollutants of concern with resulting from the regulated processes. Categorical pretreatment standards are published for each industrial category regulated under the NPDES permitting program.

Chemical Abstracts Services--a registry of over 10 million different chemical substances.

COD (Chemical Oxygen Demand)--a measure of the oxygen required to oxidize organic and oxidizable inorganic compounds in water.

Chlorination--the application of chlorine to drinking water, sewage, or industrial waste to disinfect or to oxidize undesirable compounds.

Chlorine contact chamber--a detention basic where chlorine is diffused through liquid.

Chlorine residual--the total amount of chlorine (combined and free available) remaining in wastewater at the end of a specified contact period following chlorination.

Chronic value (ChV)--the geometric mean of the lowest observed effect level and the no-observed effect level.

Clarification--clearing action that occurs during wastewater treatment when solids settle out, often aided by centrifugal action and chemically induced coagulation.

Clarifier--a settling tank where solids are mechanically removed from wastewater.

Code event--the term used by the EPA's computerized Permit Compliance System to mean a required action with a corresponding deadline.

Coliform organism--organisms found in the intestinal tract of humans and animals; their presence in water indicates pollution and potentially dangerous bacterial contamination.

Combined sewers--a system that carries both sewage and stormwater runoff. In dry weather, all flow goes through to the waste treatment plant. During a storm, only part of the flow is intercepted due to overloading. The remaining mixture of sewage and stormwater overflows untreated into the receiving stream.

Communitor--a device used to catch or shred heavy solid matter in the primary stage of waste treatment.

Compliance evaluation inspection--one of two types of EPA field inspections. Unlike a Compliance Sampling Inspection, no sample is taken. Instead, an inspector performs an overall assessment to determine compliance with the Clean Water Act and NPDES permit requirements.

Compliance inspection report--a periodic summary prepared by the EPA inspector of the status of NPDES permittee compliance. Field inspection on which the report is based are conducted by EPA personnel to detect violations and provide evidence to support enforcement actions.

Compliance sampling inspection--one of two types of EPA field inspections. Inspectors collect and analyze a sample.

Conservative pollutants--pollutants that are not biodegraded or volatilized at a wastewater treatment works.

Contact aerator--a biological unit consisting of stone, cement-asbestos, and other surfaces supported in an aeration tank, in which air is diffused up and around the surface and settled wastewater flows through the tank.

Contact stabilization process--a modification of the activated sludge process in which wastewater is aerated with a high concentration of activated sludge for a short period, usually less than 60 minutes, to obtain BOD removal. The solids are subsequently separated by sedimentation and transferred to a stabilization tank where aeration is continued, thus starving the activated sludge before returning it to the aeration basin.

Conventional pollutants--substances such as BOD, total suspended solids, fecal coliform, and pH, and other pollutants as designated by the EPA.

Co-permittee--a permittee to a NPDES permit who is only responsible for permit conditions relating to its portion of the discharge.

Detention time--the time required to fill a tank at a given flow or the theoretical time required for a given flow of wastewater to pass through a tank.

Digester--in a wastewater treatment, a closed tank where sludge is subjected to intensified bacterial action.

Digestion--process by which the components of sludge are decomposed resulting in partial gasification, liquefaction, and mineralization of pollutants.

Dilution ratio--the relationship between the volume of water in a stream and the volume of incoming waste. It can affect the ability of the stream to assimilate waste.

Direct discharge--any discernible, confined and discrete conveyance, including, but not limited to: any pipe, ditch, channel, conduit, well, discrete fissure, container, rolling stock, concentrated animal feeding operation or vessel, or other floating craft from which pollutants are or may be discharged.

Discharge monitoring report (DMR)--information that permittees must submit, at least quarterly, on their self-monitoring program to the respective NPDES permitting authority.

Dispersant--a chemical agent used to break up concentrations of organic material such as spilled oil.

Distillation--process of heating the waste effluent, removing the vapor or steam, and reconstituting the vapor/steam to nearly pure water; the pollutants remain in the concentrated residue.

Diurnal flow--flows that show marked and regular variations through the course of a day.

Effluent limitation--any restriction established by a state or EPA on quantities, discharge rates, and concentrations of pollutants or other constituents that are discharged from point sources into navigable waters, waters of the contiguous zone, or the ocean.

Enforceable requirement of the Clean Water Act--conditions of limitations of Section 402 or 404 permits that, if violated, could result in the issuance of a compliance order or initiation of a civil or criminal action under Section 309 of the act.

Environmental impact statement (EIS)--a document required of federal agencies by the National Environmental Policy Act for major projects or legislative proposals. EISs are used to determine the positive and negative effects of a proposed project.

Estuaries--areas where fresh water meets salt water (bays, mouths of rivers, salt marshes, lagoons). These brackish waster ecosystems shelter and feed marine life, birds, and wildlife.

Eutrophication--the slow aging process of a lake evolving into a marsh and eventually disappearing. During eutrophication, the lake is choked by abundant plant life. Human activities that add nutrients to a water body can speed up this action.

Excursion provisions--exemptions for unintentional violations in NPDES permits.

Existing source or existing discharger--in the NPDES program this means any source that is not a new source or new discharger.

Facultative--pond microorganism (bacteria) that can live in aerobic or anaerobic conditions.

Fecal coliform bacteria--a beneficial group of organisms found in the intestines of warm-blooded animals. If found in water, it is likely that other organisms may be present which are capable of creating dangerous bacterial contamination and subsequent health problems.

Flocculation--the process by which lumps of solids in sewage are increased in size by chemical, physical, or biological action.

Graywater--galley, bath, and shower water discharge from vessels.

Grit--the heavy mineral material present in wastewater, such as sand, egg shells, gravel, and cinders.

Groundwater runoff--groundwater that is discharged into a stream channel as spring or seepage water.

Headworks--the first part of the treatment plant, usually intake valves, flowmeters, grit chambers, flow equalization, bar screens, and communitors.

Headworks analysis--monitoring performed on the flow of wastewater into a treatment plant.

Heavy metals--metallic elements, such as mercury, chromium, cadmium, arsenic, and lead that have high molecular weights. They can damage living things at low concentrations, and tend to accumulate in the food chain.

Illicit discharge--any discharge to a municipal separate storm sewer that is not composed entirely of stormwater; exceptions include discharges made pursuant to an NPDES permit and discharges from firefighting activities.

Indirect discharger--a non-municipal, non-domestic, discharger introducing pollutants to a POTW. Such an introduction does not constitute a "discharge of pollutants."

Influent--wastewater or other raw or partially treated liquid flowing into a reservoir, basin, treatment process, or treatment plant.

Inorganic waste--waste material, such as sand, salt, iron, calcium, and other mineral materials that are not converted in large quantities by organism action. Inorganic wastes are chemical substances of mineral origin and may contain carbon and oxygen, whereas organic wastes are chemical substances of animal or vegetable origin and contain mainly carbon and hydrogen along with other elements.

Local limits--national pollutant discharge limits developed for and enforced by the POTW. Local limits address specific pollutants of concern to the POTW. They are designed to ensure compliance with the discharge standards.

Maximum allowable toxicant concentration--the concentration of toxic waste that may be present in the receiving water without causing significant harm to its productivity and uses; usually determined by long-term bioassay.

MCRT (Mean Cell Residence Time)--as it pertains to the testing of sewage sludge, is the mass of cells in the digester divided by mass of cells removed per day. The resulting number in days is related to the time an average cell spreads in the digester.

Megarad--as it pertains to the testing and treatment of sewage sludge, is a measure of the energy dose received per unit mass of the material being irradiated. One megarad is equivalent to 10 joules per grain.

National pretreatment standard--any regulation containing pollutant discharge limits promulgated by the EPA in accordance with Section 307 of the CWA, which applies to industrial users.

New discharger--any building, structure, facility, or installation that has never received a final NPDES permit for discharging pollutants at a particular site before August 13, 1979; and from which there is or may be a discharge of pollutants.

New source--any building, structure, facility or installation from which there is or may be a discharge of pollutants, and the construction of which began after the promulgation performance standards under Section 306 of the CWA, which are applicable to such source; or after the proposal of performance standards which are applicable to such source, but only if the standards are promulgated in accordance with Section 306 with 1120 days of their proposal.

Nitrification--the conversion of nitrogenous matter into nitrates by bacteria.

Non-point source (NPS)--a source of pollution which are not a point source, such as form or forest land runoff, urban stormwater runoff, mine runoff, or saltwater intrusion.

Notice of violation--a written notice from a permitting authority to an NPDES permittee that specifies the exact nature of a violation.

Overburden--material of any nature, consolidated or unconsolidated, that overlies a mineral deposit, excluding topsoil or similar naturally occurring surface materials, which is not disturbed by mining operations.

Oxidation pond--a man-made lake or body of water in which wastes are consumed by bacteria. It is used most frequently with other waste treatment processes and is basically the same as a sewage lagoon.

Permit compliance system--or PCS which is a computerized database used in EPA regional enforcement offices to monitor the NPDES. Stage permit issuers also must comply with PCS procedures.

Persistent pollutant--pollutant that is not subject to decay, degradation, transformation, volatilization, hydrolysis, or photolysis.

Pre-existing discharge--with regard to coal mining, any discharge in existence before February 4, 1987.

Pretreatment standards for existing sources--categorical standards and requirements applicable to industrial sources that began construction prior to the publication of the proposed pretreatment standards for that industrial category.

Primary treatment--the first stage of wastewater treatment; removal of floating debris and solids by screening and sedimentation.

Priority pollutants--those pollutants listed under Section 307(a) of the Clean Water Act.

Privately OTW--any device or system that is used to treat wastes from any facility of a treatment work.

Publicly OTW--a state-owned or municipally-owned device or system that is used in the treatment of municipal sewage or liquid industrial wastes.

Putrefaction--biological decomposition of organic matter with the production of ill-smelling products associated with anaerobic conditions.

Receiving water--any body of water to which untreated or treated wastes are discharged.

Saprophytic organisms--organisms living on dead or decaying organic matter. They help natural decomposition of the organic solids in wastewater.

Secondary treatment--a wastewater treatment process used to convert dissolved or suspended materials into a form more readily separated from the water being treated.

Shock load--the arrival at the plant of a waste in sufficient quantity or strength to cause operating problems, and which is toxic to organisms.

Specific oxygen uptake rate (SOUR)--the mass of oxygen consumed per unit time per unit mass of total solids and sewage sludge.

Stabilization--to convert the active organic matter in sludge into inert harmless materials.

State revolving fund--a low-interest state loan program that provides loans to local authorities for the purposes of constructing wastewater treatment works.

STORET--a database that includes water-related environmental data for all 50 states.

Tertiary treatment--advanced cleaning of wastewater that goes beyond the secondary or biological stage. It removes nutrients such as phosphorus and nitrogen and most suspended solids.

Toxic pollutants--any pollutant listed as toxic under Section 307 of the CWA.

TRI (Toxic Release Inventory)--system established under Section 313 of the 1986 Emergency Planning and Community Right-to-Know Act. TRI is a database containing information reported by industrial and commercial facilities of the amount of toxic pollutants they discharge into the environment, including discharges to POTWs.

Trace metals--metals with an amount of a chemical constituent not quantitatively determined because of minuteness.

Trickling filter--a device for the biological or secondary treatment of wastewater that consults a bed of rocks or stones to support bacterial growth. Sewage is trickled over the bed enabling the bacteria to break down organic wastes.

Upset--an exceptional incident in which there is unintentional and temporary noncompliance with technology-based permit effluent limitations because of factors beyond the reasonable control of the permittee.

Vector attraction--is the characteristics of sewage sludge that attract rodents, flies, mosquitos, and other organisms capable of transporting infectious agents.

Volatile solids--the amount of total solids in sludge lost when it is combusted at 550°C in the presence of excess air.

Waste stabilization pond--any pond, natural or artificial, receiving raw or partially treated sewage or waste, in which stabilization occurs due to sunlight, air, and microorganisms.

Wet electrostatic percipitator--an air pollution control device that uses both electrical forces and water to remove pollutants in the exit gas from a sewage sludge incinerator stack.

Wet scrubber--an air pollution control device that uses water to remove pollutants in the exit gas from a sewage sludge incinerator stack.

Whole effluent toxicity (WET)--the aggregated toxic effect of an effluent measured directly by toxicity test.

Zone of engineering control--an area under control of the owner/operator that, upon detection of hazardous waste release, can be cleared up before the groundwater or surface water is contaminated.

SUGGESTED READING

1. *Environmental Law Handbook*, twelfth ed, Government Institute Inc., p.152.
2. *Environmental Law Handbook*, twelfth ed, Government Institute Inc., p.153.
3. *Water Pollution Protection*, The Bureau of National Affairs, Inc., pp.901:111-113; p.141.
4. 33 U.S.C. 1251 section 101.
5. 40 C.F.R. 103.
6. 40 C.F.R. 149.
7. 40 C.F.R. 122.4, 122.45.
8. 40 C.F.R. 122.44 (I).
9. 40 C.F.R. 122.41 (k)(6), 122.42(a).
10. 40 C.F.R. 122.29(b).
11. 33 U.S.C. 1251, 304(b).
12. 40 C.F.R. 403.5 as amended, 55 Fed. Reg. 30082 (July 24, 1990).
13. Chicago Assoc. of Commerce and Industry v. EPA, 873 F.2d 1025 (7th Cir. 1989); Chemical Manufacturers Association v. EPA, 870 F.2d 177, 257-61 (5th Cir. 1989); Armco, Inc. v. EPA, 869 f.2d 975 (6th Cir. 1989).
14. 40 C.F.R. 403.7, 49 Fed. Reg. 31221, August 3, 1984.
15. Natural Resources Defense Council c. EPA, 790 F.2d 289 (3d Cir. 1986), cert. denied, 107 S.O 1285 (1987).
16. 40 C.F.R. Part 133.
17. 33 U.S.C. 1251 section 311(j)(5)(A).
18. 33 U.S.C. 1251 section 311(j)(5)(C).
19. 40 C.F.R. Part 116.
20. 40 C.F.R. 117.12.
21. 33 U.S.C. 1251 section 402(p)(2).
22. 55 Fed. Reg. 47990 (Nov, 16, 1990).
23. 40 C.F.R. 122.26(b)(14).
24. J. Kevin Farmer, Pharmaceutical Manufacturers Association *"Update on Current Wastewater Treatment Technologies."*
25. Pollution Engineering - *Exploring Wastewater Treatment Option*, May 94.

26. *Physicochemical Process for Water Quality Control*, Weber 1972.

27. 1994 PhRMA Engineering and Environmental Joint Spring Symposium, R.R. Bobil, "*Pharmaceutical Wastewater Pretreatment Standards and Effluent Guidelines: an Overview.*"

28. Warner Lambert Environmental Workshop, K. Tracy, "*Cleanwater Act Authorization.*"

29. The Bureau of National Affairs Inc. 1993, pp.981:201, "*Glossary of terms.*"

30. Hazmat World - *Field Applications Demonstrate Capabilities of Upflow Biological Reactors*, May 1994.

5 THE SAFE DRINKING WATER ACT

INTRODUCTION AND OVERVIEW

The Safe Drinking Water Act (SDWA) is the primary federal law with the purpose of ensuring that the public is provided with safe drinking water. In the United States, regulations governing drinking water quality date back to 1893 when the United States Congress enacted the Interstate Quarantine Act, which was developed to stop the spread of communicable diseases across national, state, and territorial boundaries. The surgeon general of the United States Public Health Service (USPHS) was empowered to promulgate and enforce these regulations. The first water related regulation, adopted in 1912, prohibited the use of the common cup on carriers of interstate commerce, such as trains.

The first federal safe drinking water standards (SDWs) were adopted in 1914. The USPHS, then a branch of the United States Treasury Department, was charged with the task of administering a health care program for sailors in the Merchant Marine. The surgeon general recommended, the U.S. Treasury Department adopted, standards that applied to water supplied to the public by interstate carriers. These standards were referred to as the "Treasury Standards" which included limits on total bacterial plate count, and B. coli (now called *Escherichia coli*). Standards for specific chemical and physical requirements were not drafted in the 1914 SDWs because the commission was unable to agree on the qualifications of the standards, therefore the 1914 SDWs only considered bacteriological quality of water.

These 1914 standards only legally applied to water supplies used by interstate carriers, but many state and local governments used them as guidelines, and because so, they were subsequently revised in 1925, 1942, 1946, and 1962.

The 1962 standards covered 28 constituents and were the most detailed pre-SDWA federal drinking water standards. These standards set limits for health related chemical and biological impurities and recommended limits for aesthetics properties such as appearance, taste, and odor. All 50 states accepted these standards, with minor modifications, either as regulations or guidelines. These regulations were federally legally binding by only 2 percent of the nation's water supply systems which consists of approximately 700 water systems that supplied common carriers in interstate commerce. This limited the use of these standards as enforcement tools for ensuring clean drinking water for the vast majority of consumers.

In 1969, initial action was taken by the USPHS to review and revise the standards of 1962. The USPHS's Bureau of Water Hygiene undertook a thorough study of water supplies in the United States known as the Community Water Supply Study (CWSS), whose objective was to determine whether the U.S. consumer's drinking water met the 1962 standards. The study included a total of 969 public water systems, serving a population of about 18.2 million people which was 12 percent of the total population served by public water systems.

The study was released in 1970 and found that 41 percent of the systems surveyed did not meet the guidelines of the 1962 standards. Many state programs were severely under-staffed and under-funded, therefore public water systems failed to receive badly needed surveillance and technical assistance which subsequently caused many deficiencies in their water systems. In some cases these deficiencies included: inadequate treatment and distribution systems resulting in poor water quality; insufficient monitoring for physical, chemical, and biological quality (in some states monitoring was not required); poor bacteriological quality (in a large number of state's water from more than one third of the water systems consistently failed the 1962 USPHS drinking water standards for microbiological quality); inadequate operation and maintenance of facilities; inadequate state surveillance programs; insufficient state resources to adequately address the problems; inadequate source protection, disinfection, clarification, system pressure, and combinations of these. Although the water served to the majority of the population was safe, the study indicated that several million people were being supplied an unsatisfactory quality of water and that 360,000 were being supplied a potentially dangerous supply of drinking water.

The results of the CWSS generated congressional interest in federal safe drinking water legislation. The first series of bills to give federal government power to set enforceable standards for drinking water were introduced in 1970. Congressional hearings on legislative proposals concerning drinking water were held in 1971 and 1972.

In 1972, a report of an investigation into the quality of water in the Mississippi River in Louisiana was published which focused on the amount of organic compounds in the effluent of the Carrollton water treatment plant in New Orleans. As a result of this report, new legislative proposals for safe drinking water laws were introduced and debated in Congress in 1973. In late 1973, the General Accounting Office (GAO) released a report investigating 446 community water systems in the state of Maryland, Massachusetts, Oregon, Vermont, Washington, and West Virginia. Only 60 systems were found to fully comply with the bacteriological and sampling requirements of the USPHS standards. Bacteriological and chemical monitoring programs of community water supplies were inadequate in five of the six states studies which lead to the conclusion that many water treatment plants needed to be expanded, replaced, or repaired.

Public awareness of organic compounds in drinking water increased in 1974 from a series of events which initiated, after more than four years of effort by the Congress, a passage of federal legislation to develop a national program to protect the quality of the nation's public drinking water systems. One of the events that led to the increased public awareness was a three-part series in *Consumer Reports* which drew attention to organic contaminants in New Orleans drinking water. Part one dealt with questions concerning "Is the Water Safe to Drink?," part two focused on "How to Make It Safe," and part three focused on "What You Can Do." As a result of this series, in July of 1974, representatives of the State of Louisiana and the City of New Orleans asked the United States Environmental Protection Agency (USEPA) to determine the identities and the quantitative concentrations of trace organic compounds that might be present in the finished water of New Orleans and the surrounding communities. The study compared death rates from communities using the lower Mississippi River as a drinking water source with those from nearby communities using groundwater sources for drinking water. The report was released on Nov. 7, 1974, and stated that people drinking treated Mississippi River water were more likely to get cancer than those who drank from neighboring sources. Further publicity and public awareness occurred on Dec. 5, 1974, when CBS aired a nationally prime time program with Dan Rather titled "Caution, Drinking Water May Be Dangerous to Your Health."

Additionally, in 1974, researchers at the USEPA and in the Netherlands discovered a class of compounds that were formed as a by-product of free chlorine in disinfection known as Trihalomethanes (THMs). On November 8, 1977, a USEPA press conference was held concerning the New Orleans study and the subsequent finding of synthetic organic compounds (SOCs) in the drinking water. As well as addressing the SOC problem, the USEPA announced that a national survey would be conducted to determine the extent of the THM problem in the United States. This survey was known as the National Organics Reconnaissance Survey, or NORS, and was completed and the findings published in November 1975.

In response to public awareness of problems concerning drinking water in the United States, especially the concern of low levels of organic compounds in drinking water systems, Congress radically altered the federal role in drinking water by passing the Public Law 93-523, the SDWA, which was signed into law by President Ford on Dec. 16, 1974. This act set up a cooperative program among local, state, and federal agencies which mandated the establishment of drinking water regulations and was designed to ensure that water that comes from the tap in the United States is fit to drink. These regulations were the first to apply to all public water systems in the United States, covering both micro-biological and chemical constituents. The SDWA requires the EPA to set national drinking water standards that must be met by the persons who deliver the water to the tap. In this regard, SDWA is somewhat different than the other major environmental statutes in that it regulates the person who supplies a product rather than the person who generates pollution.

The SDWA enacted major changes in the broad spectrum of the phases and components of drinking water systems by establishing specific roles for state and federal governments and for public water supplies. The federal government, specifically the USEPA, is authorized to set national drinking water regulations, conduct special studies and research, and oversee the implementation of the act. The state governments through their health departments and environmental agencies are expected to accept the major responsibility, called primary enforcement responsibility or primacy, for the implementation and enforcement of the act's provisions. Public water suppliers are assigned the day to day responsibility of meeting the regulation by: establishing monitoring programs with results sent directly to the regulatory agency, maintaining public awareness of violations, and correcting violations as they occur.

KEY PROVISIONS

The key provisions of the SDWA include the following:

- Definition of a public water supply.
- Process of establishing national drinking water standards.
- Procedures for obtaining variances and exemptions from the regulations.
- Provisions for public notice.
- Provisions for enforcement.
- Establishment of a program to protect underground sources of drinking water.

Many of these provisions have been substantially altered by the 1986 amendments to the act.

The SDWA applies to all "public water systems," which do not have to be publicly owned. The act defines a "public water system" for the provision of piped water for human consumption, "so long as it has" at least fifteen service connections or regularly serve at least 25 individuals.

"Water for human consumption" has been held to include not only water that is drunk but also water that is used for bathing, showering, cooking, dishwashing, and maintaining oral hygiene.

In its initial regulations implementing the SDWA in 1975, the EPA interpreted "regularly serves" as a system that "regularly serves an average of at least twenty-five individuals daily at least 60 days out of the year." This clarification was necessary to establish a broad definition for a public water system because without it, it would be possible for people to escape regulation by simply showing that there are days when they do not serve 25 individuals.

In 1975, the EPA also divided public water systems into two types: community and non-community. There are approximately 60,000 community public water systems which are those that "regularly serve at least 25 year-round residents." Non-community water systems are the remaining which follow the definition of a public water system. The division was made so that drinking water standards that are designed to protect against the adverse effects of long term or chronic exposure would apply only to those systems, serving residential populations, whose users would drink the water over long periods of time.

The act specified the process by which USEPA was to adopt national drinking water regulations. Interim regulation (National Interim Primary Drinking Water Regulations of IPRs later referred to as (NIPDWRs)) were to be adopted within six months of its enactment. Within approximately two and a half years (by March 1977), USEPA was to propose revised regulations (Revised National Drinking Water Regulations) based on a study of health effects of contaminants in drinking water conducted by the National Academy of Sciences (NAS).

Establishment of the revised regulations was to be a two-step process. First, the agency was to publish recommended maximum contaminant levels (RMCLs) for contaminants believed to have an adverse health effect based on the NAS study. These RMCLs were to be set at levels such that no known or anticipated health effects would occur including an adequate safety margin. These levels were to act only as health goals and were not intended to be federally enforceable.

Second, USEPA was to establish maximum contaminant levels (MCLs) as close to the RMCLs as the agency thought feasible. The agency was also authorized to establish a required treatment technique instead of an MCL if it was determined that a contaminant could not be accurately measured or if it was not economically or technologically feasible to determine the level of contamination. The MCLs and treatment techniques comprise the National Primary Drinking Water Regulations (NPDWRs) and are federally enforceable and are to be reviewed at least every three years.

Interim regulations were adopted December 24, 1975, and became effective on June 24, 1977. These rules were based on the 1962 USPHS standards with very little additional support. The findings of the National Organics Reconnaissance Survey (NORS) were published in November 1975. The four THMS--chloroform, bromodichloromethane, dibromochloromethane, and bromoform--were found in abundant quantities in chlorinated drinking waters in the 80 cities studied throughout the United States. This prompted USEPA to conduct the National Organics Monitoring Survey (NOMS) between 1976 and 1977 to determine the frequency of specified organic compounds in drinking water supplies. NOMS included 113 community water supplies representing different sources and treatment processes. Each of these sources was monitored three times during a 12-month period producing data which showed that THMs were the most widespread organic contaminants in drinking water found in large concentrations. NORS, NOMS, and other surveys identified more than 700 specific organic chemicals in various drinking water supplies.

On June 21, 1976, the Environmental Defense Fund petitioned the USEPA, alleging that the initial interim regulations set in 1975 did not sufficiently control organic compounds in drinking water. In response, the USEPA issued an Advance Notice of Proposed Rule Making (ANPRM) on July 14, 1976, requesting public input on how THMs and SOCs should be regulated.

On February 9, 1978, the USEPA proposed a two-part regulation for the control of organic contaminants in drinking water. The first part concerned the control of THMs. The second part concerned control of source water SOCs and propose the use of granular activated carbon (GAC) adsorption by water utilities vulnerable to possible SOC contamination.

The next day, February 10, 1978, the U.S. Court of Appeals, District of Columbia Circuit, issued a ruling in the Environmental Defense Fund case filed in June 21, 1976. The court upheld USEPA's discretion to not include comprehensive regulations for SOCs in the NIPDWRs, but as a result of new data being collected by USEPA, the court told the agency to report a plan for amending the interim regulations to control organic contaminants. The court stated: "In light of the clear language for the legislative history the incomplete state of our knowledge regarding the health effects of certain contaminants and the imperfect nature of the available measurement and treatment techniques cannot serve as justification for delay in controlling contaminants that may be harmful." The agency contested that the proposed rule published the day before satisfied the court's judgement.

Reaction to the proposed regulation on GAC adsorption treatment varied. Federal health agencies, environmental groups, and a few water utilities supported the proposed rule. Many state health agencies, consulting engineers, and most water utilities opposed it. USEPA responded to early opposition of the GAC proposal by publishing an additional statement in the July 6, 1978, *Federal Register*. Nevertheless, significant opposition continued based on several technical considerations. USEPA promulgated regulations for the control of THMs in drinking water on Nov. 29, 1979, but subsequently on March 19, 1981 withdrew its proposal to control organic contaminants by GAC.

Congress severely underestimated the time required for USEPA to develop credible regulations. USEPA's unresponsive action in regulating contaminants and its failure to require GAC treatment for organic contaminants served as a focal point for future discussion of possible revisions to the law. Minor amendments to the 1974 SDWA were made in 1977, 1979, and 1980. These amendments provided for funding reauthorization and made a number of minor changes to the SDWA.

Reports in the early 1980s of drinking water contamination by organics and other chemicals and pathogens such as Giardia lamblia rose a congressional concern over the adequacy of the SDWA. Extreme criticism by the Congress about the rate of progress of regulating contaminants prompted new legislation to amend the SDWA. The SDWA was strengthened, especially the regulation setting process and ground water protection, by Congress amending most of the 1974 SDWA in 1986. These amendments, known as the SDWA amendments of 1986, were signed into law (Public Law 99-339) June 19, 1986. The 1986 amendments significantly altered the rate at which USEPA was to set drinking water standards and clarified and added several other provisions. Since the time of the 1986 amendments, a significant increase in the number of contaminants regulated by USEPA has occurred.

The SDWA was most recently amended in 1988 when Lead Contamination Control Act (Public Law 100-572) was signed into law. This law, along with other things, stated a program to eliminate lead-containing drinking water coolers in schools by adding to the SDWA Part F-Additional Requirements to Regulate the Safety Drinking Water.

The provisions of the SDWA are shown in Table 1 and are described as follows:

PART A--is the definitions section of the act and consists of the legal meaning of key terms. Some of the key terms defined in the act are; Primary Drinking Water Regulation, Secondary Drinking Water

TABLE 1	
OUTLINE OF THE SDWA	

Section	Title

Part A -- Definitions

1401	Definitions

Part B -- Public Water Systems

1411	Coverage
1412	National Drinking Water Regulations
1413	State Primary Enforcement Responsibility
1414	Enforcement of Drinking Water Regulations
1415	Variances
1416	Exemptions
1417	Prohibition on Use of Lead Pipes, Solder, and Flux

Part C -- Protection of Underground Sources of Drinking Water

1421	Regulations for State Program
1422	State Primary Enforcement Responsibility--Underground Injection Control (UIC) Program
1423	Enforcement of Program (UIC Program)
1424	Interim Regulation of Underground Injections
1425	Optional Demonstration by States Relating to Oil or Natural Gas
1426	Regulation of State Program (UIC Program)
1427	Sole Source Aquifer Demonstration Program
1428	State Programs to Establish Wellhead Protection Areas

Part D -- Emergency Power

1431	Emergency Powers
1432	Tampering With Public Water Systems

Part E -- General Provisions

1441	Assurance of Availability of Adequate Supplies of Chemicals Necessary for Treatment of Water
1442	Research, Technical Assistance, Information, and Training of Personnel
1443	Grants for State Programs

TABLE 1 (Continued)		
OUTLINE OF THE SDWA		
Section	**Title**	

Part E -- General Provisions (continued)

1444	Special Study and Demonstration Project Grants; Guaranteed Loans	
1445	Records and Inspections	
1446	National Drinking Water Advisory Council (NDWAC)	
1447	Federal Agencies	
1448	Judicial Review	
1449	Citizens Civil Action	
1450	General Provisions	
1451	Indian Tribes	

Part F -- Additional Requirements to Regulate
the Safety of Drinking Water

1461	Definitions	
1462	Recall of Drinking Water Coolers With Lead-Lined Tanks	
1463	Drinking Water Coolers Containing Lead	
1464	Lead Contamination in School Drinking Water	
1465	Federal Assistance for State Programs Regarding Lead Contamination in School Drinking Water	

Regulation, Public Water System, Maximum Contaminant Level, Municipality, Person, Administrator, and Supplier of Water.

PART B--deals with Public Water Systems. As defined in the act Section 1411 it covers all publicly or privately owned water systems having 15 or more service connections and/or serving at least 25 people. Under the SDWA definitions, USEPA has established by regulation two types of public owned water systems: community and non-community.

A community water system (CWS) has at least 15 service connections used by year-round residents or it serves at least 25 year-round residents. Municipalities and rural water districts are typical examples. A non-community water system does not have year-round residents, but it has at least 15 service connections used by travelers or intermittent users for at least 60 days each year, or it can serve an average of 25 individuals for at least 60 days. Typical non-community systems are restaurants, campgrounds, factories, schools, and motels with their own water supplies.

Systems not covered are those that obtain water from a regulated public water system, have only distribution and storage facilities, do not resell the water, and are not an interstate carrier. Private water supplies (individual homes) are also not covered by the act.

Section 1412, National Drinking Water Regulations, empowers USEPA to establish drinking water regulations. Under this section the national interim and revised primary drinking water regulations promulgated prior to 1986 were redefined as National Primary Drinking Water Regulations (NPDWRs) and recommended MCLs published prior to 1986 be redefined as maximum contaminant level goal (MCLGs). These MCLs and MCLGs, for a given contaminant, must be proposed and promulgated simultaneously.

Congress specifically required USEPA to set NPWDRs for 83 contaminants listed in the Advance Notice to Proceed Rulemakings published March 4, 1982, and October 5, 1983, as shown in Table 2. USEPA must set regulations for these contaminants, including the seven substitutes shown in Table 2. Most aspects of the SDWA were given specific timelines in which regulatory development was to be completed. Most actions in the timeline did not meet the specified deadlines.

The administrator is given broad authority by Congress to publish MCLGs and NPDWRs for each contaminant, that by the administrator's judgement may have an adverse health impact on people associated with a particular water system. The adverse health impact of a contaminant need not be proved in order to establish a regulation as in the 1974 amendment. A complete list of all drinking water standards is given in Table 4.

A drinking water priority list (DWPL) was also developed under this section which is a list of contaminants, developed by the USEPA, known or anticipated to occur in public water systems that may require regulation under the act. The USEPA is to form an advisory group that includes members from the: National Toxicology Program, the USEPA offices of Ground Water and Drinking Water, Toxic Substances, Solid Waste, and Emergency Response; and other USEPA offices as are appropriate. Some important priorities include hazardous substances

---

TABLE 2

CONTAMINANTS REQUIRED TO BE REGULATED
UNDER THE 1986 SDWA AMENDMENTS

Volatile Organic Chemicals
  Benzene
  Carbon tetrachloride
  Chlorobenzene
  Dichlorobenzene
  1,2-Dichloroethane
  1,1-Dichloroethylene
  cis-1,2-Dichloroethylene
  trans-1,2-Dichloroethylene
  Methylene chloride
  Tetrachloroethylene
  Trichlorobenzene
  1,1,1-Trichloroethane
  Trichloroethylene
  Vinyl chloride

Microbiology and Turbidity
  Giardia lamblia
  Legionella
  Standard plate count
  Total coliforms
  Turbidity
  Viruses

Inorganics
  Aluminum
  Antimony
  Arsenic
  Asbestos
  Barium
  Beryllium
  Cadmium
  Chromium
  Copper
  Cyanide
  Fluoride
  Lead

  Mercury
  Molybdenum
  Nickel
  Nitrate
  Selenium
  Silver
  Sodium
  Thallium
  Vanadium
  Zinc

Organics
  Acrylamide
  Adipates
  Alachlor
  Aldicarb
  Atrazine
  Carbofuran
  Chlordane
  Dalapon
  Dibromochloropropane (DBCP)
  Dibromomethane
  1,2-Dichloropropane
  Dinoseb
  Diquat
  Endothall
  Endrin
  Epichlorohydrin
  Ethylene dibromide (EDB)
  Glyphosphate
  Hexachlorocyclopentadiene
  Lindane
  Methoxychlor
  Pentachlorophenol
  Phthalates
  Pichloram

TABLE 2 (Continued)

**CONTAMINANTS REQUIRED TO BE REGULATED
UNDER THE 1986 SDWA AMENDMENTS**

Polychlorinated biphenyls (PCBs)	**Removed from SDWA List of 1983**
Polynuclear aromatic hydrocarbons (PAHs)	Aluminum
Simazine	Dibromomethane
2,3,7,8-Tetrachlorodibenzodioxin (dioxin)	Molybdenum
	Silver
Toluene	Sodium
Toxaphene	Vanadium
2,4,5-TP (Silvex)	Zinc
1,1,2-Trichloroethane	
Vydate	**Substituted into SDWA List of 1983**
Xylene	Aldicarb sulfone
	Aldicarb sulfoxide
Radionuclides	Ethylbenzene
Beta particle and photo-radioactivity	Heptachlor
Gross alpha particle activity	Heptachlor epoxide
Radium-226 and radium-228	Nitrite
Radon	Styrene
Uranium	

---

under the Comprehensive Environmental Response, Compensation, and Liability Act (CERCLA), and pesticides registered under the Federal Insecticide, Fungicide, and Rodenticide Act (FIFRA). The DWPL is to be updated at three-year intervals with regulation occurring within 24 months of publication.

Maximum Contaminant Levels Goal (MCLGs) are also outlined in Section 1412 of the act. MCLGs are nonenforceable health-based goals which must be set at levels which pose no adverse effect on human health with a significant margin of safety, without concerns for the costs associated with reaching these goals. MCLs are enforceable standards which must be set as close to the MCLGs as technologically possible with costs being considered.

State Primary Enforcement Responsibility (primacy) is covered in Section 1413 which gives the USEPA the authority to delegate primacy to the state. The USEPA has developed primary requirements which states must apply for and retain primacy which are listed below.

- Adopt drinking water regulations that are no less stringent than the NPDWRs.
- Adopt and implement adequate procedures for enforcement.
- Retain records and prepare reports as USEPA may require.
- If issued, permit variances of exemptions in a manner that is not less stringent than permitted under Sections 1415 and 1416.
- Adopt and be able to implement an adequate plan for the provision of safe drinking water under emergency situations.

To maintain primacy, a state must adopt regulations that are as strict as all NPDWRs promulgated by the USEPA within an 18-month adoption period.

ENFORCEMENT

The enforcement backbone of the act is the enforcement of the Drinking Water Regulations. The enforcement provisions of the act were amended in 1986 in three keys areas:

- Civil penalties were increased.
- Public notification procedures were changed.
- USEPA was given authority to issue administrative orders.

Section 1414 empowers USEPA to bring a civil action in the appropriate U.S. direct court to require compliance with and NPDWR, and administrative order of any other schedule or other requirement imposed by a variance of exemption. The penalty that a court may impose was increased in 1986 from $5000 to a maximum of $25,000 for each day of violation. If the USEPA issues and administrative order and the total penalty does not exceed $5000, the agency can access the fine without going to court after a hearing before an administrative law judge. Any penalty over $5000 must be assessed by a district court.

In regard to public notification, the 1974 SDWA required owners or operators of community water systems to notify their customers when drinking water standards were violated. The purpose of the public notification requirement was stated as part of the House of Representatives Report on the 1974 SDWA which stated--the purpose of the requirement was to educate the public as to the extent of the inadequate performance of their public water system and to make them aware of potential or actual health hazards. Public notification provisions of the 1974 SDWA were cumbersome because all of the violations were treated the same. Section 1413 amended in 1986 directed the EPA's attention to revising public notification procedures which prompted their promulgation in October 1987. This newly promulgated regulation outlined specific criteria to be used by a water system when public notification is required.

Notice must be given to general circulation newspapers that are published in the area served by the public water system. The agency may require mailings and press releases to television and radio stations, depending on the severity of the problem. Each notice must contain a clear, concise explanation of the violation, potential health hazards, and the actions being taken to resolve the problem. A civil penalty of $25,000 may be imposed for failing to comply with the requirements of public notification.

Section 1415 deals with variances which may be issued by states, or the USEPA in states not having primacy. To receive a variance, a public water system must demonstrate that the characteristics of its water source prevent compliance with an NPDWR.

Section 1416 deals with the exemptions which can be obtained by public water systems if they show compelling factors of noncompliance with an NPDWRs. The system must be in operation as of the effective date of the particular regulation and exemption must not cause an unreasonable risk to health.

Section 1417, Prohibition of Use of Lead Pipes, Solder, and Flux, was added to the SDWA to decrease future problems of lead contamination in water systems. The section states that "any pipe, solder, or flux used after July 19, 1986, in the installation or repair of public water systems and plumbing used for drinking water must be lead free."

Public water systems were required to identify and provide notice to anyone who may be affected by lead contamination caused by lead content of materials used in the distribution or plumbing systems or caused by the corrosivity of the water, which may cause leaching of the lead. This notice was given even if the system was in compliance with the MCL for lead. States were required to start enforcement of this prohibition and the public notice requirements by June 1988. If a state refused, the USEPA could reduce that state's program grant by as much as 5 percent.

PART C--deals with the Protection of Underground Sources of Drinking Water. Section 1421 of this division requires the administrator to establish regulations for state underground injection control programs. Minimum requirements include assurance that injections will not endanger ground injection of brine or other fluids brought to the surface in connection with oil or natural gas production. Underground injection is the subsurface emplacement of fluids by well injection. Section 1421 specifies that underground injection endangers drinking water sources if: such injection may result in the presence of any contaminant in underground water which supplies or can reasonably be expected to supply any public water system.

States as well as Indian tribes may assume primacy for the underground injection program under Section 1422. These states must meet requirements set by the USEPA to retain primacy.

The enforcement of this program is explained in Section 1423 in which the administrator is authorized to take action against any person who is subject to underground injection control regulation and is found in violation of any requirement. A notice of violation must be issued to the person violating the requirement and primacy agent. If after the thirteenth day after the notice of violation the state has not taken enforcement action, the administrator may issue an administrative order or commence a civil action in the appropriate U.S. district court. Any person found in violation of any underground injection control program requirement of an administrative order is subject to a civil penalty of not more than $25,000 for each day of violation. In the case of willful violations, the violator may be imprisoned for no more than three years or receive additional fines in addition to or in place of a civil penalty.

Aquifer protection is part of Section 1427 which establishes procedures for development, implementation, and assessment of demonstration programs designed to protect critical aquifer protection areas designated as sole or principal source aquifers. A critical aquifer protection area is defined as either all or part of an area located within an area for which an application of designation as a sole or principal source aquifer has been approved by the administrator; or all or part of an area that is within a aquifer designated as a sole source aquifer under an approved area with a groundwater quality protection plan under Section 208 of the Clean Water Act.

Applications must include a comprehensive management plan that is designed to maintain the quality of groundwater in the critical protection area in a manner reasonably expected to protect human health, the environment, and groundwater resources.

The 1986 amendments added Section 1428, which provided for state programs intended specifically to protect groundwater supplying public water system wells. The programs protect wellhead areas from contaminants that could effect public health. States were required to submit a program to the USEPA by June, 1989. The program must:

- Outline the roles of state, local, and public water supply agencies in carrying out the program.
- Determine wellhead protection areas for each well of wellfield supplying a public water system.
- Identify all potential anthropogenic contamination sources within each wellhead protection area.
- Describe a plan to protect the wells from contamination which includes technical assistance, training, control measures, or financial assistance.
- Include contingency plans to provide safe drinking water in the event of a supply becoming contaminated.
- Require that potential sources of contamination be investigated before a new well is constructed.

A wellhead protection area is defined as the surface and subsurface area surrounding a well or wellfield supplying a public water supply through which contaminants could reach the well(s). The size of a particular area will be determined by the state using the USEPA guidelines. Hydrogeological factors such as radius of influence, which should be considered, will be outlined in these guidelines.

States that submit programs were eligible to receive grants to cover from 50 to 90 percent of costs needed to develop and implement the program. No funding would be available for programs submitted after

June 1989. If a state chose not to submit a program, the USEPA would not implement a federal wellhead protection program. This differs from the public water supply and underground injection programs.

PART D--emergency powers Section 1431 of the 1986 SDWA, expanded the USEPA's authority to act when an imminent threat to health is present. The USEPA can now issue an order to anyone causing or contributing to contamination of a public water system and endangering public health. The order can require the provision of alternative, safe water supplies until the threat is over. Violation of an order may lead to a penalty of up to $5000 per day.

Section 1432 was added in 1986, making it a federal offense to tamper with a public water system. Tamper means introducing a contaminant into a public water system with harmful intentions.

PARTS E & F--Part E consists of general provisions of the SDWA. The sections of this part deal with many generalities that range from the assurance of availability of adequate supplies of chemicals necessary for water treatment, research, technical assistance, information, training personnel, grants, record keeping, and provisions for Indian tribes.

Part F is the final section of the 1986 amendment and involves the regulation of lead in drinking water. The sections of this part includes the recall of drinking water coolers with lead-lined tanks, the lead contamination in school drinking water, and federal assistance guidelines for state programs regarding lead contamination in school drinking water.

The Safe Drinking Water Act is complex and very broad. Funding for such a broad piece of legislation is important and is necessary in order for the USEPA to implement the many provisions of the act. Congress appropriates different amounts of money each year which is the basis of how much the SDWA can be implemented.

The SDWA is only one of several environmental laws impacting water utilities. As shown in Table 3 there are many other that need to be considered along with the SDWA. Complying with regulations establish under the SDWA may cause violation of other environmental laws. Conversely, constraints imposed by other environmental laws and regulation may restrict or hinder a utility from complying with SDWA regulation. Hence, water utilities must pursue compliance using an integrated approach to ensure that full compliance with all applicable

TABLE 3
FEDERAL ENVIRONMENTAL LAWS IMPACTING WATER UTILITIES

Law	Last Amended	Agency*	Primary Utility Impact						
			Water Supply Development	Source Water Quality	Treatment	Residuals Disposal	Distribution	Operations	Drinking Water Quality
Clean Air Act (CAA)	1990	USEPA				X			
Clean Water Act (CWA)	1987	USEPA COE	X	X		X	X	X	
Coastal Zone Management Act (CZMA)	1991	NOAA USEPA	X						
Comprehensive Environmental, Response, Compensation, Liability Act (CERCLA)	1980	USEPA		X		X			
Endangered Species Act (ESA)	1979	DOC DOI	X						
Federal Insecticide, Fungicide, and Rodenticide Act (FIFRA)	1988	USEPA		X					
Federal Power Act (FPA)		FERC	X						
Food, Agriculture, Conservation, and Trade Act	1990	DOA	X	X					
Food Security Act (FSA)	1985	DOA		X					
Hazardous Materials Transportation Act (HMTA)	1988	DOT				X			

TABLE 3 (Continued)
FEDERAL ENVIRONMENTAL LAWS IMPACTING WATER UTILITIES

Law	Last Amended	Agency*	Primary Utility Impact						
			Water Supply Development	Source Water Quality	Treatment	Residuals Disposal	Distribution	Operations	Drinking Water Quality
Laws Related to National Parks, Forests, and Public Lands†		DOA DOI	X						
National Environmental Policy Act (NEPA)	1975	CEQ USEPA	X						
National Historic Preservation Act (NHPA)		DOI	X						
Ocean Dumping Act (ODA)	1988	USEPA				X			
Resource Conservation and Recovery Act (RCRA)	1984	USEPA				X		X	
Rivers and Harbors Act (RHA)	1899	COE	X						
Safe Drinking Water Act (SDWA)	1988	USEPA	X	X	X	X	X	X	X
Solid Waste Disposal Act (SWDA)	1965	USEPA				X			
Superfund Amendments and Reauthorization Act (SARA)	1986	USEPA		X		X			

Surface Mining Control and Reclamation Act (SMCRA)	1977	DOI	X						
Toxic Substances Control Act (TSCA)	1988	USEPA		X	X			X	
Water Resources Development Act (WRDA)	1990	COE	X						
Wild and Scenic Rivers Act (WSRA)		DOI	X						
TOTALS			13	8	2	9	3	3	1

*Agency: USEPA–U.S. Environmental Protection Agency; COE–Corps of Engineers (U.S. Department of the Army); NOAA–National Oceanic and Atmospheric Administration; DOC–U.S. Department of Commerce; DOI–U.S. Department of Interior; FERC–Federal Energy Regulatory Commission; DOA–U.S. Department of Agriculture; DOT–U.S. Department of Transportation; CEQ–Council on Environmental Quality. For more information about these government agencies see *The United States Government Manual*, U.S. Government Printing Office, Washington, DC 20402.

†Organic Act of 1897; Federal Land Policy and Management Act of 1976; Reclamation Act of 1902; Reclamation Reform Act of 1982; Forest and Range Land Renewable Resources Planning Act; National Forest Management Act; Multiple-Use Sustained Yield Act of 1960; Wilderness Act of 1964; Land and Water Conservation Statutes located in various sections of Title 16.

environmental regulation is maintained. Understanding the SDWA and its regulator requirements is the focus of the report. The SDWA has evolved over several decades into a comprehensive law intended to ensure safe drinking water for over 267 million people served by public water systems in the United States. The adequacy of the law will continue to be questioned as new contaminants are discovered, technology improves, and financial resources become more limited. These and other pressures will provide additional incentives for Congress to consider future amendments to the law. Possible changes to the SDWA are constantly being thought of, some of which will be promulgated in the future. Tables 4 through 8 at the end of this chapter are a compilation of the Safe Drinking Water Effluent Standards.

CHRONOLOGICAL LISTING OF *FEDERAL REGISTER* NOTICES FOR DRINKING WATER REGULATIONS

1962

1. U.S. Public Health Service (USPHS); Drinking water standards. *Fed. Reg.*, 2152-2155 (Mar. 6, 1962).
    USPHS standards of 1962 that served as the basis for the U.S. Environmental Protection Agency's (USEPA's) interim drinking water regulations.

1975

2. National Interim Primary Drinking Water Regulations. *Fed. Reg.*, 40:248:59566-59588 (Dec. 24, 1975).
    Promulgates national interim primary drinking water regulations.

3. Drinking Water Regulations; Public Notification; Final Rule. *Fed. Reg.*, 40:59570 (Dec. 24, 1975).
    Promulgates general public notification requirements.

1976

4. National Primary Drinking Water Regulations Implementation; Primary Enforcement Responsibility; Final Rule. *Fed. Reg.*, 41:2917 (Jan. 20, 1976).
    Promulgates state primacy requirements.

5. National Interim Primary Drinking Water Regulations; Promulgation of Regulations on Radionuclides. *Fed. Reg.*, 41:133:28402-28409 (July 9, 1976).

   Promulgates interim regulations for radionuclides.

6. Organic Chemical Contaminants; Control Options in Drinking Water. *Fed. Reg.*, 41:136:28991-28998 (July 14, 1976).

   Advance notice of the development of regulations for controlling organics in drinking water.

1977

7. Drinking Water and Health; Recommendations of the National Academy of Sciences. *Fed. Reg.*, 42:132:35764-35779 (July 11, 1977).

   Presents findings of National Academy of Sciences study required by the 1974 SDWA.

1978

8. Interim Primary Drinking Water Regulations; Control of Organic Chemical Contaminants in Drinking Water, Proposed Rule. *Fed. Reg.*, 43:28:5756-5780 (Feb. 9, 1978).

   Proposes regulations for trihalomethanes and a treatment technique requiring use of granular activated carbon for systems serving > 75,000 people.

9. Interim Primary Drinking Water Regulations; Control of Organic Chemicals; Notice of Availability. *Fed. Reg.*, 43:130:29135 (July 6, 1978).

   Announces additional information and discusses issues associated with the proposed rule to control organic contaminants.

1979

10. National Secondary Drinking Water Regulations; Final Rule. *Fed. Reg.*, 44:231:42195-42202 (July 19, 1979).

    Promulgates secondary drinking water regulations.

11. National Interim Primary Drinking Water Regulations; Control of Trihalomethanes in Drinking Water; Final Rule. *Fed. Reg.*, 44:231:68624-68707 (Nov. 29, 1979).

    Promulgates regulations for trihalomethanes.

1980

12. Interim Primary Drinking Water Regulations; Amendments. *Fed. Reg.*, 45:168:57332-57357 (Aug. 27, 1980).

    General amendments to the interim regulations.

1981

13. Interim Primary Drinking Water Regulations; Control of Organic Chemicals in Drinking Water; Notice of withdrawal. *Fed. Reg.*, 46:53:17567 (Mar. 19, 1981).

    Withdraws proposed treatment technique rule requiring the use of granular activated carbon.

1982

14. National Revised Primary Drinking Water Regulations; Volatile Synthetic Organic Chemicals in Drinking Water; Advanced Notice of Proposed Rulemaking. *Fed. Reg.*, 47:43:9350-9358 (Mar. 4, 1982).

    Advanced notice of the development of regulations for volatile organic chemicals.

15. National Interim Primary Drinking Water Regulations; Trihalomethanes. *Fed. Reg.*, 47:44:9796-9799 (Mar. 5, 1982).

    Proposes best available technology for meeting trihalomethane regulations.

1983

16. National Interim Primary Drinking Water Regulations; Trihalomethanes; Final Rule. *Fed. Reg.*, 48:40:8406-8414 (Feb. 28, 1983).

    Promulgates best available technology for trihalomethanes.

17. National Revised Primary Drinking Water Regulations; Advanced Notice of Proposed Rulemaking. *Fed. Reg.*, 48:194:45502-45521 (Oct. 5, 1983).

    Advanced notice of the development of regulations for organic, inorganic, microbial, and radionuclide contaminants.

1984

18. National Primary Drinking Water Regulations; Volatile Synthetic Organic Chemicals; Proposed Rulemaking. *Fed. Reg.*, 49:114:24330-24355 (June 12, 1984).

    Proposes recommended maximum contaminant levels for 8 volatile organic chemicals.

1985

19. Fluoride; National Primary Drinking Water Regulations; Proposed Rule. *Fed. Reg.*, 50:93:20164-20175 (May 14, 1985).

    Proposes a recommended maximum contaminant level for fluoride.

20. National Primary Drinking Water Regulations; Volatile Synthetic Organic Chemicals; Proposed Rule. *Fed. Reg.*, 50:219:46880-46901 (Nov. 13, 1985).

    Promulgates recommended maximum contaminant levels for 8 volatile organic chemicals.

21. National Primary Drinking Water Regulations; Volatile Synthetic Organic Chemicals; Proposed Rule. *Fed. Reg.*, 50:219:46902-46933 (Nov. 13, 1985).

    Proposes regulations for 8 volatile organic chemicals.

22. National Primary Drinking Water Regulation; Synthetic Organic Chemicals, Inorganic Chemicals, and Microorganisms; Proposed Rule. *Fed. Reg.*, 50:219:46936-47022 (Nov. 13, 1985).

    Proposes recommended maximum contaminant levels for synthetic organic chemicals, inorganic chemicals, and microbials.

23. National Primary Drinking Water Regulations; Fluoride; Final Rule. *Fed. Reg.*, 50:220:47142-47155 (Nov. 15, 1985).
    Promulgates a recommended maximum contaminant level for fluoride.

24. National Primary Drinking Water Regulations; Fluoride; Proposed Rule. *Fed. Reg.*, 50:220:47156-47171 (Nov. 14, 1985).
    Proposes regulations for fluoride.

1986

25. National Primary and Secondary Drinking Water Regulations; Fluoride; Final Rule. *Fed. Reg.*, 51:63:11396-11412 (Apr. 2, 1986).
    Promulgates regulations for fluoride.

26. Water Pollution Control; National Primary Drinking Water Regulations; Radionuclides; Advanced Notice of Proposed Rulemaking. *Fed. Reg.*, 51:189:34836-34862 (Sept. 30, 1986).
    Advance notice of the development of regulations for radionuclides.

1987

27. Water Pollution Control; National Primary Drinking Water Regulations; Volatile Synthetic Organic Chemicals; Paradichlorobenzene; Propose Rule. *Fed. Reg.*, 52:74:12876-12883 (Apr. 17, 1987).
    Proposes regulations for paradichlorobenzene.

28. National Primary Drinking Water Regulations; Synthetic Organic Chemicals; Monitoring for Unregulated Contaminants; Final Rule. *Fed. Reg.*, 52:130:25690-25717 (July 8, 1987).
    Promulgates regulations for 8 volatile organic chemicals (Phase I).

29. Drinking Water; Proposed Substitution of Contaminants and Proposed List of Additional Substances Which May Require

Regulation Under the Safe Drinking Water Act. *Fed. Reg.*, 52:130:25720-25734 (July 8, 1987).

Proposes seven substitute contaminants and the first drinking water priority list.

30. Drinking Water Regulations; Public Notification; Final Rule. *Fed. Reg.*, 52:208:41534-41550 (Oct. 28, 1987).

Promulgates regulations for public notification.

31. National Primary Drinking Water Regulations; Filtration and Disinfection; Turbidity; *Giardia lamblia*, Viruses, *Legionella*, and Heterotrophic Bacteria; Proposed Rule. *Fed. Reg.*, 52:212:42178-42222 (Nov. 3, 1987).

Proposes the Surface Water Treatment Rule.

32. Drinking Water; National Primary Drinking Water Regulation; Total Coliforms; Proposed Rule. *Fed. Reg.*, 52:212:42224-42245 (Nov. 3, 1987).

Proposes the Total Coliform Rule.

1988

33. Drinking Water; Substitution of Contaminants and Drinking Water Priority List of Additional Substances Which May Require Regulation Under the Safe Drinking Water Act. *Fed. Reg.*, 53:14:1892-1902 (Jan. 22, 1988).

Announces the seven substitute contaminants and the first drinking water priority list.

34. National Primary Drinking Water Regulations; Filtration and Disinfection; Turbidity, *Giardia lamblia*, Viruses, *Legionella*, and Heterotrophic Bacteria; Total Coliforms; Notice of Availability; Close of Public Comment Period; Proposed Rule. *Fed. Reg.*, 53:88:16348-16358 (May 6, 1988).

Solicits specific data, offers additional regulatory options for comment, and clarifies and corrects statements made in the proposed Surface Water Treatment Rule.

35. National Primary Drinking Water Regulations; Synthetic Organic Chemicals; Monitoring for Unregulated Contaminants; Correction; Final Rule. *Fed. Reg.*, 53:127:25108-25111 (July 1, 1988).
Corrects errors and clarifies the preamble to the Phase I rule.

36. National Primary Drinking Water Regulations Implementation; Primary Enforcement Responsibility; Proposed Rule. *Fed. Reg.*, 53:29194-29207 (Aug. 2, 1988).
Proposed changes to state primacy regulations.

37. Drinking Water Regulations; Maximum Contaminant Level Goals and National Drinking Water Regulations for Lead and Copper; Proposed Rule. *Fed. Reg.*, 53:160:31516-31553 (Aug. 18, 1988).
Proposes regulations for lead and copper.

1989

38. Drinking Water Regulations; Public Notification; Final Rule; technical amendment. *Fed. Reg.*, 54:72:15185-15188 (Apr. 17, 1989).
Amends public notification requirements.

39. National Primary and Secondary Drinking Water Regulations; Proposed Rule. *Fed. Reg.*, 54:97:22062-22160 (May 22, 1989).
Proposes regulations for 30 synthetic organic chemicals and 8 inorganic chemicals (Phase II).

40. Drinking Water; National Primary Drinking Water Regulations; Filtration, Disinfection; Turbidity, *Giardia lamblia*, Viruses, *Legionella*, and Heterotrophic Bacteria; Final Rule. *Fed. Reg.*, 54:124:27486-27541 (June 29, 1989).
Promulgates Surface Water Treatment Rule.

41. Drinking Water; National Primary Drinking Water Regulations; Total Coliforms (Including Fecal Coliforms and *E. coli*); Final Rule. *Fed. Reg.*, 54:124:27544-27568 (June 29, 1989).
Promulgates Total Coliform Rule.

42. National Primary Drinking Water Regulations; Analytical Techniques; Coliform Bacteria; Final Rule. *Fed. Reg.*, 54:135:29998-30002 (July 17, 1989).

   Approves the MMO-MUG test for detecting total coliforms.

43. National Primary Drinking Water Regulations Implementation; Primary Enforcement Responsibility; Final Rule. *Fed. Reg.*, 54:243-52126-52140 (Dec. 20, 1989).

   Amends state primacy requirements.

1990

44. National Primary and Secondary Drinking Water Regulations; Fluoride; Request for Information. *Fed Reg.*, 55:2:160-161 (Jan. 3, 1990).

   Requests information bearing on the current standards for fluoride in drinking water.

45. National Primary Drinking Water Regulations; Analytical Techniques; Coliform Bacteria; Proposed Rule. *Fed. Reg.*, 55:106:22752-22756 (June 1, 1990).

   Proposes 3 analytical methods for detecting *E. coli*.

46. National Primary and Secondary Drinking Water Regulations; Synthetic Organic and Inorganic Chemicals; Proposed Rule. *Fed. Reg.*, 55:143:30370-30448 (July 25, 1990).

   Proposes regulation for 18 synthetic organic chemicals and 6 inorganic chemicals (Phase V).

47. Variances and Exemptions for Primary Drinking Water Regulations; Unreasonable Risk to Health Guidance. *Fed. Reg.*, 55:191:40205 (Oct. 2, 1990).

   Announces availability of draft guidance on determining unreasonable risks to health.

48. National Primary Drinking Water Regulations Implementation; Primary Enforcement Responsibility; Notice of proposed

rulemaking. *Fed. Reg.*, 55:229:49398-49399 (Nov. 28, 1990).
Proposed changes to state primacy requirements.

1991

49. National Primary Drinking Water Regulations; Analytical
    Techniques; Coliform Bacteria; Final Rule. *Fed. Reg.*, 56:5:636-
    643 (Jan. 8, 1991).
    Approves 2 analytical methods for detection of *E. coli*.

50. Priority List of Substances Which May Require Regulation Under
    Safe Drinking Water Act; Notice. *Fed. Reg.*, 56:9:1470-1474
    (Jan. 14, 1991).
    Announces updated drinking water priority list.

51. Drinking Water; National Primary Drinking Water Regulations;
    Total Coliforms; Partial Stay of Certain Provisions of Final Rule.
    Fed. Reg., 56:10:1556-1557 (jan. 15, 1991).
    Stays the no variance provisions of the Total Coliform Rule,
    allowing states to issue variances under limited conditions.

52. National Primary Drinking Water Regulations; Synthetic Organic
    Chemicals and Inorganic Chemicals; Monitoring for Unregulated
    Contaminants; National Primary Drinking Water Regulations
    Implementation; National Secondary Drinking Water Regulations;
    Final Rule. *Fed. Reg.*, 56:20:3526-3597 (Jan. 30, 1991).
    Promulgates regulations for 26 synthetic organic chemicals and
    7 inorganic chemicals.

53. National Primary Drinking Water Regulations; Monitoring for
    Synthetic Organic Chemicals; MCLGs and MCLs for Aldicarb,
    Aldicarb Sulfoxide, Aldicarb Sulfone, Pentachlorophenol, and
    Barium; Proposed Rule. *Fed. Reg.*, 56:20:3600-3614 (Jan. 30,
    1991).
    Proposes changes in monitoring requirements for 8 volatile
    organic compounds and regulations for aldicarb, aldicarb sulfoxide,
    aldicarb sulfone, pentachlorophenol, and barium.

54. National Primary Drinking Water Regulations Implementation; Primary Enforcement Responsibility; Final Rulemaking. *Fed. Reg.*, 56:106:254046-25050 (June 3, 1991).
   Promulgates changes to state primacy requirements.

55. Maximum Contaminant Level Goals and National Primary Drinking Water Regulations for Lead and Copper; Final Rule. *Fed. Reg.*, 56:110:26460-26564 (June 7, 1991).
   Promulgated regulations for lead and copper.

56. National Primary Drinking Water Regulations; Monitoring for Volatile Organic Chemicals; MCLGs and MCLs for Aldicarb, Aldicarb Sulfoxide, Aldicarb Sulfone, Pentachlorophenol, and Barium. Final Rule. *Fed. Reg.*, 56:126:30266-30281 (July 1, 1991).
   Revises monitoring requirements for 8 volatile organic compounds and promulgates regulations for aldicarb, aldicarb sulfoxide, aldicarb sulfone, pentachlorophenol, and barium.

57. Drinking Water Regulations; Maximum Contaminant Level Goals and National Primary Drinking Water Regulations for Lead and Copper; Final Rule; Correction. *Fed. Reg.*, 56:135:32112-32113 (July 15, 1991).
   Corrects errors in the effective date and the text of the Lead and Copper Rule.

58. National Primary Drinking Water Regulations; Radionuclides; Proposed Rule. *Fed. Reg.*, 56:138:33050-33127 (July 18, 1991).
   Proposes new rules for radionuclides in drinking water.

59. National Primary Drinking Water Regulations; Analytical Techniques; Coliform Bacteria; Notice of Availability. *Fed. Reg.*, 56:188:49153-49154 (Sept. 27, 1991).
   Announces the availability of two reports regarding the performances of the MMO-MUG test for detecting *E. coli*.

60. Drinking Water; National Primary Drinking Water Regulations; Synthetic Organic Chemicals and Inorganic Chemicals; Notice of

Availability With Request For Comments. *Fed. Reg.*, 56:230: 60949-60956 (Nov. 29, 1991).

Requests comments on new information the USEPA will consider in developing the final Phase V rule.

1992

61. National Primary Drinking Water Regulations; Analytical Techniques; Coliform Bacteria; Final Rule. *Fed. Reg.*, 57:10:1850-1852 (Jan. 15, 1992).

    Approves transfer of cultures from total coliform-positive MMO-MUG test for *E. coli* detection.

62. Drinking Water; National Primary Drinking Water Regulations; Aldicarb, Aldicarb Sulfoxide, and Aldicarb Sulfone; Notice of Postponement of Certain Provisions of Final Rule. *Fed. Reg.*, 57:102:22178-22179 (May 27, 1992).

    Postpones the effective date of the regulations for aldicarb, aldicarb sulfoxide, and aldicarb sulfone.

63. National Primary Drinking Water Regulations; Analytical Techniques; Coliform Bacteria; Final Rule. *Fed. Reg.*, 57:112: 24744-24747 (June 10, 1992).

    Approves the use of the MMO-MUG test for the detection of *E. coli*.

64. Drinking Water Regulations; Maximum Contaminant Level Goals and National Primary Drinking Water Regulations for Lead and Copper; Final Rule; Correcting Amendments. *Fed. Reg.*, 57:125: 28785-28789 (June 29, 1992).

    Corrects errors in the text of the Lead and Copper Rule and in the Phase II rule.

65. National Primary Drinking Water Regulations; Synthetic Organic Chemicals and Inorganic Chemicals; Final Rule. *Fed. Reg.*, 57:138:31776-31849 (July 17, 1992).

    Promulgates regulations for 18 synthetic organic chemicals and 5 inorganic chemicals, known as the Phase V rule.

66. Draft Groundwater Disinfection Rule Available for Public Comment; Notice of availability and review. *Fed. Reg.*, 57:148:33960 (July 31, 1992).

    Announces availability of draft Groundwater Disinfection Rule.

67. Intent to Form an Advisory Committee to Negotiate the Drinking Water Disinfection By-Products Rule and Amendment of Public Meeting; Notice of Intent. *Fed. Reg.*, 57:179:42533-42536 (Sept. 15, 1992).

    Announces intent to form an advisory committee to negotiate a Proposed Disinfectant-Disinfection By-products Rule.

68. Establishment and Open Meeting of the Negotiated Rulemaking Advisory Committee for Disinfection By-products; Establishment of FACA Committee and Meeting Announcement. *Fed. Reg.*, 57:220:53866 (Nov. 13, 1992).

    Announces formation of the advisory committee to negotiate a Proposed Disinfectant-Disinfection By-products Rule.

69. National Primary and Secondary Drinking Water Regulations; Fluoride; Update of Ongoing Review of National Primary and Secondary Drinking Water Regulations for Fluoride. *Fed. Reg.*, 57:226:54957-54958 (Nov. 23, 1992).

    Presents an update of the USEPA's ongoing review of the fluoride drinking water standard.

1993

70. National Primary Drinking Water Regulations; Analytical Techniques; Trihalomethanes; Final rule. *Fed. Reg.*, 58:147: 41344-41345 (Aug. 3, 1993).

    Approves USEPA methods 502.2 and 524.2 for trihalomethane compliance monitoring.

71. National Primary and Secondary Drinking Water Regulations; Analytical Methods for Regulated Drinking Water Contaminants; Proposed Rule. *Fed. Reg.*, 58:239:65622-65632 (Dec. 15, 1993).

Proposes several new analytical methods and updates previously approved methods for several regulated chemical, microbial, and physical contaminants in drinking water.

72. Drinking Water Maximum Contaminant Level Goal; Fluoride; Notice of Intent Not to Revise Fluoride Drinking Water Standards. *Fed. Reg.*, 58:248:68826-68827 (Dec. 29, 1993).

    Announcement of USEPA's decision not to revise the MCLG for fluoride.

1994

73. National Primary Drinking Water Regulations; Monitoring Requirements for Public Drinking Water Supplies; *Cryptosporidium, Giardia*, Viruses, Disinfection By-products, Water Treatment Plant Data and Other Information Requirements; Proposed Rule. *Fed. Reg.*, 59:28:6332-6444 (Feb. 10, 1994).

    Proposes Information Collection Rule to require microbial monitoring, disinfection by-product monitoring, and data reporting for water system serving $> 10,000$ people.

74. National Primary Drinking Water Regulations; Monitoring Requirements for Public Drinking Water Supplies. Proposed rule; Notice of extension of public comment period. *Fed. Reg.*, 59:50:11961-11962 (Mar. 15, 1994).

    Extends public comment period for the Information Collection rule until Mar. 28, 1994.

75. Drinking Water; Maximum Contaminant Level Goals, and National Primary Drinking Water Regulations for Lead and Copper. Final Rule; Technical corrections. *Fed. Reg.*, 59:125:33860-33864 (June 30, 1994).

    Amends the Lead and Copper Rule to correct typographical errors, clarify language, and restore special primacy requirements inadvertently deleted by the Phase II rule.

76. Drinking Water; National Primary Drinking Water Regulations; Synthetic Organic Chemicals and Inorganic Chemicals; National Primary Drinking Water Regulations Implementation; Monitoring

for Unregulated Contaminants. Final Rule; Technical amendments. *Fed. Reg.*, 59:126:34320-34321 (July 1, 1994).

Amends Phase I/II/V regulations to correct typographical errors, clarify language that was unclear, and correct mistakes where the preamble correctly indicated the Agency's intent but the language of the regulation was in error.

77. National Primary and Secondary Drinking Water Regulations; Analytical Methods for Regulated Drinking Water Contaminants; Notice of data availability. *Fed. Reg.*, 59:134:35891-35893 (July 14, 1994).

Announces availability of data demonstrating that the Colisure test is at least as effective for total coliform and *E. coli* as currently approved test. Also invites public comment on approval of the Colisure test, approval of 4 updated analytical methods, and withdrawal of 27 outdated analytical methods.

SUGGESTED READING

1.   McDermott, J.H. Federal Drinking Water Standards-Past, Present and Future. *Journal Environmental Engineering Division-ASCE*, EE4:99:469 (August 1973).

2.   American Water Works Association. *Water Quality and Treatment*, McGraw Hill Book Co., New York (3rd ed. 1971).

3.   American Water Works Association. *The SDWA Advisor,* 1994.

4.   Harris, R.H. & Brecher, E.M. Is the Water Safe to Drink? Part I: The Problem. Part II: How to Make it Safe. Part II: What You Can Do. *Consumer Reports*, 436, 538, 623 (June, July, August 1974).

5.   The Safe Drinking Water Act of 1974. P.L. 92-523, 93rd Congress (Dec. 16, 1974).

6.   Government Institutes Inc., *Environmental Law Handbook*, 12th ed. (1993).

7.   USEPA. Organic Chemical Contaminants; Control Options in Drinking Water. *Fed. Reg.*, 41:136:289991 (July 14, 1976).

8.   USEPA. Control of Organic Chemicals in Drinking Water. Proposed Rule. *Fed. Reg.*, 43:28:5756 (Feb. 9, 1978).

9. Symons, J.M. A History of the Attempted Federal Regulation Requiring GAC Adsorption for Water Treatment. *Jour. AWWA,* 76:8:34 (August 1984).

10. USEPA. Control of Organic Chemicals in Drinking Water. Notice of availability. *Fed. Reg.,* 43:130:29135 (July 6, 1978).

11. USEPA. Control of Trihalomethanes in Drinking Water. Final Rule. *Fed. Reg.,* 4:231:68624 (Nov. 29, 1979).

12. USEPA. Control of Organic Chemicals in Drinking Water. Notice of withdrawal. *Fed. Reg.,* 46:53:17567 (Mar. 19, 1981).

13. Public Law 99-339. Safe Drinking Water Act Amendments of 1986 (June 19, 1986).

14. Public Law 100-572. Lead Contaminated Control Act of 1988 (Oct. 31, 1988).

TABLE 4
USEPA DRINKING WATER STANDARDS--ORGANICS

Contaminant	Drinking Water Health Effects	MCLG. mg/L	MCL mg/L	Status	Effective Date	Contaminant Sources
Acrylamide	Probable cancer	Zero	0.005% dosed at 1 mg/L	Final	July 30, 1992	Water treatment chemicals (polymers)
Alachlor (Lasso)	Probable cancer	Zero	0.002	Final	July 30, 1992	Herbicide
Aldicarb (Temik)	Nervous system toxicity	0.001	0.003	Final	Delayed	Pesticide, herbicide; restricted in some areas
Aldicarb sulfone	Nervous system toxicity	0.001	0.002	Final	Delayed	Pesticide, herbicide; restricted in some areas
Atrazine (Altranex, Crisazina)	Nervous system, liver, heart effects	0.003	0.003	Final	July 30, 1992	Herbicide
Benzene	Cancer	Zero	0.005	Final	Jan. 9, 1989	Fuel (leaking tanks), solvent commonly used in manufacture of industrial chemicals, pharmaceuticals, pesticides, paints, and plastics
Benzo(a)pyrene	Probable cancer	Zero	0.002	Final	Jan. 17, 1994	Fossil fuel and wood burning; coal tar, forest fires
Carbofuran (Furadan 4F)	Nervous system, reproductive effects	0.04	0.04	Final	July 30, 1992	Pesticide, herbicide
Carbon tetrachloride	Cancer	Zero	0.005	Final	Jan. 9, 1989	Cleaning agents, industrial wastes from manufacture of coolants
Chlordane	Nervous system, probable cancer	Zero	0.002	Final	July 30, 1992	Pesticide, herbicide; most uses banned in 1980
2,4-D (Formula 40)	Liver, kidney effects	0.07	0.07	Final	July 30, 1992	Herbicide
Dalapon	Liver, kidney effects	0.02	0.02	Final	Jan. 17, 1994	Herbicide for fruit trees, corn, cotton
Di(2-ethylhexyl) adipate	Possible cancer; liver, reproductive system effects	0.5	0.5	Final	Jan. 17, 1994	Plastics
Di(2-ethylhexyl) phthalate	Probable cancer	Zero	0.006	Final	Jan. 17, 1994	Plastics
Dibromochloropropane (DBCP)	Probable cancer	Zero	0.0002	Final	July 30, 1992	Pesticide; canceled in 1977
p-Dichlorobenzene	Possible cancer	0.075	0.075	Final	Jan. 9, 1989	Used in insecticides, moth balls, air deodorizers

TABLE 4 (Continued)
USEPA DRINKING WATER STANDARDS--ORGANICS

Contaminant	Drinking Water Health Effects	MCLG. mg/L	MCL mg/L	Status	Effective Date	Contaminant Sources
o-Dichlorobenzene	Nervous system, lung, liver, kidney effects	0.6	0.6	Final	July 30, 1992	Industrial solvent, pesticide
1,2-Dichloroethane	Probable cancer	Zero	0.005	Final	Jan. 9, 1989	Used in manufacture of insecticides, gasoline
1,1-Dichloroethylene	Liver, kidney effects	0.007	0.007	Final	Jan. 9, 1989	Used in manufacture of plastics, dyes, perfumes, paints, SOCs
cis-1,2-Dichloroethylene	Nervous system, liver, kidney effects	0.07	0.07	Final	July 30, 1992	Extraction solvent, dyes, perfumes, pharmaceuticals, lacquers
trans-1,2-Dichloroethylene	Nervous system, liver, kidney effects	0.1	0.1	Final	July 30, 1992	Extraction solvent, dyes, perfumes, pharmaceuticals, lacquers
Dichloromethane (methylene chloride)	Probable cancer	Zero	0.005	Final	Jan. 17, 1994	Solvent
1,2-Dichloropropane	Probable cancer	Zero	0.005	Final	July 30, 1992	Pesticide, solvent
Dinoseb	Thyroid, reproductive organ effects	0.007	0.007	Final	Jan. 17, 1994	Herbicide
Diquat	Liver, kidney, GI tract effects; cataract formation	0.02	0.02	Final	Jan. 17, 1994	Herbicide; defoliant
Endothall	Liver, kidney, GI tract, reproductive system effects	0.1	0.1	Final	Jan. 17, 1994	Herbicide; defoliant
Endrin	Liver, kidney, heart effects	0.002	0.002	Final	Aug. 17, 1992	Pesticide; insecticide
Epichlorohydrin	Probable cancer	Zero	TT* (0.01% dosed at 20 mg/L)	Final	July 30, 1992	Water treatment chemicals (polymers)
Ethylbenzene	Liver, kidney effects	0.7	0.7	Final	July 30, 1992	Manufacture of styrene
Ethylene dibromide (EDB)	Probable cancer	Zero	0.00005	Final	July 30, 1992	Gasoline additive, soil fumigant solvent; most pesticide uses restricted in 1984
Glyphosate (Rodeo, Roundup)	Liver, kidney effects	0.7	0.7	Final	Jan. 17, 1994	Herbicide

Name	Health effects	MCLG	MCL	Status	Date	Uses/comments
Heptachlor (H-34, Heptox)	Probable cancer	Zero	0.0004	Final	July 30, 1992	Insecticide; most uses restricted in 1983
Heptachlor-epoxide	Probable cancer	Zero	0.0002	Final	July 30, 1992	Insecticide; most uses restricted in 1983
Lindane	Neurological, liver, kidney effects	0.0002	0.0002	Final	July 30, 1992	Insecticide to control fleas, lice, ticks; some uses restricted in 1983
Methoxychlor (DMDT, Marlate)	Central nervous system effects	0.04	0.04	Final	July 30, 1992	Insecticide
Monochlorobenzene	Respiratory, nervous system, liver, kidney effects	0.1	0.1	Final	July 30, 1992	Solvent, pesticide
Oxamyl (Vydate)	Kidney effects	0.2	0.2	Final	Jan. 17, 1994	Pesticide for potatoes and tomatoes
Pentachlorophenol	Organ, central nervous system effects; cancer	Zero	0.001	Final	Jan. 1, 1993	Wood preservative; nonwood uses banned in 1987
Picloram	Liver, kidney effects	0.5	0.5	Final	Jan. 17, 1994	Herbicide
Polychlorinated biphenyls (PCBs)	Probable cancer; reproductive effects	Zero	0.0005	Final	July 30, 1992	Transformers, capacitors; production banned in 1987
Simazine	Possible carcinogen; circulatory system effects	0.004	0.004	Final	Jan. 17, 1994	Herbicide
Styrene	Possible cancer; liver, central nervous system effects	0.1	0.1	Final	July 30, 1992	Manufacture of polystyrene plastic
2,3,7,8-TCDD (Dioxin)	Probable carcinogen	Zero	3×10^{-8}	Final	Jan. 17, 1994	By-product in manufacturing process of some chlorinated herbicides; pulp and paper mill effluent
Tetrachloroethylene	Probable cancer	Zero	0.005	Final	July 30, 1992	Dry-cleaning solvent
Toluene	Nervous system, lung, liver effects	1	1	Final	July 30, 1992	Solvent, gasoline additive

TABLE 4 (Continued)
USEPA DRINKING WATER STANDARDS--ORGANICS

Contaminant	Drinking Water Health Effects	MCLG. mg/L	MCL mg/L	Status	Effective Date	Contaminant Sources
Toxaphene	Probable cancer	Zero	0.003	Final	July 30, 1992	Pesticide, herbicide; most uses canceled in 1977
2,4,5-TP (Silvex)	Liver, kidney effects	0.05	0.05	Final	July 30, 1992	Herbicide; canceled in 1983
1,2,4-Trichloro-benzene	Liver, kidney effects	0.07	0.07	Final	Jan. 17, 1994	Manufacture of herbicides; dye carrier
1,1,1-Trichloroethane	Nervous system effects	0.2	0.2	Final	Jan. 9, 1989	Manufacture of food wrappings, synthetic fibers
1,1,2-Trichloroethane	Possible cancer; kidney, liver effects	0.003	0.005	Final	Jan. 17, 1994	Solvent manufacture of vinylidene chloride
Trichloroethylene	Probable cancer	Zero	0.005	Final	Jan. 9, 1989	Waste from disposal of dry-cleaning materials and manufacture of pesticides, paints, waxes, and varnishes, paint stripper, metal degreaser
Total trihalomethanes	Probable cancer		0.1	Final	Varied, depending on system size	Disinfection by-products
Vinyl chloride	Cancer	Zero	0.002	Final	Jan. 9, 1989	Industrial waste from manufacture of plastics and synthetic fiber
Xylenes (total)	Central nervous system effects	10	10	Final	July 30, 1992	Solvent; used to manufacture paint, dyes, adhesives, detergents; fuel additive

*TT: treatment technique.

TABLE 5

USEPA DRINKING WATER STANDARDS--INORGANICS

Contaminant	Drinking Water Health Effects	MCLG. mg/L	MCL mg/L	Status	Effective Date	Contaminant Sources
Antimony	Decreases growth and longevity	0.006	0.006	Final	Jan. 17, 1994	Geological; manufacture of flame retardants, ceramics, glass, pesticides, and tin-antimony solder
Arsenic	Dermal, nervous system toxicity effects; possible cancer		0.05	Final	June 24, 1977	Geological; pesticide residues, industrial wastes, smelter operations
Asbestos	Benign tumors	7 million fibers/L (>10μm)	7 million fibers/L (>10 μm)	Final	July 30, 1992	Geological; asbestos–cement pipe
Barium	Circulatory system effects	2	2	Final	Jan. 1, 1993	Geological
Beryllium	Possible cancer, damage to bones and lungs	0.004	0.004	Final	Jan. 17, 1994	Geological; manufacture of high-thermal-conductivity materials
Cadmium	Kidney effects	0.005	0.005	Final	July 30, 1992	Geological, mining, smelting, and corrosion of galvanized pipe
Chromium (total)	Gastrointestinal effects	0.1	0.1	Final	July 30, 1992	Geological
Copper	Stomach and intestinal distress; Wilson's disease	1.3	TT*	Final	Dec. 7, 1992	Corrosion of interior household and building pipe
Cyanide	Spleen, brain, liver effects	0.2	0.2	Final	Jan. 17, 1994	Used in electroplating, steel processing, plastics, synthetic fibers, mining, fertilizers, and farm products
Fluoride	Dental and skeletal fluorosis	4	4	Final	Oct. 2, 1987	Geological; additive to drinking water, toothpaste, foods processed with fluoridated water

TABLE 5 (Continued)
USEPA DRINKING WATER STANDARDS--INORGANICS

Contaminant	Drinking Water Health Effects	MCLG. mg/L	MCL mg/L	Status	Effective Date	Contaminant Sources
Lead	Central and peripheral nervous system damage; kidney effects; highly toxic to infants and pregnant women	Zero	TT*	Final	Dec. 7, 1992	Leaches from lead pipe and lead-based solder pipe joints
Mercury	Kidney effects	0.002	0.002	Final	July 30, 1992	Used in manufacture of paint, paper, vinyl chloride; used in fungicides; geological
Nickel	Heart, liver effects	0.1	0.1	Final	Jan. 17, 1994	Geological; used in electroplating, battery production, ceramics, and glass coloration
Nitrate (as N)	Methemoglobinemia ("blue baby syndrome")	10	10	Final	July 30, 1992	Fertilizer, sewage, feedlots
Nitrite (as N)	Methemoglobinemia ("blue baby syndrome")	1	1	Final	July 30, 1992	Fertilizer, sewage, feedlots
Nitrate plus nitrite (as N)	Methemoglobinemia ("blue baby syndrome")	10	10	Final	July 30, 1992	Fertilizer, sewage, feedlots
Selenium	Neurological effects	0.05	0.05	Final	July 30, 1992	Geological; mining
Sulfate	Gastroenteritis	400/500	400/500	Proposed		Geological; steel and metal industries; fungicide manufacture
Thallium	Kidney, liver, brain effects	0.00005	0.002	Final	Jan. 17, 1994	Geological; electronics industry; alloys and glass manufacturing

*TT: treatment technique.

TABLE 6
USEPA DRINKING WATER STANDARDS--RADIONUCLIDES

Contaminant	Drinking Water Health Effects	MCLG. mg/L	MCL mg/L	Status	Effective Date	Contaminant Sources
Alpha emitters	Probable cancer	Zero	15 pCi/L	Final	June 24, 1977	Naturally occurring and synthetic
		Zero	15 pCi/L	Proposed	June 24, 1977	
Beta-particle and photon emitters	Probable cancer	Zero	4 mrem	Final	June 24, 1977	Naturally occurring and synthetic
Radium 226 + 228	Probable cancer	Zero	4 mrem	Proposed		
Radium 226	Probable cancer	Zero	5 pCi/L	Final	June 24, 1977	Naturally occurring
Radium 228	Probable cancer	Zero	20 pCi/L	Proposed		Naturally occurring
Radon	Probable cancer	Zero	20 pCi/L	Proposed		Naturally occurring
Uranium	Probable cancer, kidney toxicity	Zero	300 pCi/L	Proposed		Naturally occurring
		Zero	20 µg/L	Proposed		Naturally occurring

TABLE 7
USEPA DRINKING WATER STANDARDS--MICROBIALS

Contaminant	Drinking Water Health Effects	MCLG. mg/L	MCL mg/L	Status	Effective Date	Contaminant Sources
Giardia lamblia	Giardiasis (stomach cramps, intestinal distress)	Zero	TT*	Final	Dec. 30, 1990	Human and animal fecal matter
Heterotrophic plate count (HPC)	Not necessarily disease-causing; HPC bacteria can be indicators of other disease-causing organisms	N/A†	TT*	Final‡	Dec. 30, 1990	Naturally occurring
Legionella	Legionnaires' disease (pneumonia), Pontiac fever	Zero	TT*	Final‡	Dec. 30, 1990	Water aerosols
Total coliforms	Not necessarily disease-causing; coliforms can be indicators of organisms that can cause gastroenteric infections and other diseases	Zero	No more than 5% of the samples per month may be positive. For systems collecting fewer than 1 sample per month may be positive.	Final	Dec. 30, 1990	Human and animal fecal matter

Turbidity	Not necessarily disease-causing; turbidity can indicate the presence of organisms that cause disease; interferes with disinfection	N/A†	PS§	Final	Dec. 30, 1990	Erosion, runoff, discharges	12
Viruses	Gastroenteritis	Zero	TT*	Final‡	Dec. 30, 1990	Human and animal fecal matter	12

*TT: treatment technique.
†N/A: not applicable.
‡Final for systems using surface water; also being considered for groundwater systems.
§PS: performance standard.

TABLE 8
BAT FOR REGULATED CONTAMINANTS

Contaminant	Conventional Processes	Specialized Processes	Regulation
Organics			
Acrylamide	Polymer addition practices		Phase II
Alachlor		GAC	Phase II
Aldicarb		GAC	Phase II
Aldicarb sulfone		GAC	Phase II
Aldicarb sulfoxide		GAC	Phase II
Atrazine		GAC	Phase I
Benzene		GAC; PTA	Phase V
Benzo(a)pyrene		GAC	Phase II
Carbofuran		GAC	Phase I
Carbon tetrachloride		GAC; PTA	Phase II
Chlordane		GAC	Phase II
2,4-D		GAC	Phase V
Dalapon		GAC	Phase V
Di(2-ethylhexyl) adipate		GAC; PTA	Phase V
Di(2-ethylhexyl) phthalate		GAC	Phase II
Dibromochloropropane (DBCP)		GAC; PTA	Phase I
p-Dichlorobenzene		GAC; PTA	Phase II
o-Dichlorobenzene		GAC; PTA	Phase I
1,2-Dichloroethane		GAC; PTA	Phase I
1,1-Dichloroethylene		GAC; PTA	Phase II
cis-1,2-Dichloroethylene		GAC; PTA	Phase II
trans-1,2-Dichloroethylene		GAC; PTA	Phase V
Dichloromethane methylene chloride		PTA	Phase II
1,2-Dichloropropane		GAC; PTA	Phase V
Dinoseb		GAC	Phase V
Diquat		GAC	Phase V
Endothall		GAC	Phase V
Endrin		GAC	Phase II
Epichlorohydrin	Polymer addition practices		

TABLE 8 (Continued)

BAT FOR REGULATED CONTAMINANTS

Contaminant	Conventional Processes	Specialized Processes	Regulation
Organics			
Ethylbenzene		GAC; PTA	Phase II
Ethylene dibromide (EDB)		GAC; PTA	Phase II
Glyphosate		GAC	Phase V
Heptachlor		GAC	Phase II
Heptachlor epoxide		GAC	Phase II
Hexachlorobenzene		GAC	Phase V
Hexachlorocyclopen-tadiene		GAC; PTA	Phase V
Lindane		GAC	Phase II
Methoxychlor		GAC	Phase II
Monochlorobenzene		GAC; PTA	Phase II
Oxamyl (Vydate)		GAC	Phase V
Pentachlorophenol		GAC	Phase II
Picloram		GAC	Phase V
Polychlorinated biphenyls (PCBs)		GAC	Phase II
Simazine		GAC	Phase V
Styrene		GAC; PTA	Phase II
2,3,7,8-TCDD (Dioxin)		GAC	Phase V
Tetrachloroethylene		GAC; PTA	Phase II
Toluene		GAC; PTA	Phase II
Toxaphene		GAC	Phase II
2,4,5-TP (Silvex)		GAC	Phase II
1,2,4-Trichloro-benzene		GAC; PTA	Phase V
1,1,1-Trichloroethane		GAC; PTA	Phase I
1,1,2-Trichloroethane		GAC; PTA	Phase V
Trichloroethylene		GAC; PTA	Phase I
Total trihalomethanes*	AD; PR; discontinue pre-Cl$_2$		Interim
Vinyl chloride		PTA	Phase I
Xylenes (total)		GAC; PTA	Phase II

TABLE 8 (Continued)

BAT FOR REGULATED CONTAMINANTS

Contaminant	Conventional Processes	Specialized Processes	Regulation
Inorganics			
Antimony	C-F†	RO	Phase V
Arsenic			Interim
Asbestos (fibers/1 > 10μm)	C-F†DF; DEF; CC		Phase II
Barium	LS†	IX; RO	Phase II
Beryllium	C-F†; LS†	AA; IX; RO	Phase V
Cadmium	C-F†; LS†	IX; RO	Phase II
Chromium (total)	C-F†; LS (Cr III)†	IX; RO	Phase II
Copper	CC; SWT		Lead and copper
Cyanide	Cl$_2$	IX; RO	Phase V
Fluoride		AA; RO	Fluoride
Lead	CC; PE; SWT; LSLR		Lead and copper
Mercury	C-F (influent <10μg/L)†; LS†	GAC; RO (influent ≤ μg/L)	Phase II
Nickel	LS†	IX; RO	Phase V
Nitrate (as N)		IX; RO	Phase II
Nitrite (as N)		IX; RO	Phase II
Nitrate + nitrite (both as N)		IX; RO	Phase II
Selenium	C-F (Se IV)†; LS†	AA; RO	Phase II
Sulfate (proposed)		IX; RO	Sulfate
Thallium		AA; IX	Phase V
Radionuclides			
Beta-particle and photon emitters (proposed)		IX; RO	Interim Rads
Alpha emitters (proposed)		RO	Rads
Radium-226 (proposed)	LS†	IX; RO	Rads
Radium-228 (proposed)	LS†	IX; RO	Rads
Radon (proposed)		Aeration	Rads
Uranium (proposed)	C-F†; LS†	AX; LS	Rads

TABLE 8 (Continued) BAT FOR REGULATED CONTAMINANTS			
Contaminant	Conventional Processes	Specialized Processes	Regulation
Microbials			
Giardia lamblia	C-F; SSF; DEF; DF; D		SWTR
Legionella	C-F; SSF; DEF; DF; D		SWTR
Standard plate count	C-F; SSF; DEF; DF; D		SWTR
Total coliforms	D		TCR
Turbidity	C-F; SSF; DEF; DF		SWTR
Viruses	C-F; SSF; DEF; DF; D		SWTR

Note: Abbreviations used in this table: AA--activated alumina; AD--alternative disinfectants; AX--anion exchange; CC--corrosion control; C-F--coagulation-filtration; Cl₂--chlorination; D--disinfection; DEF--diatomaceous earth filtration; DF--direct filtration; GAC--granular activated carbon; IX--ion exchange; LS--lime softening; LSLR--lead service line removal; PE--public education; PR--precursor removal; PTA--packed-tower aeration; RO--reverse osmosis; SWT--source water treatment.

*The sum of the concentrations of bromodichloromethane, dibromochloromethane, tribromomethane, and trichloromethane.

†Coagulation-filtration and lime softening are not BAT for small systems for variances unless treatment is already installed.

6 THE COMPREHENSIVE ENVIRONMENTAL RESPONSIBILITY, COMPENSATION, AND LIABILITY ACT

INTRODUCTION

The Comprehensive Environmental Response, Compensation, and Liability Act of 1980, commonly known as the "Superfund" act or CERCLA, was passed by Congress in response to a growing national concern about release of hazardous substances to the environment. These concerns were directed primarily at inactive sites, but also from actively managed facilities and vessels which are subjected to the Resource Conservation and Recovery Act (RCRA).

The key purpose of CERCLA is to establish a mechanism of response for the immediate cleanup of hazardous waste contamination from accidental spills or from abandoned hazardous waste disposal sites that may result in long term environmental damage.

The following is a summary of the CERCLA legislation.

LIABILITY

Any person who at the time of disposal of any hazardous substance owned or operated any facility where such hazardous substances were disposed of, any person by contract arranged with a transporter for transit for disposal or treatment of hazardous substances, any person who accepts or accepted any hazardous substances for transport to disposal or treatment facilities or site selected by that person, any owner or operator of a vessel or facility from which there is a release, or "threatened release which could acquire a response cost," of a hazardous substance, will be liable to the following: "(a) all costs of removal or remedial

213

action incurred by the United States government or a state not inconsistent with the National Contingency Plan; (b) any other necessary costs of response incurred by any other person consistent with the National Contingency Plan; (c) damages for injury to, destruction of, or loss of natural resources, including the reasonable costs of assessing such injury, destruction, or loss resulting from such a release." However, a person will be found not liable if he can prove without a doubt that the release was caused only by an act of God, an act of war, or an act of omission by a third party other than an employee of the defendant and that the defendant took all necessary precautions to prevent a release.

If the defendant is found guilty of a release for reasons other than the ones listed immediately above he will face fines not to exceed $300 per gross ton or $5,000,000 which ever is greater for any vessel which carries any hazardous substance as cargo or residue. For another vessel the fines will be $300 for gross or $500,000 whichever is greater. For any motor vehicle, aircraft, or pipeline the fines will be $50,000,000 or a lesser amount determined by the President, but in no case less than $5,000,000 or $8,000,000 for release in navigable waters. Fines shall take into account size, type, location, storage, handling capacity or other matters related to the likelihood of a release and the economic impact of such a release. For any other facility not mentioned above the fines will be the total cost of response plus $50,000,000 for any damages.

Aside from the limitations mentioned above the responsible person may be liable for the total costs of response and damages if the release or threat of a hazardous substance was the result of misconduct or negligence, the cause of the release was a violation of safety, construction, or operating standards, or the responsible person fails to provide all reasonable cooperation and assistance requested by a public official in connection with response activities.

A person may be liable to the United States for punitive damages in an amount at least equal to and not more than three times the amount of costs incurred by the fund if he fails without sufficient cause to properly provide removal or remedial action upon order of the President. The President can also begin civil action against a person to recover the punitive damages. Any money received by the United States from this will be deposited in the fund. The person shall not be liable for damages as a result of actions taken or omitted while doing the cleanup, with respect to an incident creating a danger to public health or welfare or environment, as a result of a hazardous substance release. However, this does not preclude liability for damages as a result of negligence or intentional misconduct.

If there is destruction or loss of natural resources liability will be to the United States government and to any state for natural resources within the state. The President, or the authorized representative of any state, will act as trustee of such natural resources to recover the damages. The money recovered will be available for use to restore, rehabilitate, or acquire the equivalent of natural resources by the appropriate agencies. The amount of money collected will not be limited by the sum which can be used to restore or replace the resources.

For an operator of a hazardous waste disposal facility which received a permit under Subtitle C of the Solid Waste Disposal Act, the liability will be transferred to the Post-Closure Liability Fund, if the owner of the facility has complied with the requirements of Subtitle C of the Solid Waste Disposal Act, and the facility and surrounding area has been monitored for the time required in the permit. The transfer will take place ninety days after the owner of the facility notifies the Administrator of the Environmental Protection Agency that the conditions of the permit have been satisfied, if before ninety days are up the Administrator of the Environmental Protection Agency or the state determines that the facility has not complied with the conditions imposed or that insufficient information has been provided to show compliance, the Administrator or the state will notify the owner of the facility and the administrator of the Post-Closure Fund. If this happens the owner of the facility will continue to be liable for the facility until the Administrator and the state determine that the facility has met all the conditions of the permit.

FINANCIAL RESPONSIBILITY

Any owner of a vessel over three hundred gross tons that uses any port in the United States or navigable waters of any shore facility will establish and maintain evidence of financial responsibility of three hundred dollars per gross ton, or for a vessel carrying hazardous substances as cargo three hundred dollars per gross ton or $5,000,000 whichever is greater. Financial responsibility will be established by any of the following ways: insurance, guarantee, surety bond, or quali- fication as a self insurer. If the owner operates or charters more than one vessel, evidence of financial responsibility needs to be established to meet the maximum liability applicable to the largest vessel.

The Secretary of the Treasury may revoke the clearances for any vessel that does not have certification from the President that the financial responsibility mentioned above have been met. The Secretary of Transportation may deny entry to any port in the United States or navigable waters to the United States. He may also detain a vessel from leaving a port if the owner of the vessel does not show certification from the President that financial responsibility has been complied with.

Under the order of the President classes of facilities must maintain evidence of financial responsibility consistent with the degree of risk associated with the production, transportation, treatment, storage, or disposal of hazardous substances. The financial responsibility will be made accordingly to protect against the level of risk which the President in his discretion believes is appropriate based on payment experience of the Fund, commercial insurers, count settlement and judgements, and voluntary claims satisfaction.

When a facility is owned by more than one person evidence of financial responsibility is maintained by one of the owners in consolidation form by the two or more owners. The proportional share of each participant shall be shown when the evidence of financial responsibility is in consolidation form. A statement authorizing the applicant to act in behalf of each participant in submitting and maintaining the evidence of financial responsibility must accompany the evidence of financial responsibility.

After notification and an opportunity for a hearing, any person who is found to have failed to comply with the preceding requirements or any denial or detention order shall be liable to the United States for a civil penalty not to exceed $10,000 for each day of violation.

NATIONAL CONTINGENCY PLAN

After public comment the President revised the National Contingency Plan, originally prepared and published pursuant to Section 311 of the Federal Water Pollution Control Act, for removal of oil and hazardous substances to reflect the responsibilities and powers created by the act. One revision was to include a section of the plan to be known as the National Hazardous Substance Response Plan which will establish procedures and standards for responding to releases of hazardous substances, pollutants, and contaminants. The plan included the following:

1. "Methods for discovering and investigating facilities at which hazardous substances have been disposed of or otherwise come to be located.

2. Methods for evaluating, including analysis of relative cost and remedying any releases or threats of releases from facilities which pose substantial danger to public health and the environment.

3. Methods and criteria for determining the appropriate extent of removal remedy, and other measures authorized by this act.

4. Appropriate roles and responsibilities for federal, state, and local government and for nongovernment entities in effectuating the plan.

5. Provisions for identification, procurement, maintenance, and storage of response equipment and supplies.

6. A method for and assignment of responsibility for reporting the existence of such facilities which may be located on the federally owned or controlled properties and any releases of hazardous substances from such facilities.

7. Means of assuring the remedial action measures are cost effective over the period of potential exposure to the hazardous substances or contaminated materials.

8. Criteria for determining priorities among releases or threatened release throughout the United States for the purpose of taking remedial action and, to the extent practicable taking into account the potential urgency of such action for the purpose of taking removal action.

9. Specified roles for private organizations and entitled in preparation for response and in responding to releases of hazardous substances, including identification of appropriate qualification and capacity therefore."

The President shall determine the priorities based upon risk or danger to the public health or welfare of the environment. He will take into account the population at risk, the hazard potential of the hazardous substance, the potential for contamination of drinking water supplies, the potential for human contact, the potential for destruction to sensitive ecosystems, state ability to assume state costs and responsibilities, among other factors. Based on these factors, the President shall make a list of known releases or threatened releases in the United States, and he shall revise the list no less than once a year. Each state will submit a list to the President every year of releases or potential releases in the state. At least four hundred of the highest priority facilities will be designated individually and will be known as the top priority among known response targets, and if at all possible the top one hundred highest priority facilities will include at least one facility from each state, that the state feels posses the greatest danger to public health or the environment.

The plan will include procedures, techniques, materials, equipment, and methods to be used in identifying, removing, or remedying releases of hazardous substances comparable to those required in the Federal Water Pollution Control Act. All responses and actions to minimize damage from hazardous substances releases shall be done in accordance with the national contingency plan. The President from time to time may revise and republish the national contingency plan. Table 1 provides a list of NPL sites in New Jersey. Refer also to Figure 1.

TABLE 1			
NATIONAL PRIORITIES LIST			
New Jersey Facilities on the National Priorities List			
A. O. Polymer	Sparta Township	Bridgeport Rental & Oil Service	Bridgeport
Amerivan Cyanamid	Bound Brook	Brook Industrial Park	Bound Brook
Asbestos Dump	Millington	Burnt Fly Bog	Marlboro
Bog Creek Farm	Howell Township	Caldwell Trucking Co.	Fairfield
Brick Township Landfill	Brick Township		

TABLE 1 (Continued)

NATIONAL PRIORITIES LIST
New Jersey Facilities on the National Priorities List

Chemical Control	Elizabeth	Montgomery Township Housing Development	Montgomery Township
Chemical Insecticide Corp.	Edison Township	Myers Property	Franklin Township
Chemcal Leeman Tank Lines Inc.	Bridgeport	Nascolite Corp.	Millville
		NL Industries	Pedricktown
Chemsol Inc.	Piscataway	Pepe Field	Boonton
Ciba-Geigy Corp.	Toms River	Pijak Farm	Plumstead Township
Cinnamonson Township (Block 702)	Cinnamonson Township	PJP Landfill	Jersey City
		Pohatcong Valley Ground Water	Warren County
Combe Fill North Landfill	Mount Olive Township		
Combe Fill South Landfill	Chester Township	Pomana Oaks Residential Wells	Galloway Township
Cosden Chemical Coating Corp.	Beverly	Price Landfill	Pleasantville
CPS/Madison Industries	Old Bridge Township	Radian Technology, Inc.	Rockaway Township
Curicio Scrap Metal, Inc.	Saddles Brook Township	Reich Farms	Pleasant Plains
D'Impetio Property	Hamlton Township	Renora, Inc.	Edison Township
Dayco Corp./L.E. Carpenter Co.	Wharton Borough	Ringwood Mines/Landfill	Ringwood Borough
		Rockaway Borough Well Fields	Rockaway Township
De Rewal Chemical Co.	Kingwood Township		
Deliah Road	Egg Harbor Township	Rockaway Township Wells	Rockaway
Denzer & Schafer X-Ray Co.	Bayville	Rocky Hill Municipal Wells	Rocky Hills Borough
Diamond Alkali Co.	Newark	Roebling Steel Co.	Florence
Dover Municiple Well #4	Dover Township	Sayreville Landfill	Sayreville
Ellis Property	Evasham Township	Scientific Chemical Processing	Carlstadt
Evor Phillips Leasing	Old Bridge Township	Sharkey Landfill	Parsippany/Troy Hills
Fair Lawn Well Field	Fair Lawn	South Jersey Clothing Co.	Minotola
Florence Land Recountoring Landfill	Florence	Spence Farm	Plumstead Township
		Swope Oil & Chemical Co.	Pennsauken
Fried Industries	East Brunswick Township		South Kearney
Garden State Cleaners	Minotola	Syncon Resins	Tabernacle Township
GEMS Landfill	Gloucester Township	Tabermacle Drum Dump	Orange
Glen Ridge Radium Site	Glen Ridge	U.S. Radium Corp.	East Orange
Global Sanitary Landfill	Old Bridge Township	Universal Oil Products (Chemical Div.)	
Goose Farm	Plumstead Township		
Helen Karmer Landfill	Mantua Township	Upper Deerfield Township Sanitary Landfill	Upper Deerfield Township
Hercules, Inc.	Gibbstown		
Higgins Disposal	Kingston	Ventron/Veisicol	Wood Ridge Borough
Higgins Farm	Franklin Township	Vineland State School	Vineland
Hopkins Farm	Plumstead Township	Waldwick Aerospace Devices, Inc.	Wall Township
Imperial Oil Co./ Champion Chemicals	Morganville		
Industrial Latex Corp.	Wallington	White Chemical Corp.	Newark
Jackson Township Landfill	Jackson Township	Williams Property	Swainton
		Willison Farms	Plumstead Township
JIS Landfill	Jamesburg/S. Brunswich	Witco Chemical Co.	Oakland
Kaufman & Minteer Inc.	Jobstown	Woodland Route 532 Dump	Woodland Township
Monroe Township Landfill	Monroe Township		
		Woodland Route 72 Dump	Woodland Township
Montclair/West Orange Radium Site	Montclair/W. Orange		

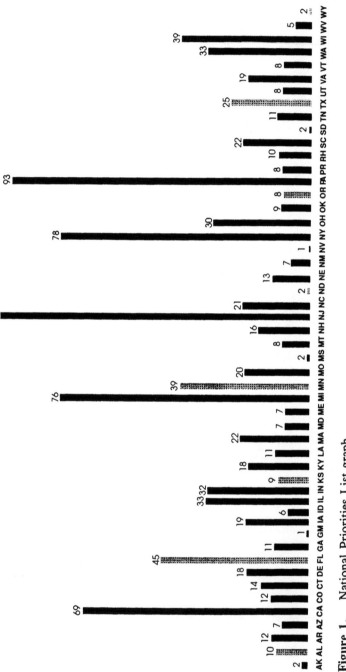

Figure 1. National Priorities List graph.

ABATEMENT ACTION

Along with any other action taken by a state or local government, if the President feels that there may be an immanent and substantial endangerment to the public health or welfare to the environment due to a release or threatened release of a hazardous substance from a facility, he may tell the Attorney General of the United States to get relief as necessary to abate such danger. The district court of the United States in the district in which the threat occurs shall have jurisdiction to grant such relief as the public interest. After notifying the state the President may take other action including issuing such orders as may be necessary to protect public health and welfare and the environment.

If any person refuses to comply with any order of the President under abatement action, he may have action taken against him by the United States district court in which he may be fined up to $5000 per day in which the violation occurs or for as long as he fails to comply.

After meeting with the Attorney General, the Administrator of the Environmental Protection Agency published guidelines for using the immanent hazard enforcement and emergency response authorities and other existing statutes administered by the Administrator of the Environmental Protection Agency to carry out the responsibilities created by CERCLA. The guidelines are consistent with the national hazardous substance response plan. The plan includes "the assignment of responsibility for coordinating response actions with the assurance of administrative orders, enforcement of standards and permits, the gathering of information and other imminent hazard and emergency powers authorized by the Federal Water Pollution Control Act, Solid Waste Disposal Act, Safe Drinking Water Act, Clean Air Act, and Toxic Substance Control Act."

REPORTABLE QUANTITIES

The Administrator will revise as needed the regulations designating as hazardous substances in addition to those mentioned before, such elements, compounds, mixtures, solutions, and substances, that when released may present a danger to the public health or welfare or the environment, and he will announce regulations establishing that quantity of a hazardous substance that when released needs to be reported to the

National Response Center. This will be known as the Reportable Quantity. The Administrator may also determine that one quantity may be the reportable quantity for a hazardous substance no matter what the medium of the hazardous substance.

As soon as a person has knowledge of a release of a hazardous substance from his vessel, off-shore facility, or on-shore facility, that is equal or greater than the reportable quantity he must immediately notify the National Response Center. The National Response Center will notify all appropriate government agencies, including the governor of any affected state.

Other than a federal permitted release, any person in charge of a vessel which a hazardous substance is released into navigable water of the United States, or any person in charge of a vessel which has a hazardous substance released that may affect natural resources belonging to the United States, or any person in charge of a facility which a hazardous substance is released in a quantity equal to or greater than the reportable quantity who fails to notify the appropriate agency of the United States government after finding out about the release can be fined up to $10,000 and sentenced to one year in prison or both if convicted. If notification is made immediately no misuse of the information can be used against any person in a criminal case, except a prosecution for perjury or for giving a false statement.

Any person who owns and operates or who owned or operated a facility where hazardous waste was stored, disposed of, transferred from, or any one who arranges for transport of hazardous waste to such a facility, must notify the Administrator of the Environmental Protection Agency of the existence of such a facility unless the facility has a permit issued under Subtitle C of the Solid Waste Disposal Act. The person must specify the amount and type of hazardous substance to be found there and any known or suspected releases of the substance. In most cases the Administrator may require greater information in the notice. The Administrator will then inform the state agency, or any department the government of that state has designated to receive such notice. If a person knows of such a facility and fails to inform the Administrator, that person can be fined $10,000 or imprisoned up to one year or both, if convicted. Also, any person who fails to report such a facility will not be entitled to any limitations on liability or to any defense to liability mentioned in the liability section of this chapter.

For fifty years after the establishment of a record or any earlier time as a waiver is obtained, it is illegal for any person to destroy, mutilate, erase, dispose of, conceal, or otherwise make unavailable or unreadable or falsify any records of a release. If convicted of this, a person could be fined up to $20,000 or imprisoned or both.

A person may apply to the administration of the Environmental Protection Agency for a waiver to declare the release is a federally permitted release. The Administrator is authorized to give such a waiver, if in his discretion, the waiver will not interfere with the attainment of the provisions of CERCLA. The Administrator will implement rules to inform the parties of the proper application procedures and conditions for approval of a waiver.

The Administrator may require a person to retain records of releases for a longer period of time than mentioned above, if he determines it to be necessary to protect the public health and welfare.

No notification is needed for a release of a hazardous substance if it is to be reported under Subtitle C of the Solid Waste Disposal Act and which has been reported to the National Response Center, or the release is a continues release stable in quantity and rate from a facility which notification has been given to the Administrator of the Environmental Protection Agency. Notification for this type of release must be made once a year or when there is an increase in the quantity of any hazardous substance above that previously reported.

RESPONSE AUTHORITIES

Whenever any hazardous substance is released or there is a threat of release into the environment, the President is authorized to remove or arrange for the removal of, and provide remedial action relating to such hazardous substance and take any response measures consistent with the National Contingency Plan, unless the President feels that the removal or remedial action will be done properly by the owner or operator of the facility.

Whenever the President is authorized to act according to this section, he may undertake an investigation, monitoring, surveys, testing, and other gathering as he feels is necessary to identify the existence and extent of the release. He may also investigate into the extent of the danger to public health or welfare to the environment. In addition, the

President may undertake such planning, legal, fiscal, and other studies or investigations as he may feel necessary or appropriate to plan and direct response action, and to enforce the provisions of this act.

Unless the President finds that continued response actions are immediately required to prevent any further damage or release, assistance will not otherwise be provided. The President shall consult with the affected state or states before determining any remedial action to be taken. He will not provide any remedial action unless the state in which the release occurs first signs a contract or a cooperate agreement with the President assuring the state will assure all future maintenance of the removal and remedial actions provided for the life of actions, or assure the availability of a hazardous waste disposal facility in compliance with the requirements of Subtitle C of the Solid Waste Disposal Act for any off site treatment. The state will also agree to pay for a determined amount for costs of remedial action, including all future maintenance.

The President will select appropriate remedial actions determined to be necessary to carry out the action which provide for cost-effective response and provide for a balance between the need for protection of public health and the availability of amounts from the Fund.

If the President determines that a state has a capacity to carry out any or all of the remedial actions, he may make a contract or cooperate agreement with the state to be reimbursed for the reasonable response costs from the fund.

In rewarding contracts to a company engaged in response actions the President or the state, shall require compliance with federal health and safety standards of this act by contractors as a condition of such contracts. All laborers employed by contractors involved in the remedial action will be paid wages that are not less than those of projects of a similar nature. The President will not approve any funding without first getting assurance that required labor standards will be maintained upon the construction work.

Under this act there was established the Agency for Toxic Substance and Disease Registry, which will report directly to the Surgeon General. The Administrator of the registry will in cooperation with the United States, maintain a national registry of serious diseases and illness and a national registry of persons exposed to toxic substances. He will in cooperation with the state and other agencies of the federal government, maintain a complete list of areas closed to the public or otherwise restricted in use because of toxic substances contamination. The registry will provide medical care and testing to exposed individuals. The Administrator will conduct periodic survey and screening programs to determine relationships between exposure of toxic substances and illness. In cases of public health emergencies, exposed persons shall be eligible for admission to hospitals, and other facilities and services operated or provided by the Public Health Service.

EMPLOYEE PROTECTION

Under CERCLA no person will be fired or discriminated against in any way for providing information to a state or federal government that causes enforcement of the provisions of CERCLA.

If a person feels that he was fired or discriminated against by his employer can before thirty days apply to the Secretary of Labor for a review of such firing of discrimination. If the Secretary of Labor finds that there has been a violation, after a hearing, he may have the employee rehired and compensated. If he finds that there was no such violation, he shall issue an order denying the application. Where there is a decision under this section to abate such a violation, at the request of the applicant a sum equal to the total amount of all costs and expenses, including attorney fees determined by the Secretary of Labor to have been reasonably incurred by the applicant, shall be assessed against the person committing such violation. This section will have no application to any employee who acts without discretion from his employer deliberately violates any requirements of the act.

USE OF FUNDS

Section 11 of CERCLA describes the use of the Hazardous Substance Response Trust Fund. The schedule the payment of government response costs incurred under the Response Authorities Section of this act, as well as these costs incurred by any other person who acts under the National Contingency Plan, provided that these costs are approved and certified by the responsible federal official.

Another use of the funds money is the payment of claims made by the President in his capacity as trustee of the natural resources of the United States. These claims may be payable, but are not satisfied under the Clean Water Act for injury to, destruction, or loss of natural resources.

The Response Trust Fund is also to be used to cover the assessment costs of short- and long-term injury, destruction, or loss of any natural resource due to the release of a hazardous substance. It also covers the costs of restoring, rehabilitating, or replacing these resources.

Also payable from the fund is the cost of implementing a program to identify, investigate, and take action against hazardous substance spills.

The trust fund pays for the cost of creating the Agency for Toxic Substance in order to study long-term health effects of exposure of hazardous substance to carry out epidemiologic studies and for provisions of diagnostic services.

Another use of the fund is the compensation of the costs of supplementary equipment and services in addition to that provided through non-federal sources; and of the establishment and maintenance of damage assessment capability for federal agencies and other responsible teams.

The fund is also used to offset the costs of establishing a program to protect the health and safety of high risk workers, that is those involved in the response and removal actions.

This section of the act states that no claims will be paid for damages to natural resources sustained before CERCLA was enacted, or those sustained as a result of long-term exposure to ambient concentrations of air pollution.

If any claims which exceed the money in the fund are made, those claims will not be paid until there is sufficient money in the fund. If total claims exceed the amount in the fund, then they will be paid in full in the order they were determined.

Regarding regulations, the section authorizes the President to promulgate certain rules and regulations. These include the designation of federal and certain state officials to obligate money in the fund to be used according to this section. These regulations also address the notification of potential injury parties in the event of a hazardous substance release from a vessel or facility. This notice must be provided by the responsible owner or operator. However, in the case of release from a public vessel the President becomes responsible for providing notice to potential injured parties.

As far as damages for injury to, destruction of, or loss of natural resources as a result of a spill, this section states that these damages must be assessed by President-appointed federal officials and that assessment will have the force and effect of a rebuttable presumption on behalf of any claimant in any administrative proceeding.

The trust fund may not be used for the restoration, rehabilitation, or replacement of any natural resources until a plan has been proposed, and then adopted by the involved state government or federal agency, and public notice and comment have taken place. The only exception to this is if there is a need for emergency action, for example to prevent or reduce continuing damage to a natural resource.

The money in the Post-Closure Liability Fund shall be used by the President for the same purpose as the Hazardous Substance Response Trust Fund, except that it will be done with respect to a hazardous waste disposal facility whose liability has been transferred to the former fund under the liability section of CERCLA.

The Inspector General of each department or agency which obligates money for the trust fund must provide an audit review to audit all uses of the fund such as payments and reimbursement. This is done to ensure proper administration of the fund and to ensure that claims are considered in an appropriate and timely manner.

Finally foreign claims may be asserted to the same extent as nature claims if the release occurs in navigable waters or territorial seas of the foreign country; if the claimant is not otherwise compensated; and if recovery is authorized by an agreement between the United States and the foreign country, or if that country provides similar compensation for the United States claims.

CLAIMS PROCEDURE

Section 112 of the Title I of CERCLA describes the claim procedure. In order to make a claim against the fund, the claim must first be asserted against the owner or operator of the vessel or facility responsible for the release of a hazardous substance; or to any other person who may bear liability. If this initial claim is not satisfied within 60 days, then the claimant has two choices. He may either start court proceedings against the alleged responsible party, or he may file a claim against the trust fund.

Once a claim has been filed and received all known affected parties must be notified by the President. He will then try to settle the claim between these parties. If an agreement is reached, then no further claims can be made against the fund.

If the liable party is unknown, a settlement against the trust fund will be made between the claimant and the President.

In order to implement this procedure, the President may also use the services of private insurance and claims adjusting organizations or state agencies. Under special circumstances, the services of federal organizations may be used as well.

Private organizations which are employed shall be paid under contract. If a state agency has provided its services, then no compensation can be made or any claim filed by that state or its agencies, unless the President approves the payment.

If a settlement is not reached within 45 days of filing, the President may make and pay an award of that claim. If the claimant is not satisfied he may appeal the award. If the President chooses not to make an award, then the claim is submitted to the Board of Arbitrators for a decision.

This is implemented by the establishment of the Board of Arbitrators after 90 days of enactment of CERCLA. Selection of the board member will be made using the procedures of the American Arbitration Association. However, no regular employee of the President or federal personnel can act as a member of the board.

The hearing will be public and informal and the claimant will have to prove his claim. Once a decision has been made, it must be rendered in writing within 90 days of the submission. This time unit may be extended either by the President, or if all involved parties agree to it. The decisions are considered final and binding but appeals may be made within 30 days.

If no appeal is made during the allowed time or if a final judgement has been made on an appeal, then the award must be paid out of the fund within 20 days.

Once a claim has been paid by the fund, the United States government has the right to cover the costs of removal and damages for which the claimant has been compensated. As a result, the President may direct the Attorney General to initiate action against the liable party. If any other person compensates the claimant, then that person is subrogated the above mentioned rights.

The time limit for asserting a claim or commencing action is 3 years from the date the loss is discovered, or the date CERCLA is enacted, whichever is later.

Finally, this section states that no person who presents a claim against the fund has waived his right to assert another claim which has the same origin, but which is not covered by the fund.

LITIGATION JURISDICTION AND VENUE

Exclusive original jurisdiction over all claims is held by the United States district court. The venue for all controversies shall be the district court in which the incident occurred, or in which the defendant may be found. The only exception to this concerns application for the review of any promulgated regulations under this act. This application must be made in the Circuit Court of Appeals of the United States for the District of Columbia, within 90 days of the promulgation date.

The residence of the Hazardous Substance Response Trust Fund may be considered to be the District of Columbia.

These provisions do not apply to the assessment of tax collection under Title II of this act, or the review of regulations promulgated under the Internal Revenue Code of 1954.

This act does not make moot any litigation concerning hazardous substances released, which was initiated before the enactment of CERCLA.

RELATIONSHIP TO OTHER LAWS

This act does not prevent any state from imposing additional liability regarding spills within that state.

This section also states that if a claimant receives compensation under this act, then compensation for the same removal costs or damages may not be sought under another state or federal law or vice versa.

With a few exceptions, no one has to contribute a fund which will compensate claimants if those costs are compensable under CERCLA. However, states may use general revenues to set up such a fund or may collect taxes in order to prepare for responding to spills in that state.

AUTHORITY TO DELEGATE ISSUE REGULATIONS

The President has the authority to delegate any of his assigned duties and to promulgate regulations which are necessary for the implementation of Title I of CERLCA.

TITLE II - HAZARDOUS SUBSTANCE RESPONSE REVENUE ACT OF 1980

Subtitle A of this Title II concerns the imposition of taxes on petroleum and certain chemicals.

Section 4611 of Subchapter A describes the tax on petroleum. It imposes a tax of $0.79 per barrel of crude oil entering a United States refinery, as well as on petroleum products entering the United States for the purpose of consumption, use, or warehousing. This same tax is also imposed on domestic crude oil which is used in or exported from the United States, and which had not been previously taxed.

The only exception to this applies to crude oil which is used for the extraction of oil or natural gas on the premises where the crude oil has been produced. The liability for paying taxes fall on:

1. The operator of the United States refinery receiving crude oil.

2. The person importing petroleum products into the United States for consumption, use, or warehousing.

3. The person who uses or exports the previously untaxed crude oil.

The imposed taxes where slated to terminate on September 30, 1985, except that if on September 30, 1983 or September 30, 1984, the unobligated balance in the fund exceeds $900,000,000 and it has been determined that the balance will exceed $500,000,000 one year later if no tax is imposed during that time, then those taxes under Sections 4611 and 4661 will not be imposed for the following calendar year.

Section 4661 of Subchapter B describes the tax on certain chemicals.

In general a tax is imposed on any taxable chemicals which is sold by its manufacturer, producer, or importer.

The amount of tax imposed is determined in accordance with a certain given table, and it ranges from $0.22 per ton of potassium hydroxide to $4.87 per ton of benzene.

The imposition of this tax terminates at the same time as the imposition of the petroleum tax Section 4611. (In 1986, the Superfund Amendments and Reauthorization Act (SARA) renewed CERCLA, and raised the funding from $1,400,000 to $8,500,000).

Under Section 4662, a taxable chemical is any substance which is listed in the above mentioned table and which is manufactured or produced in the United States, or is exported into the United States for consumption, use, or warehousing. Exceptions to this rule include:

1. Methane or butane used as a fuel.

2. Substances used in the production of fertilizer.

3. Sulfuric acid produced as a by-product of air pollution control.

4. Coal-derived substances.

The provisions of this section were effective from April 1, 1981. Refer to Table 2.

SUBTITLE B - ESTABLISHMENT OF HAZARDOUS SUBSTANCE TRUST FUND

Section 221 of Subtitle B CERCLA creates the Response Trust Fund and describes the transfers to, and expenditures from the fund. Money is transferred to the fund from the Treasury in the following amounts: (1) that received under Sections 4611 or 4661 of the Internal Revenue Code of 1954; (2) that recovered for this trust fund under CERCLA; (3) that recovered under section 311(b)(B) of the Clean Water Act; and (4) that collected as punitive damages under the liability section of CERCLA.

TABLE 2	
TAX ON CERTAIN CHEMICALS	
In the case of:	The tax is the following amount per ton
Acetylene	$4.87
Benzene	4.87
Butane	4.87
Butylene	4.87
Butadiene	4.87
Ethylene	4.87
Methane	3.44
Naphthalene	4.87
Propylene	4.87
Toulene	4.87
Xylene	4.87
Ammonia	2.64
Antimony	4.45
Antimony Trioxide	3.75
Arsenic	4.45
Arsenic Trioxide	3.41
Barium Sulfide	2.30
Bromine	4.45
Cadmium	4.45
Chlorine	2.70
Chromium	4.45
Chromite	1.52
Potassium Dichromate	1.69
Sodium Dichromate	1.87
Cobalt	4.45
Cupric Sulfate	1.87
Cupric Oxide	3.59
Cuprous Oxide	3.97
Hydrochloric Acid	0.29
Hydrogen Fluoride	4.23
Lead Oxide	4.14
Mercury	4.45
Nickel	4.45
Phosphorus	4.45

TABLE 2 (Continued)

TAX ON CERTAIN CHEMICALS

In the case of:	The tax is the following amount per ton
Stannous Chloride	$2.85
Stannic Chloride	2.12
Zinc Chloride	2.22
Zinc Sulfate	1.90
Potassium Hydroxide	0.22
Sodium Hydroxide	0.28
Sulfuric Acid	0.26
Nitric Acid	0.24

Authorization is also made for appropriations of $44,000,000 for each fiscal year from 1981 to 1985. In addition, if the total authorized amount has been collected before October 1, 1984, then the required balance shall also be appropriate in 1985.

If there is an unobligated balance remaining in the fund of Section 311 of the Clean Water Act before CERCLA is enacted, than one half of these monies must be transferred to the Response Trust Fund.

Finally, transfers to the Response Trust Fund also include money collected under Section 504(b) of the Clean Water Act during any fiscal year.

Expenditures from the Response Trust Fund include:

1. Costs of response to releases or threats of releases of hazardous substances.

2. Claims which are assessed and can be compensated, but not under Section 311 of the Clean Water Act.

3. Claims for injuries to or destruction or loss of natural resources.

4. Related costs as described in Section 111(c) (uses of fund) of CERCLA.

Section 222 of CERCLA states that all claims assessed against the Response Trust Fund may be paid only out of the fund, and not from any other source. These claims are paid in full according to the order in which they were finally determined.

ADMINISTRATIVE PROVISIONS

The monies which were previously mentioned in Section 221 are transferred from the treasury to, the Response Trust Fund on a monthly basis. If an incorrect amount is transferred then the appropriate adjustment is made to subsequent transfers.

The Secretary of the Treasury is the trustee of the Response Trust Fund and must submit a report to Congress for each fiscal year ending on or after September 30, 1980. As well as the current condition and operations of the fund, the report must include expectations for the next five fiscal years.

The Secretary is also required to invest a portion of the trust fund in public debt securities. Any income thus generated, will become a part of the fund.

In addition monies are required to further the objectives of the trust fund, provisions have been made which authorize the appropriation of these sums as repayable advances. However, repayable advances may not exceed the estimated amount which will be transferred to the trust fund in the following year.

After March 31, 1983, advances are not made to the trust fund in order to pay for response costs. The only exception in the case is a catastrophic spill. If costs other than these have to be paid by an advance, those amounts may not exceed a third of the estimated transfer to the fund in the following 12 months.

SUBTITLE C - POST-CLOSURE TAX AND TRUST FUND

Section 281 Subchapter C describes the tax in hazardous wastes. Section 4681 imposes a tax on hazardous waste that is received at a qualified waste disposal facility. The amount of the tax is $2.13 per dry weight ton of hazardous waste. This tax is imposed on the owner or operator of the facility.

Section 4682 defines hazardous as those wastes which have the characteristics identified under Section 3001 of the effective Solid Waste Disposal Act; or which are subject to the reporting and recovery requirements of Sections 3002 and 3004 of that act. Also defined is "qualified hazardous waste disposal facility"--this is a facility which has received a permit, or is enjoying interim status under Section 3005 of the Solid Waste Disposal Act.

The tax does not apply to hazardous waste which will be removed from the disposal facility once it is closed. This tax affects hazardous waste that is received after September 30, 1983, unless the Post-Closure Liability Trust Fund contains on unobligated balance of more than $200,000,000 on September 30, of any subsequent calendar year. In this case the tax will not be paid during the following 12 months.

POST-CLOSURE LIABILITY TRUST FUND

Section 232 creates the Post-Closure Liability Trust Fund which is available for those purposes described in the "Liability" and "Uses of Fund" section of CERCLA.

The provisions regarding repayable advances to the Response Trust Fund also apply to the Liability Trust Fund, except that those advances must not exceed $20,000,000.

TITLE III - MISCELLANEOUS PROVISIONS

Section 301: This section called for the President to submit a report to Congress which cover:

1. The effectiveness of the act and fund in helping the government to respond and remediate hazardous substance spills.

2. A summary of past receipt and expenditures.

3. A projection of future funding needs and the risk posed by projected releases which create these needs.

4. Recovery of fund disbursements from liable parties.

5. A record of state participation in the implementing of CERCLA.

6. An assessment of the feasibility and desirability of a schedule of taxes.

7. Recommendations for legislative changes in order to improve the effectiveness of CERCLA.

8. An exemption from or an increase in the substance or amount of taxes imposed on certain metallic oxides under Section 4661 of the Internal Revenue Code of 1954, and also on substances utilized in fertilizer production.

Also due at the same time is another report to Congress by the Administrator of the EPA. The report is to identify additional waste which have been designated as hazardous under Section 3001 of the Solid Waste Disposal Act, and recommend appropriate tax rates for those wastes for the Liability Trust Fund.

The report must have been submitted within three years of the enactment of CERCLA. The report described the necessity for the adequacy of the revenue raised as related to projected future requirements of the Liability Trust Fund.

The President shall also conduct a study on private insurance protection, its adequacy, and competitiveness of the market. Results and recommendations were submitted two years after the enactment of CERCLA.

The President shall promulgate regularities for assessing damages for injury to, or loss or destruction of natural resources arising from a spill, under CERCLA and Section 311(f)(4) and (5) of the Federal Water Pollution Control Act.

These regulations shall specify standards procedures and alternative protocols for conducting assessments and shall identify best available procedure. The regulations shall be reviewed and revised every two years.

The Administrator of the EPA shall conduct a study and report on the issues, alternatives, and policy considerations involved in selecting locations for hazardous waste treatment, storage, and disposal facilities.

A study shall be conducted to review the adequacy of existing common laws and statutory remedies in compensating for harm to man and environment as the result of a hazardous substance release. A report shall be submitted to Congress describing the result and offering recommendations.

Lastly the President shall conduct a study and shall modify the nation contingency plan to provide for the protection of the health and safety of employees involved in response actions.

EFFECTIVE DATES, SAVINGS PROVISION

Unless otherwise listed, all provisions become effective on the date that CERCLA was enacted.

Any regulations under Section 311 of the Clean Water Act which is effective before CERCLA is enacted, but is repeated or superseded by that act, become effective under CERCLA and remains in effect until it is superseded by new regulations.

This act does not affect a person's obligation of liability under other federal or state laws regarding hazardous substance releases.

EXPIRATION, SUNSET PROVISION

Section 303 states that unless reauthorized by Congress, CERCLA will no longer have the authority to collect taxes after September 30, 1985, or when a total of $1,380,000,000 is collected under Sections 4611 and 4661 of the Internal Revenue Code of 1954 (See Section 4661 of Subtitle B).

CONFORMING AMENDMENTS

Section 304 of CERCLA repeats subsection (b) of Section 504 of the Federal Water Pollution Control Act.

All funds which have been collected under this section as well as half of the unobligated balance under Section 311(k) of the Federal Water Pollution Control Act shall be transferred to the fund created under the Hazardous Substance Revenue Act of 1980.

The provisions of CERCLA apply in any conflict between those provisions and those under Section 311 of the Federal Water Pollution Control Act.

LEGISLATIVE VETO

If a regulation is being promulgated or repromulgated, the head of the involved department or agency must submit a copy to the Senate and the House of Representatives. The rule or regulation will not become effective if:

1. Both the House of Congress adopt a concurrent resolution and disapprove the rule or regulation. This must occur within 90 calendar days of continuous session after the promulgation date.

2. Within 60 calendar days, one House of Congress adopts a concurrent resolution and transmits this to the other house, and that resolution is not disapproved by the other House within 30 calendar days of continuous session after the transmittal.

If no House of Congress reports or is discharged from further consideration of concurrent resolution which disapproves the rule or regulation and neither adopts such a resolution within 60 calendar days after promulgation, then the rule or regulation goes into immediate effect.

In the opposite case, the rule or regulation may go into effect not sooner than 90 calendar days of continuous session, unless otherwise disapproved.

Congressional inaction on, or rejection of, a resolution or disapproval does not imply approval of the rule or regulation.

TRANSPORTATION

All hazardous substances which are listed under CERCLA shall be listed as a hazardous material under the Hazardous Material Transportation Act.

If there is a release of hazardous substance during its transportation prior to listing under CERCLA or Hazardous Material Transportation Act, then the common or contract carrier shall bear liability unless he can prove no actual knowledge of the identity or nature of the substance released.

Violation of the regulations will result in civil penalties not exceeding $20,000.

SEPARABILITY

Section 308 states that even if one provision of CERCLA is invalid when applied to a certain person or circumstances that provision still applies to other persons or circumstances, and remainder of the act is not affected.

SUPERFUND REFORM ACT OF 1994

The Comprehensive Environmental Response, Compensation, and Liability Act if not reauthorized by Congress, will expire in 1995. With the deadline approaching and with all criticism against the effectiveness of the existing statute, the Clinton administration and the 103rd Congress have undertaken a major revision of this legislation. Referred to as the Superfund Reform Act of 1994, H.R. 3800 and S. 1834 have emerged as the main means for amending the Superfund Law. On June 14, 1994, the House Subcommittee on transportation and Hazardous Materials adopted amendments to H.R. 3800, followed on May 18, 1994, by the Committee on Energy and Commerce. On June 14, 1994, The Senate Environmental and Public Works Subcommittee on Superfund, Recycling, and Solid Waste Management approved moving S. 1834 out of Subcommittee for consideration by full Senate Environmental and Public Works Committee. The changes presented in H.R. 3800 and S. 1834 are as follows:

Government Enforcement and Response Authority

CERCLA authorizes the government to respond to releases of hazardous substances that pose an eminent endangerment to human health or the environment. Under the proposed bills individual removal actions conducted by the EPA may extend for up to two years, and cost up to $4 million (an increase of one year and $2 million). Under the current law, sites that are potential candidates for Superfund are ranked using a scoring system developed by the EPA (HRS II). H.R. 3800 calls for the EPA to revise the manner in which some sites are scored using HRS II. Multiple sites may be scored as if they are a single site where potential exposure or releases from the site overlap and effect the same population. This change will allow the EPA to target groups of sites within heavily industrialized areas for inclusion on the NPL.

Liability Exemption and De Minimis Parties

In attempt to address the concerns of joint and several liability on small business and on those parties responsible for small quantities of hazardous substances at a site, H.R. 3800 and S. 1834 created an exemption from liability for those who are responsible for 55 gallons of liquid or less, or less than 100 pounds of solid at a site, referred to as de micromis parties. Also, certain small businesses and small non-profit organizations are shielded from liability as are those who acquire contaminated property through donation, as long as they do not contribute to or exacerbate the release.

For those persons who do not qualify for the above described exemptions, but whose contribution to a site are relatively small there are expedited settlement opportunities if they qualify as a de minimis PRPs. De minimis PAPS are defined as those whose "volumetric contribution of materials constituting hazardous substance is minimal in comparison to the total volumetric contributions of the material containing hazardous substances at the facility; such individual contribution at the facility, unless the administrator identifies a different threshold based on site-specific factors and the potentiality responsible party's hazardous substances do not prevent toxic or other hazardous effects that are significantly greater than those of the other hazardous substances at the facility."

Allocation System

If adopted H.R. 3800 and S. 1834 would significantly limit the scope of joint and several liability and potentially responsible parties now face at multi-party sites. The proposed approach will allow the PRPs for a site to select the person, through a ballet process, who will allocate responsibility amongst them. Allocators must be drawn from an approved EPA list. Others not on the list may be accepted at the discretion of the EPA.

After gathering all the necessary information, the allocator will make a non-binding apportionment based upon the following factors:

- The amount of hazardous substance contributed by each allocation part.
- The degree of toxicity of hazardous substances contributed by each allocation party.
- The mobility of hazardous substance contributed by each allocation party.
- The degree of involvement of each allocation party in generation, transportation, treatment, storage, or disposal of the hazardous substance.
- The degree of care exercised by each allocation party with respect to the hazardous substance, taking into account the characteristic of the hazardous substance.
- The cooperation of each allocation party in contributing to the response action and in providing complete and timely information during the allocation process.

The allocator will assign percentages of responsibility to each PRP and identify orphan shares (related to the portion of the hazardous substance for whom a responsible party cannot be assigned). The orphan shares will be distributed amongst the PRPs according to their respective share of liability. Cost of conducting the allocation will be considered orphan shares.

The EPA and the Department of Justice may reject the allocation only if it is unreasonable in the light of all the facts, was illegally or fraudulently prepared, or if substantial procedural error occurred during the allocation process. If it is rejected, a second allocator will be appointed who will make a new determination based upon the record generated during the first attempt.

While the allocation is non-binding on PRPs, if PRP makes a settlement offer within 90 days and in accordance with the final allocation percentages the EPA is not allowed to seek a higher percentage from the PRP. The EPA may add a specific percentage as a premium on top of the allocated percentage to cover the government's litigation risk with respect to PRPs who have not settled. Non-setting PRPs will be jointly and severally liable for all response costs not covered through settlements with other persons.

Under this new allocation system, the EPA will absorb orphan shares and be responsible for costs of the allocation process. H.R. 3800 authorizes $300 million per year for the EPA to pay orphan shares.

The non-binding process creates opportunities for PRPs to bring in as much information as available on other PRPs and allows parties to review this information in a confidential setting. The cost of considering parties ultimately determined to have no liability will be apportioned to party who brought them into the process.

Municipal Liability

Small quantities of household hazardous waste included within municipal solid waste have in many instances, caused municipalities to be drawn into Superfund litigation, where their waste have been commingled with other hazardous substances at the industrial landfills or dumps. In these instances, municipalities have faced significant liability due to the volume of waste disposed.

H.R. 3800, in an effort to limit municipal liability, places a 10 percent cap on the amount of liability attributable to municipal solid waste generators at any given site. That is all generator of municipal solid waste as a site will together be liable for no more than 10 percent of the liability allocated to municipal solid waste generator that exceed the 10 percent cap becomes orphan shares to be paid by the government.

H.R. 3800 also exempts from liability these persons who arrange for the transport, disposal, or treatment of only municipal solid waste or sewage sludge and who are the owner, operator, lessee of residential property, a small non-profit organization, or a small business.

Remedy Selection

Under H.R. 3800 the EPA will establish Protective Concentration Levels that will serve as national goals for human health expressed as a single numerical level. To do this, the EPA will define the acceptable exposure concentrations for carcinogens and non-carcinogens. These exposure concentrations will provide the basis for Protection Concentration Levels unless the achievement of such concentrations is technically infeasible or unreasonably costly. H.R. 3800 requires that the Protective Concentration Levels will result in final protection at the 90th exposure percentile of the affected population while the Senate bill proposes to have the EPA establish national risk levels that will provide final protection at approximately the 90th exposure percentile.

Once Protective Concentration Levels are established, a risk assessment is conducted to factor in the exposure pathways and expected points of exposure. The risk assessment will determine whether selected remedies will meet the Protective Concentration Levels. To achieve consistency in risk assessment methodology, these bills direct the EPA to establish a national risk protocol for conducting risk assessment.

Under the existing statute and EPA policy, remedy selection is driven largely by a preference for remedies that reduce the toxicity and mobility of hazardous substances through treatment. Many people feel that contaminated-based remedies H.R. 3800 and S. 1834 now direct that remedies must be cost effective, and that contamination remedies may be appropriate. The legislation does call for treatment of toxic "hot spots" but will even allow for interim contaminant of hot spots pending development or availability of technology that will effectively address the hot spots at some future date.

Public Participation, Community Work Groups, and Environmental Justice

H.R. 3800 has various provisions which call for the EPA and the states to create more opportunity for public involvement in remedial decision making activities. The bill calls for enhancement to the Technical Assistance Grant (TAG) program, more public meetings during Superfund site investigation, remedy selections, and remedy implementation activities, and creation of Community Working Groups and Citizen Information and Access Offices.

Technical assistance grants are available to affected citizen groups to obtain technical assistance in interpreting information with regard to the nature of the hazard at the facility, the remedial investigation and feasibility study, the record of decision; the selection design, and the construction of the remedial action, operation and maintenance, and removal activities at a facility. H.R. 3800 expands the site at which

TAG grants may be awarded to include State Registry sites, and elaborates on acceptable uses of these funds.

Upon either the EPA's determination or by petition of local citizens, a representative public forum, known as Community Working Groups (CWGs) may be established in NPL communities. CWGs will provide an official forum through which local citizens may convey information and views on the Superfund process. The EPA is directed to consult with CWGs on a regular basis throughout the remedy selection process regarding the reasonable anticipated future use of land at the facility and institution controls required to assure that land use determines remain in effect.

H.R. 3800 directs the Citizens Information and Access Offices (CIAO) be established in each state to act as a clearinghouse for information relating to State Registry and NPL sites within the state.

In Section 105(i) of H.R. 3800, the EPA is directed to evaluate major urban areas and any other areas environmental justice concerns may warrant special attention and identify 5 facilities in each region of the EPA that are likely to warrant inclusion on the NPL. These facilities shall accord a priority in evaluation for NPL listing and scoring, and shall be evaluated for listing within 2 years after enactment.

Voluntary Response Programs

Recognizing that voluntary as opposed to government mandated cleanings are preferable in terms of cost efficiency and speed, H.R. 3800 calls for the EPA to establish a program to provide technical, financial, and other assistance, including grants to states to establish and expand voluntary response programs. This program will be designed to provide opportunities for public participation in selecting response action, streamlines procedures, and some degree of government oversight to ensure that cleanups are protective of human health and environment. State programs satisfying the EPA standards will be deemed Qualified State Programs.

Cleanups performed pursuant to a qualified state program shall be presumed to be constant with the NCP for the purpose of private cost recovery claims under CERCLA and will not be subject to listing on the federal NPL. The Senate version also established a state voluntary cleanup program, it goes further in creating an Economic Redevelopment Credit Assistance Program designed to encourage lenders to loan funds for investigation and remediation of contaminated sites. The Voluntary Response Program represents Congressional efforts to further shift cleanup oversight and programs implementation to the state and away from federal government.

State Delegation and Authorization

While states may conduct or oversee pre-medial and response actions, under CERCLA, only the EPA may select remedies and require response actions. H.R. 3800 and S. 1834 propose to allow states to take or require pre-medial and response actions via contracts or cooperate agreements with the EPA. Remedies and enforcement must be selected and conducted in accordance with all the procedures and requirements set forth in Sections 117 and 112 of this act, the NCP and other relevant regulations and guidelines adopted by the EPA This bill allows the EPA to delegate certain responsibilities and certain sites to the state. The Senate bill lists those authorities that the state will have upon delegation and makes clear that the state will not be delegated authority relating to the government's authority to require actions to abate an imminent and substantial endangerment, and state will not have the authority to add or delete sites from the NPL.

Environmental Insurance Resolution Fund

Disputes between insurers over the availability of insurance coverage for response cost incurred at Superfund sites has generated an enormous volume of expensive and complex litigation. Because of widely varying results and with prospects for, litigation continuing into the future, people within the insurance industry and amongst PRPs have called for a legislative solution to this situation. H.R. 3800 offers a solution in the form of the Environmental Insurance Resolution Fund. This fund supported through a tax on commercial insurance premiums will be available to claimants who can demonstrate eligible cost incurred at NPL facilities or during removal actions under CERCLA and NCP.

To account for the variability in state court law pertaining to coverage, the state proposes to allow varying rates of recovery equalling 20, 40, and 60 percent depending upon whether the designated litigation venue in a state where state law favor insurers, claimants, or neither. The statute classifies Florida, Maine, Maryland, Massachusetts,

Michigan, New York, North Carolina, and Ohio as states favorable to insurers and establishes a state percentages category of 20 percent. California, Colorado, Georgia, Illinois, New Jersey, Washington, West Virginia, and Wisconsin are classified as generally favoring the insurers and are designated as a 60 percent category. All other states fall within the 40 percent category.

Phase I Environmental Site Assessment and Environmental Professional Standards

Under existing law, a person acquiring property may be exempt from liability if they can demonstrate that they are an innocent landowner. This term comes from the liability provisions under Section 107(b)(3) of CERCLA and the definition of contracted relationship at section 35 of CERCLA, that exempt from liability those who at the time of acquisition conduct an appropriate inquiry into the previous ownership and users of property. Persons seeking to qualify as innocent landowners have typically undertaken what is referred to as a Phase I Environmental Site Assessment to identify potential liabilities associated with a propriety acquisition. The content and elements of Phase I ESAs, however, have never have been defined by the EPA and therefore, uncertainty remains as to whether any given Phase I ESA will qualify as an appropriate inquiry rendering the landowner innocent.

H.R. 3800 authorizes the EPA to establish standards, or adopt standards developed by others, for conducting appropriate injury and satisfying the conditions of the innocent landowner defense. Section 612 of the proposed legislation further directs the EPA to establish a model state program for organization that train and certify individuals to perform Phase I ESAs.

If enacted H.R. 3800 and S. 1834 will tackle many of these issues that the regulated community has long considered particularly troublesome under existing Superfund program. The new legislation will not only address these areas of concern, but will authorize $9.6 billion for Superfund activities during the next five years, expend the sites subject to inclusion on the NPL, and enhance the EPA's information gathering and response authorities. The Superfund Reform Act of 1994 will also encourage state and public involvement in the Superfund process.

For almost two years, representatives from industry, municipal government, environmental groups, public interest groups, and state and federal government have been involved in discussions aimed at improving the Superfund program. The proposed H.R. 3800 reflects these negotiations and the comprises which have resulted from those negotiations. Not everyone is fully satisfied with all the aspects of the proposed changes to the Superfund process; however, it appears there is a commitment within Congress to reauthorize the law before it expires.

SUGGESTED READING

1. Pollution Law Handbook, Sidney M. Wolfe, Quorum Books, Washington, D.C., 1988.
2. Superfund Reform Act of 1994, H.R. 3800, section 104-130.
3. Superfund Reform Act of 1994, S. 1834
4. United States vs. Fleet Factors Corp., 901, F.2d 1550, 1990.
5. 1994 Book of Lists, Government Institutes, Washington, D.C., 1994.
6. 1980 Public Law 96-510 42 USC 9608.
7. 1986 Public Law 99-499 42 USC 9602.
8. 40 CFR 261.4 - 261.35.

7 THE SUPERFUND AMENDMENTS AND REAUTHORIZATION ACT

HISTORICAL PERSPECTIVE

The Comprehensive Environmental Response, Compensation, and Liability Act (CERCLA), better known today as Superfund, was created out of the need to address past disposal problems. Prior to Superfund there was no national legislation to address past disposal transgressions. Awareness of the problem was peaking in the late 1970s with the disastrous events at the Love Canal in Niagara Falls, N.Y. The Hooker Chemical and Plastic Corporation had used an old canal bed in this area as a chemical dump from the 1930s to 1952. The filled land was then given to the state and subsequently homes and a school were built on it. Years later, toxic liquids including carcinogens leaked out of the dump. Extremely high birth defect and miscarriage rates developed along with liver cancer and nervous disorder diseases. The state paid $10 million to buy the homes and another $10 million in an attempt to clean up the site. In order to prevent further Love Canal incidents, CERCLA was passed into law by Jimmy Carter on December 11, 1980.

SUPERFUND LEGISLATION

To better understand and discuss SARA, a review of the Superfund legislation is required. Unlike most federal environmental laws, Superfund does not focus on compliance with a regulatory system that governs current activities. Rather Superfund addresses site contamination created by past activities. "Superfund established a program to identify sites where hazardous substances have been or might be, released into the environment; to ensure that they are cleaned up by responsible parties or the government; to evaluate damage to natural resources; and to create a claims procedure for parties who have cleaned up sites or spent money to restore natural resources." In general CERCLA was designed to give the federal government the authority to take action to respond to releases or the threats of releases of hazardous substances, pollutants, or contaminants. With the development of the Hazardous Ranking System, CERCLA developed a comprehensive program to prioritize hazardous waste sites nationally. CERCLA also sought to identify and compel potentially responsible parties to cleanup or pay for clean up of sites. A $1.6 billion Hazardous Substance Response Fund (Superfund) was set up to make money available to finance cleanups where the principle responsible parties could not be found or were unable to pay for cleanup. And finally CERCLA was designed to advance scientific and technological capabilities in all aspects of hazardous waste management, treatment, and disposal. Superfund was founded on the premise that the polluter must pay for any problems created by the polluter. So CERCLA was specifically designed to first ensure that the polluter was responsible and paid for cleanup. Only if the principal responsible party could not be found or could not pay were trust fund monies used.

Monies for the trust came from several sources. Monies were created from general tax revenues, interest earned by the trust fund, cost recoveries from principal responsible parties (PRPs) and a tax on the petroleum and chemical industries.

There are seven phases in the Superfund remedial process. The first phase is the Preliminary Assessment which is the first action after site notification and includes a site visit and paper review of site. Site inspection is then performed if deemed necessary and includes groundwater and soil tests to determine the extent of contamination. The site is then ranked using the Hazard Ranking System. This system uses a structured value analysis approach to scoring sites. The Hazard Ranking System score is the primary criterion the EPA uses to determine whether a site is to be placed on the National Priorities List (NPL). The system assigns numerical values to factors that relate to or indicate risk based on the conditions at the site. The factors are grouped into three categories; observed release/route characteristics, waste characteristics, and targets and were combined to obtain category scores. Each category and component factor had a maximum score.

These category scores were then multiplied together within each of the migration pathways (groundwater, surface water, and air) and normalized to obtain a pathway score. The final Hazard Ranking System score was the square root of the sum of the squares of the pathways scores divided by a factor 1.73, which put all the scores on a scale of 1-100. Typically scores greater than 28.6 make the site eligible for the NPL. Sites on the NPL are eligible for long term remedial action.

If only one migration pathway score was high, the Hazard Ranking System (HRS) score could be relatively high. This was certainly an important requirement since some extremely dangerous sites pose threats through only one migration pathway (such as buried drums contaminating only drinking water wells).

The next step in the process is the Remedial Investigation/Feasibility Study. During this stage the nature and extent of the problem is evaluated along with recommended cleanup alternatives. The Remedial Investigation and Feasibility Study are usually concurrent and interactive activities. This stage represents the first phase of remediation once the Superfund site has been placed on the NPL. During this stage the nature and extent of the problem is evaluated along with recommended cleanup alternatives. At the conclusion of the feasibility study alternative remedial actions have been developed, screened, and evaluated. The final remedial alternatives should pass the following nine criteria:

1. Overall protection of human health and environment.

2. Compliance with other Applicable and Relevant Appropriate standards.

3. Long-term effectiveness and permanence.

4. Reduction of toxicity, mobility, or volume through treatment.

5. Short-term effectiveness.

6. Implementability.

7. Cost.

8. State (support agency) acceptance.

9. Community acceptance.

The Record of Decision signifies a significant milestone in the Superfund process. The   purpose of the Record of Decision is to formally document the selected remedy at the site.   The Record of Decision is a legal, technical, and public information document.   After establishing the ROD, the Remedial Design phase can begin.   During this stage, all construction and design plans are finalized for remedial action. During the Remedial Action state, actual construction takes place at the site.   The final phase is operation and maintenance which includes any post action such as maintenance/monitoring etc.

"During the first five year of Superfund several facts became evident: the problem of uncontrolled hazardous substance sites was more extensive than originally thought; and its solution would be more complex and time consuming."

SARA LEGISLATION

The Superfund program was reauthorized by the Superfund Amendments and Reauthorization Act of 1986 on October 17, 1986. These amendments were based largely on part of the EPA's experiences in administering CERCLA during the first five years of its existence. "One of the primary features of the SARA amendments was that the new trust fund was increased to $8.5 billion, which represented approximately a five time increase in the amount of money previously in the fund."   In addition, SARA created a separate one-half billion dollar fund for cleanup of underground storage tanks containing petroleum.

The major source of revenue under the original law was a tax on petroleum and 42 listed chemicals. Revised funding sources are:

- Taxes on petroleum and chemicals which are similar to taxes under the original Superfund.
- A new "environmental" tax on corporations.
- $1.25 billion from general appropriations.
- Costs recovered from responsible parties.
- Punitive damages and penalties under Superfund.
- Earned interest.

SARA also increased the limits on removal actions financed by the trust fund from $1 million over six months to $2 million over one year.

Any removal actions were required to be consistent with any long-term remedial action. SARA stresses the utilization of permanent solutions wherever practical and cost effective and a preference was given to solutions that reduced the toxicity, mobility, and/or volume of waste. Off-site land disposal without treatment was the least desirable alternative.

During the preliminary years of the CERCLA legislation, the EPA realized that settlements with the responsible parties proved to be the most cost-effective methods of cleaning up hazardous sites, so SARA sought to promote settlements between the responsible parties and the government. In order to promote the accurate reporting of releases of hazardous substances, "SARA increases the criminal penalties for failure to report releases and made it a criminal offense to provide false or misleading information regarding these releases." Specifically if you knowingly fail to give facility notice, you personally could be faced with the following criminal sanctions upon conviction:

- "A fine of not more than $10,000, and/or,
- Imprisonment of not more than 1 year."

Penalties for violation of spill reporting and record retention requirements are even more severe. It is considered a violation if you fail to report a release that you know of, report false or misleading information, fail to retain records, or make records unavailable. If convicted, you may face imprisonment of up to three years for the first conviction and up to five years for subsequent convictions (versus one year under the original act) and/or a maximum fine of $250,000 for an individual or $500,000 fine as a corporation. SARA even offered up to $10,000 for information leading to arrests and convictions on CERCLA violations. As an additional enforcement mechanism, civil penalties of up to $25,000 per violation per day for each violation continues could also be imposed. In summary, the SARA amendments greatly increased the penalties for non-compliance.

SARA expanded the government's authority to obtain information concerning a site. The EPA was also given the authority to enter and inspect your property and take any response action deemed necessary. "Refusal to allow the EPA access to property or information could result in fines of up to $25,000 for each day of non-compliance."

SARA promoted public participation and notification during the development of action plans concerning Superfund actions. Technical assistance grants were even permitted to allow local citizens to hire experts to explain the intricacies of the cleanup and potential hazards of the release. Records concerning the site information were required to be made accessible to the public.

Major disasters like that in Bhopal, India, in December 1984, which resulted in 2000 deaths and over 200,000 injuries, may be rare. However, reports of hazardous materials spills and releases are increasing. Thousands of new chemicals are being developed each year. Citizens and communities are becoming more aware and more concerned about accidents involving hazardous chemicals occurring in their community. Communities need to prepare themselves to prevent hazardous incidents from occurring and responding to accidents once they occur. During the initial stages of an accident, local governments will be on their own in the first stages of almost any hazardous materials incident.

In response to these concerns SARA Title III, The Emergency Planning and Community Right-to-Know Act of 1986 was instituted (EPCRA). This legislation was designed to allow communities to be prepared in the event of a hazardous chemical emergency situation and also to increase the public's general knowledge of the threat of hazardous chemicals. SARA Title III established a four-part program as outlined below.

1. Define emergency planning structures at the state/local level and develop local emergency response plans.

2. Require emergency notification of chemical releases.

3. Require notification of chemical use, storage, and production activities.

4. Report annual emissions requirements.

Figure 1 illustrates the various activities that are part of the emergency planning process.

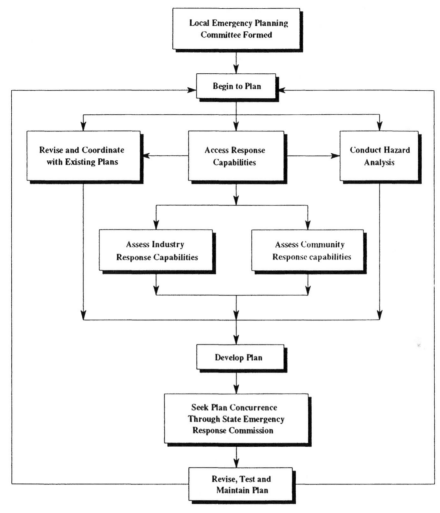

Figure 1.   Overview of planning process.

"The emergency planning sections are designed to develop state and local government emergency preparedness and response capabilities through better coordination and planning, especially at the local level." The state governors were required to set up a state emergency response commission (SERC). This commission should have representation from public agencies, transportation, natural resources, public health, emergency management, etc. The SERC is also delegated with the responsibility of appointing local emergency planning committees (LEPCs). The LEPCs should be represented by local fire, police, public health professionals, transportation, etc. LEPCs must develop an emergency response plan that includes the identification of facilities and hazardous material transportation routes, emergency site procedures, emergency notification procedures, evacuation plans, and a description of training programs along with methods of determining the occurrence of a release and the probable effected population.

Emergency Notification

Any facility that produces, stores, or uses one or more hazardous chemicals must notify the LEPC and the SERC if there is a release of a listed hazardous substance in reportable quantities. The emergency notification must include the following information: chemical's name, an estimate of quantity released into the environment, the time and duration of the release, the medium in which the release occurred, any known acute or chronic health effects associated with the emergency, and the name and telephone number of a contact person.

A follow up notice is also required outlining the response actions taken along with updated information on health risks due to the release.

Table 1 shows examples of the type of facilities where Certain Extremely Hazardous Substances might be found in quantities greater than their threshold planning quantities.

Community Right-To-Know

Any facilities under the Standard Industrial Classification (SIC) Codes 20-39 must make available Material Safety Data Sheets (MSDSs) and submit copies to the LEPC, the SERC, and the local fire department. A revised MSDS must be submitted if significant new information is

	TABLE 1				
	EXTREMELY HAZARDOUS SUBSTANCES (TPQ*)				
Type of Facility	**Ammonia** (100 lbs.)	**Chlorine** (100 lbs.)	**Sulfuric Acid** (500 lbs.)	**Phosgene** (10 lbs.)	**Aldicarb** (100 lbs/ 10k lbs)
Blueprinting Facilities	X		X		
Bulk Storage Facilities	X		X		
Farms	X				X
Frozen Food Processing	X				
Pesticide Distributors					X
Processing Plants	X	X	X	X	
Plumbing, Heating & A/C Companies	X				
Pulp & Paper Plants		X	X		
Retail Stores	X				
Swimming Pools		X			
Warehouses	X		X		X
Water Treatment Facilities		X	X		

*TPQ is 100 lbs. for fine powders and solutions and 10,000 lbs. otherwise.

discovered regarding a chemical.

Toxic Chemicals Release Reporting

Subject facilities must submit a toxic chemical release form for specified chemicals in order to inform both government officials and the public about the release of toxic chemicals into the environment. "This reporting requirement is required of any owner or operator of a facility employing ten or more full-time employees, has a SIC Code of 20-39, and that manufactures, processes, or otherwise uses a listed toxic chemical in excess of specified threshold limits."

IMPACT OF SARA ON CERCLA

Under SARA, the EPA was required to revise and update the Hazardous Ranking System (HRS) in order to more clearly identify which sites on the list needed further investigation and which should be given the highest attention.

Superfund imposes liability for most response costs which can include cleanup, governmental administrative costs, consultants and attorney's fees, and damage to natural resources, including the costs of assessing the injuries. SARA added two additional sources of liability. "One is the cost of any health assessment or health effects study done for the Agency for Toxic Substances and Disease Registry (ATSDR). The other is for interest on all recoverable amounts, from the date payment of a specified amount is demanded in writing, or the date of the expenditure, whichever is later."

Also under SARA, it is now allowable for one PRP to sue other PRPs for cost recovery. This is possible under the following conditions:

- If you voluntarily undertake a response action.
- If you settle with the government, and you paid more than your fair share of costs.
- If the government sues some but not all PRPs at a site for cleanup costs, those PRPs who pay must seek contribution from the PRPs who did not pay.
- If another party (not necessarily another PRP) sues you to recover cleanup cost, you may counterclaim for contribution.

SARA requires that the EPA develop guidelines for preparing non-binding preliminary allocations of responsibility among the parties settling with the government. SARA also identified certain factors that are to be considered when dividing up liability, such as volumes, toxicity, mobility of hazardous substances, strength of evidence, availability to pay, and litigation risks.

"Under SARA the government was given discretion to allow responsible parties to conduct RI/FS and response actions." When a responsible party will perform a response action properly and promptly, the government may allow the party to carry out the action. But the government was not obligated to let responsible parties conduct response actions even if they could do so properly. If the government determined that you are qualified to conduct an RI/FS, you would be able to do so only if:

- The government arranges for a qualified person to oversee and review the RI/FS.
- You agree to reimburse the fund for any costs the government incurs for the oversight arrangement.

Under SARA, it was made easier for citizens to sue PRPs for personal injury. SARA allowed the health assessment information to be used by plaintiffs and also gave the citizen more time to bring action to state court for personal injury or property damage caused by a hazardous substance.

To shield certain landowners from liability, SARA provided for the "innocent landowner" defense. This defense may be used to protect the owner against liability for releases from property that was acquired after a hazardous substance came to be located there. This defense may apply if the owner exercised due care and took precautions against foreseeable acts or omissions of third parties and could establish at least one of the following:

- The property was acquired without reason to know of any hazardous substances was there.
- The owner is a government entity that acquired the property involuntarily or through eminent domain.
- The property was inherited.

The proper inquiries must have been made about the property's previous ownership and use. A successful innocent landowner defense will be rare because the owner must prove that the history of the site was thoroughly investigated. "Experience to date has shown that the innocent landowner defense is very difficult to qualify for and is narrowly applied."

The SARA amendments added statutes of limitation for the government's cost of recovery actions. In general for a removal action (short-term remedy) initial action for cost recovery must begin within three years after the action is completed. For longer term remedial actions, the time is increased to six years.

SARA has somewhat limited the government's response to actual or threatened releases. "The government cannot take removal or remedial action under CERCLA in response to a release or threat of release:

- Of naturally occurring substance from a location where it is naturally found.
- From products that are part of a structure of and result in exposure within, residential buildings or businesses or community structures.
- Into public or private drinking water supplies due to deterioration of the system through ordinary use."

The aforementioned limitations do not apply if the government deems a release or threat of release, a public or environmental emergency, and no other qualified person is able to respond to the emergency quickly. This exemption allows the EPA to act in an emergency. SARA directs the government to give primary attention to releases that may present a public health threat.

CERCLA's RELATIONSHIP TO OTHER LEGISLATION

The EPA's role is to protect human health and the environment. Many environmental laws have been enacted to address release, or threats of releases, of hazardous substances. An understanding of this law is necessary to see where CERCLA, or the Superfund program, fits into the national environmental protection program established by Congress. Each environmental statute has its own particular focus, whether it is to control the level of pollutants introduced into a single environmental medium (i.e., air, soil, water) or to address a specific area of concern, such as pesticides or waste management.

The legislation that serves as the basis for managing hazardous wastes can be divided into three categories:

1. The central statutory authorities are CERCLA and RCRA. The former authorizes cleanup of releases of hazardous substances. The latter creates a management system for current and future hazardous and solid wastes, and authorizes cleanup at hazardous waste management facilities.

2. Several statutes are media-specific and limit the amount of wastes introduced into the air, waterways, oceans, and drinking water.

3. Other statutes directly limit the production of chemical substances and products that may contribute to the nation's waste.

Resource Conservation and Recovery Act

The Resource Conservation and Recovery Act (RCRA), an amendment to the Solid Waste Disposal Act, was enacted in 1976 to address a problem of enormous magnitude--how to safely dispose of huge volumes of hazardous and non-hazardous municipal and industrial waste generated nationwide and to ensure the prevention of future releases. The term "solid waste" by definition, includes traditional non-hazardous wastes such as municipal; refuse and liquid, semi-solid, or gaseous material from industrial, commercial, and mining operations as well as hazardous waste.

The goals set by RCRA are:

- To protect human health and the environment.
- To reduce waste, and conserve energy and natural resources.
- To reduce or eliminate the generation of hazardous waste as expeditiously as possible.

RCRA originally provided regulatory authority to address hazardous waste management, but had limited authority to require cleanup. CERCLA was enacted in 1980 to fill the apparent gap in RCRA and the Clean Water Act authority for remedying past mismanagement of hazardous substances.

The 1984 Hazardous and Solid Waste Amendments (HSWA) significantly expanded the scope and requirements of RCRA. Regulations have been developed and continue to expand based on the HSWA provisions, e.g., Land Disposal Restrictions. In addition, HSWA expanded the EPA's authorities to address releases of hazardous waste or cleanup of wastes released from RCRA hazardous waste facilities. Furthermore, a new program for regulating underground storage tanks was developed under RCRA Subtitle 1. RCRA established four distinct, yet interrelated, regulatory programs (see Figure 2).

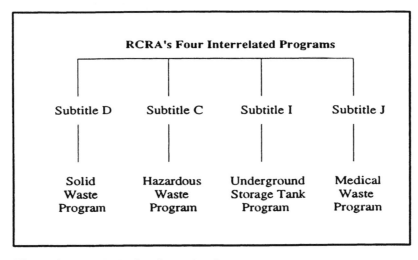

Figure 2. RCRA's four interrelated programs.

1. The **Subtitle D** *Solid Waste Management Program* sets national standards for the management of solid waste (e.g., municipal solid waste landfills).

2. The **Subtitle C** *Hazardous Waste Management Program* sets national standards for hazardous waste management, provides for oversight of state implementation of RCRA, and includes corrective action authorities to address releases to the environment.

3. The **Subtitle I** *Underground Storage Tank (UST) Program* is designed to protect ground water from leaking underground storage tanks.

4. The **Subtitle J** *Medical Waste Program* establishes a two-year demonstration program to track medical waste from generation to disposal.

5. CERCLA is most impacted by RCRA Subtitle C. The RCRA Subtitle C standards for managing hazardous waste affect many CERCLA response decisions, such as which off-site disposal facility to use or which regulatory requirements to consider in implementing on-site response actions.

How RCRA and CERCLA Overlap

RCRA and CERCLA follow roughly parallel procedures in responding to releases. In both, the first step after discovery of a release is an examination of available data to see if an emergency action is warranted. Both programs allow for short-term measures to abate the immediate adverse effects of a release. In RCRA, short-term measures may occur after the investigations. Investigations and formal study of long-term cleanup options are conducted once an emergency has been addressed. When these analyses are completed, both provide the basis for the formal selection of a remedy.

RCRA regulatory requirements are potentially applicable or relevant and appropriate requirements (ARARs) for CERCLA response actions. Thus, many CERCLA response actions must meet the applicable or relevant and appropriate RCRA requirements for on-site actions, unless a waiver is justified under the circumstances. For example, the RCRA Land Disposal Restrictions (LDRs), established under HSWA, may be applicable to a CERCLA response action involving the placement of hazardous waste in a land disposal unit. In order to determine their applicability, the EPA has issued a series of Superfund LDR Guides (LDR Guides 1-8). These guides summarize the major components of the LDRs, such as treatment standards and minimum technical requirements in respect to CERCLA response actions.

In accordance with CERCLA Section 121(d)(3) all wastes shipped off-site for treatment, storage, or disposal must be sent to facilities that have been determined by the EPA to be "acceptable." In order to be acceptable, a facility cannot have any relevant violations of applicable federal or state requirements such as RCRA or TSCA and cannot have any relevant releases.

Imminent Hazards Under RCRA and CERCLA

Both CERCLA and RCRA contain provisions that allow the EPA to require persons contributing to an imminent hazard to take the necessary actions to cleanup releases. Under CERCLA Section 106, the EPA has the authority to abate an imminent or substantial danger to public health or the environment that results from a hazardous substance release. The authority under RCRA Section 7003 is essentially the same, except that RCRA's imminent hazard provision addresses non-hazardous as well as hazardous solid waste releases. In an enforcement action, the CERCLA and RCRA imminent hazard provisions may be used in tandem.

How RCRA and CERCLA Differ

RCRA and CERCLA have the common goal to protect human health and the environment from the dangers of hazardous waste. However, these statutes address the hazardous waste problem from two fundamentally different approaches:

- RCRA has a largely regulatory approach. RCRA regulates the management of wastes from the moment of generation until final disposal, and provides some corrective action authority for investigating and cleaning up contamination at or from RCRA Subtitle C facilities.
- CERCLA has a response approach. CERCLA authorizes cleanup actions whenever there has been a breakdown in the waste management system (i.e., a threatening release of a hazardous substance occurs). Also, CERCLA addresses the problems of hazardous waste encountered at inactive or abandoned sites of those resulting from spills that require emergency response.

How RCRA Regulations Affect CERCLA Remedy Selection

In assessing cleanup remedies, the EPA takes into account the long-term uncertainties, standards, criteria, or limitations under federal or more stringent state environmental laws, including RCRA, that are determined to be ARARs. Furthermore, the NCP provides that removal actions attain ARARs whenever practicable. This means, for example, that whenever a remedial action involves on-site treatment, storage, or disposal of hazardous waste, the action must meet RCRA's technical standards for treatment, storage, or disposal. The EPA interprets CERCLA to mean that Superfund sites are not required to comply with administrative requirements, but that RCRA technical requirements may apply as ARARs. The National Oil and Hazardous Substances Pollution Contingency Plan (NCP), the blueprint for the Superfund program, details the application of ARARs to Superfund remedial actions cited in Section 300.435(b)(2).

As noted earlier, once hazardous substances, contaminants, or pollutants are transported from a Superfund site, they are subject to CERCLA's off-site requirement that they go to a facility that the EPA has determined to receive CERCLA wastes. CERCLA wastes that are RCRA hazardous wastes must go to a Subtitle C facility acceptable under the CERCLA off-site policy. Each regional office has an off-site contact who makes the acceptability determination prior to each off-site shipment of CERCLA wastes.

Finally, as of October 1989, the EPA may not take or fund remedial actions in a state unless the state ensures the availability of hazardous waste treatment and disposal capacity. This hazardous waste capacity must be adequate to manage the waste generated for a period of 20 years and for facilities that are in compliance with RCRA Subtitle C requirements.

RCRA Corrective Action vs. CERCLA Response

RCRA authorizes the EPA to require corrective action whenever there is, or has been, a release of hazardous waste or constituents. Further, RCRA allow the EPA to require corrective action beyond the facility boundary, the EPA interprets the term "corrective action" to cover the full range of possible actions, from investigations, studies, and interim measures to full cleanups. Anyone who violates the corrective action order can be fined up to $25,000 per day of noncompliance and runs the risk of having their operating permit suspended or revoked.

The general distinction between RCRA and CERCLA is as follows: RCRA focuses on waste management and corrective action, while CERCLA focuses on cleanup activities. However, the two programs overlap. For example, RCRA standards are considered ARARs and are central to selecting remedies under CERCLA. Moreover, the RCRA corrective action and CERCLA repines action programs use parallel (but not identical) procedures.

Oil Pollution Act of 1990

The Oil Pollution Act of 1990 (OPA) amends Section 311 of the Clean Water Act (CWA). Section 311 prohibits the discharge of oil and certain hazardous substances in quantities that may be harmful to public health or welfare. OPA established the Oil Spill Liability Trust Fund to pay for federal responses to oil spills. Section 311 also authorizes the Oil Spill Prevention, Control, and Countermeasures (SPCC) program.

OPA is a comprehensive statute designed to: (1) expand the federal role in response activities; (2) increase trust funds available for cleanup costs and other damages; (3) improve preparedness and response capabilities of federal agencies and owners of operators of vessels and facilities; (4) ensure that responsible parties pay for damages from spills that do occur; (5) increase vessel safety through requirements for double hulls; and (6) establish an expanded oil pollution research and development effort.

Some of the most significant provisions of OPA include the following:

- **Expanded federal role in response**--Under revised Section 311(c) of the CWA, the federal government is required to direct responses to releases that pose a substantial threat to the public health or welfare, and has the discretion to direct responses to any discharges threatening public health or welfare.
- **Oil Spill Liability Trust Fund**--OPA increases a new $1 billion trust fund that is available for cleanup costs and other damages.

The USCG administers the fund, which is used to pay for removal costs and damages resulting from an oil discharge. Fund monies are supplied by a five-cent-per-barrel fee on oil.

- **Contingency planning**--New Section 311(j) of the CWA requires the EPA and the USCG to enhance the existing National Response System by designating Area Committees to develop Area Contingency Plans to help ensure among other things the removal of a worst-case spill from a vessel or facility in or near the area covered by each plan. Also, OPA added a new requirement in CWA Section 311(j) that owners or operators of individual vessels and facilities (except onshore facilities that are not expected to cause substantial environmental harm from oil discharges) prepare response plans from worst-case oil and hazardous substances discharges. The statute also requires amendments to the NCP, including development of a Fish and Wildlife Plan.

- **Increased liability for spills**--OPA increases the liability of tanker owners and operators, responsible parties at onshore facilities and deepwater ports, and holders of leases or permits for offshore facilities in the event of a spill. OPA broadens liability to cover not only removal costs and natural resource damages, but also the provision of spill-related health and safety services by state and local governments and losses of private property, revenues, subsistence use, and profits.

- **Double hulls**--Under OPA, newly constructed tankers over certain size limits must have double hulls or other double contaminant systems. Existing tankers without double hulls are to be phased out by size, age, and design beginning in 1995. Tankers over a certain size limit without double hulls are banned after 2015.

- **Research and development**--OPA mandates the establishment of an interagency committee to coordinate efforts to improve oil spill response technology.

Primary federal responsibility for implementing OPA rests with the United States Coast Guard (USCG) and the EPA. The USCG is responsible for administering the trust fund, responding to oil coastal spills, reviewing contingency plans for vessels and transportation-related facilities, and coordinating research and development efforts along with other requirements. EPA's responsibilities include reviewing contingency plans for certain onshore facilities, responding to discharges occurring in the inland zone, and revising the NCP.

As with the Superfund program, the NCP serves as the regulatory blueprint which guides federal response to oil spills.

OPA amends the CWA and includes a number of provisions regarding the prevention, control, and CWA hazardous substances. The NCP provides the framework for CERCLA and CWA Section 311 response. CWA Section 311 requires facilities storing oil and CWA hazardous substances to develop contingency plans, and to penalize facilities for non-compliance. CWA Section 311(b) authorizes more stringent penalties for unauthorized spills of oil and/or hazardous substances and violations of the regulations. OPA provides that liability includes the cost of the response and damages to natural resources, property, and subsistence use of natural resources. The provisions are independent of CERCLA.

Clean Water Act

The Clean Water Act (CWA) was enhanced to regulate and clean up polluted waters in the United States. It is designed to ensure that the nation's waters are safe to the public and support fish and other aquatic life. Specifically, the CWA is designed to restore and maintain the chemical, physical, and biological integrity of the nation's waters.

The CWA was one of the major environmental laws passed by Congress in the 1970s. It provides the EPA with two types of authority:

- **Regulatory**--to prevent and control discharges of pollutants into waters of the U.S.
- **Response**--to respond to releases of pollutants into waters of the U.S. Prior to CERCLA, the EPA and USCG worked under the CWA to clean up releases of oil and hazardous substances into the navigable waters of the U.S.

The previous section of the OPA describes the authorities and provisions of CWA Section 311. This section describes some of the major authorities and provisions of the other sections of the CWA.

The CWA requires that all direct discharges to surface water simply with technology-based discharge standards. These standards require the use of best practicable control technology (BPCT) for conventional pollutants and best available technology economically achievable (BAT) for toxic and non-conventional pollutants. The EPA has published effluent guidelines for specific categories of industries. These guidelines are translated into specific effluent requirements in discharge permits.

The CWA requires as permit for any discharge into the nation's waterways. For waste water, only two discharge options are allowed:

- **Direct discharge** into surface water pursuant to a National Pollution Discharge Elimination System (NPDES) permit.
- **Indirect discharge**, which means that the waste water is first sent to a publicly owned treatment works (POTW), and then after treatment by the POTW, discharged into surface water pursuant and NPDES permit.

If the direct discharge option is chosen, the generator of the waste water cannot simply transfer the pollutants to a POTW. Rather, the waste water must satisfy applicable preferment standards, where they exist.

Section 304 of the CWA directs the EPA to publish water quality criteria for specific pollutants. The EPA develops two types of criteria: one for the protection of human health and another for protection of aquatic life. The EPA has published a total of 82 water quality criteria. These criteria are non-enforceable guidelines used by states to set water quality standards for surface water. Section 303 requires states to develop water quality standards, based on federal water quality criteria, to protect existing or attainable uses (e.g., recreation, water supply) of surface waters.

How the CWA and CERCLA Interact

The CWA-designated hazardous substances are incorporated into the CERCLA definition of hazardous substances. The CWA Section 311 has authority for responding to discharges of oil into U.S. waters. In addition, CERCLA provides response authority for responding to discharges of other hazardous substances, pollutants, and contaminants into the environment. The NCP, the blueprint for managing responses to releases, governs both CWA and CERCLA responses. The previous section on OPA provides a more detailed discussion of how CWA Section 311 and CERCLA interact.

On-site CERCLA responses must comply with or waive substantive requirements of federal and state environmental laws that are determined to be ARARs. CWA and state discharge and water quality standards may be ARARs for on-site remedial action at Superfund sites. The application of CWA and state ARARs is determined on a case-by-case basis. CERCLA responses conducted entirely on-site do not require CWA permits.

Clean Air Act

The Clean Air Act (CAA) was the first major environmental law passed by Congress. The CAA was enacted to limit the emission of pollutants into the atmosphere to protect human health and the environment from the effects of airborne pollution. The CAA authorizes the EPA to achieve this objective by setting air quality standards and regulating emissions of pollutants into the air. The EPA has established emission standards for mobile and stationary sources of pollutant emission. These are implemented through federal, state, and local programs.

For six pollutants, EPA has established National Ambient Air Quality Standards (NAAQS). Regulation of these six pollutants affords the public some protection from toxic six pollutants. Primary responsibility for meeting the requirements of the CAA rests with states, who must submit State Implementation Plans (SIPs) to achieve and maintain the NAAQS. Pursuant to the SIP, new or modified stationary sources of air emissions must undergo pre-construction review to determine whether the facility will interfere with attainment or maintenance of NAAQS. In addition, in some areas that do not attain NAAQS, SIPs must contain regulatory strategies to control emissions from existing stationary sources. SIPs, not NAAQS, are potential ARARs. Of chief concern to Superfund are the requirements that apply to sources of volatile organic compounds (VOCs) and other toxic air pollutants (e.g., heavy metals).

Section 112 of the CAA directs EPA to identify hazardous air pollutants and to establish emission standards for sources that emit the pollutants. These standards, known as National Emission Standard for Hazardous Air Pollutants (NESHAPs), apply to new as well as existing sources. Additionally, under Section 112(r), the accidental release provisions of the CAA, facilities are required to provide information on the ways they manage risk posed by certain substances listed by EPA and indicate what they are doing to minimize risk to the community from those chemicals.

The CAA and CERCLA interact in the following two ways:

- The CAA hazardous air pollutants are included as CERCLA hazardous substances.
- CAA emissions limitations provide substantive standards for CERCLA responses.

CERCLA provides federal response authority to address releases of air pollutants that threaten human health or the environment. CAA requirements may apply to CERCLA responses.

The accidental release provisions of CAA requires the establishment of a list of at least 100 regulated substances and thresholds under section 112. Sixteen of these substances were identified in the CAA for inclusion on the list. The rest of the list may be drawn from, but not necessarily limited to, the list of extremely hazardous substances under SARA Title III.

CAA hazardous air pollutants, identified under Section 112, are CERCLA hazardous substances by definition. Other CAA air pollutants, identified under Sections 109 and 111, are not covered by the CERCLA definition of hazardous substances but may be covered by the CERCLA definition of "pollutant or contaminant."

CAA emissions limitations provide substantive standards for CERCLA responses in two ways. CAA emissions limitations provide triggers for Superfund action. And, these limitations provide cleanup standards to attain in addressing unremediated conditions, and emission standards for certain cleanup technologies employed. CAA emission standards may be ARARs for on-site response actions at Superfund sites. The applications of CAA standards as ARARs is determined on a case-by-case basis.

CERCLA responses need not comply with CAA permit requirements.

Safe Drinking Water Act

The Safe Drinking Water Act (SDWA) was enacted in 1974 to protect human health by protecting the quality of the nation's drinking water supply. It protects drinking water sources by regulating facilities or systems that inject fluids into the ground, and protects public drinking water consumers by regulating the quality of water distributed by public water systems. These goals are achieved by authorizing the establishment of:

- Drinking water standards.
- A permit program for the underground injection of wastes.
- Resource planning programs.

The SDWA imposes requirements on persons who own or operate a system which has at least 15 service connections or 25 consumers, and provides piped water for human consumption. The regulations which implement these requirements are entitled the National Primary Drinking Water Regulations (NPDWR). All water suppliers must periodically sample the water delivered to users and record and report their findings to EPA or the state, whichever is appropriate.

The Underground Injection Control (UIC) program protects underground sources of drinking water from contamination by injection of waters or wastes into injection wells. A permit program limits substances that may be injected and how they may be injected.

EPA currently administers the SDWA public water system program in only two states, Indiana and Wyoming. In all other states, EPA oversees state implementation, but retains independent enforcement authority.

Provisions of the SDWA apply to CERCLA site discharges to public drinking water sources. SDWA provides such as Maximum Contaminant Levels (MCLs) may be applicable to CERCLA cleanup of water that may be used for drinking.

Toxic Substances Control Act

The Toxic Substances Control Act (TSCA), signed into law in October 1976, provides EPA with broad authority to regulate chemicals and chemical substances whose manufacture, processing, distribution in commerce, use, or disposal may present an unreasonable risk of injury to health or the environment. TSCA was enacted to keep harmful chemicals out of the environment and to fill the gaps in existing environmental laws in the areas of toxic substances.

TSCA deals with all chemical substances planned for production, produced, imported, or exported from the country. TSCA applied primarily to manufacturers, distributors, processors, and importers of chemicals. TSCA provides authorities to control the manufacture and sale of certain chemical substances. These authorities include:

- Testing of chemicals currently in commercial production or use.
- Pre-market screening and regulatory tracking of new chemical products.
- Controlling unreasonable risks once chemical substance is determined to have an adverse effect on health or the environment. These powers include:
  -- Prohibiting the manufacture or certain uses of the chemical.
  -- Requiring labeling with specific markings or warnings.
  -- Limiting volume of production or concentration.
  -- Requiring recordkeeping about production.
  -- Controlling disposal methods.

The only exceptions to these authorities are pesticides, tobacco or tobacco products, source material by-products or special nuclear material, food additives drugs, and cosmetics.

TSCA and CERCLA commonly interact if polychlorinated biphenyls (PCBs) are involved in a CERCLA response. PCB disposal regulations under TSCA may apply, as ARARs, at Superfund sites. PCBs are the only chemical identified by Congress by name for direct regulation under TSCA. TSCA regulations of other chemicals may also present possible ARARs, depending on the type of hazardous substances at a Superfund site.

Hazardous Materials Transportation Uniform Safety Act

The Hazardous Materials Transportation Uniform Safety Act of 1990 (HMTUSA), Section 117, evolved from the emergency preparedness proposal developed by DOT, FEMA, EPA, DOL, and DOE, and presented to the Congress during the legislative process to reauthorize the Hazardous Materials Transportation Act of 1975. The requirements of the HMTUSA were designed to allow the federal government or provide national direction and guidance to enhance hazardous materials emergency preparedness activities at the state and local levels. This will be accomplished by ensuring comprehensive, integrated, and coordinated planning, training, and technical assistance programs. Section 117, "Public Sector Training and Planning," was specifically crafted to build upon and enhance the existing framework and working relationships established within CERCLA Superfund for the National Response Team (NRT), Regional Response Teams (RRTs), and the Title III State Emergency Response Commission.

HMTUSA builds on existing programs and relationships and, in fact, specifically requires grant money to be submitted to LEPCs as established under SARA Title III. Specifically, HMTUSA provides for:

- Planning grants to develop, improve, and implement SARA Title III local emergency response plans and to determine the need for regional hazardous materials emergency response teams.
- Training grants for delivery of training to public sector employees in hazardous materials response. This grant could be used for hazardous material waste and emergency response and other training activities. However, in order to qualify for the training grants, states/tribes must certify they are in compliance with Sections 301 and 303 SARA Title III.

FUTURE DIRECTIONS OF THE SUPERFUND PROGRAM

During the first ten years of the Superfund program, EPA successfully developed a program that brought the United States to a new level of understanding about hazardous substances and how they can be treated. The Superfund program is comprehensive, yet flexible and innovative. Its mission is both immediate and long range; its focus is specific enough to handle individual site cleanups with precision, yet broad enough to encourage advances in a relatively new scientific and technical field.

Superfund already has resulted in permanent solutions to many hazardous waste problems. However, after the first 10 years of experience, it is apparent to the Superfund program participants that the program faces a workload stretching well into the next century. The hazardous waste problem in the United States remains large, complex, and long term.

EPA is looking ahead to project a program for the future. Long-term planning is important because, for example, EPA estimates indicate that the cleanup of sites on the NPL, as of 1991, is expected to cost an additional $19 billion beyond the amount already obligated. Also, EPA expects the number of NPL sites to grow from 1200 to 2000 by the end of the century.

"Superfund 200" represents EPA's strategy for responding to long-term needs. As part of this concept, EPA is conducting studies of the possible universe of sites to be cleaned up by Superfund or other parties. One study involves the development of a liability model to help estimate possible future cleanup cost under different scenarios. EPA also is looking at past remedy selection decisions and evaluation patterns that may indicate future useful technologies. In addition, opportunities for greater program integration, particularly between the Superfund and Resource Conservation and Recovery Act (RCRA) programs, are being assessed.

In keeping with EPA's goal of increasing multimedia enforcement efforts, EPA is examining the future role of responsible parties and state and local governments in the Superfund program. All these studies and activities will ensure that an integrated, practical, viable, and results-oriented Superfund program will continue to evolve.

In February of 1992, the EPA Administrator signed a plan aimed at moving sites more quickly through the Superfund process to cleanup and redefine the way progresses are measured. This new plan, called the **Superfund Accelerated Cleanup Model (SACM)**, is designed to include substantial, prioritized risk reduction in shorter time frames and better communication of program accomplishments to the public.

As outlined in this book, the current system for Superfund cleanups is based on two discrete programs: removal and remedial. Under SACM, this distinction would be retained, but EPA would view both removal and remedial actions as Superfund actions. Rather than viewing these two entities as separate programs, they are viewed as separate legal authorities with different, but complimentary, application at Superfund sites.

An integral part of SACM is the combined site assessment. The *single site assessment function* would address, in a coordinated fashion, requirements for removal assessments, preliminary assessments/site inspections (PA/SIs), remedial investigations/feasibility studies (RI/FSs), and risk assessments. Discovered sites could be screened once and, if they are considered to have a serious level of contamination, go directly to the remedial investigation and risk assessment phase of cleanup. Such a change could cut the current process by several years.

During the assessment process, a **Regional Decision Team** would decide to place a site on either an **"Early Action List"** or a **"Long-Term Remediation List"** or both. Early actions are short-term, quickly implemented cleanups that would be completed in three to five years. Early actions will include time-critical and non-time-critical removal activities, as well as remedial actions, and will be designed to address all short-term threats to public health and safety. Under SACM, such actions would be combined immediately with public participation sites and would be combined enforcement actions. Long-term remediation sites would only include sites requiring cleanup over many years.

SACM introduces the concept of Regional Decision Teams that would combine the cross-programmatic skills and experience of on-scene coordinators, remedial project managers, and community relations coordinators. As explained above, the Regional Decision Team would be responsible for expediting sites onto the Early Action List and scoring long-term restoration actions for inclusion on the NPL.

A key objective of SACM is to count the totality of risk reduction ration than focus on NPL site selections. This would be a fundamentally new way for the Superfund program to measure it success, and would show the public how Superfund is achieving appropriated cleanup at a large number of sites.

Regional pilot tests are underway. The model is being further refined, and EPA is developing protocols and guidance that will expedite implementation of SACM. The steps for the new Superfund process are illustrated in Figure 3.

EPA is proud of its hard-won accomplishments in the Superfund program, and will continue to use new management and technological approaches to significantly improve human health and the environment, accelerate the pace of cleanup, expand its efficiency and activity, improve the quality of the program over time, and build public confidence. There are no miracle cures for the hazardous waste problem, but EPA has a clear strategy for meeting this challenge.

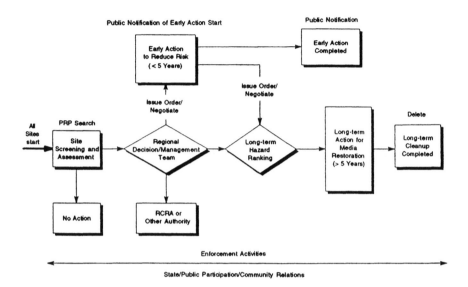

Figure 3. The SACM Process.

SUGGESTED READING

1. Sidley and Austin, et al. 1989, *Superfund Handbook, Third Edition*, ENSR Corporation, p. 12-31.
2. *CERCLA/Superfund Orientation Manual*, Technology Innovation Office, EPA/542/R-92/005, 1992, II-XIII.
3. Technical Guidance for Hazards Analysis: *Emergency Planning for Extremely Hazardous Substances*, U.S. Environmental Protection Agency, Federal; Emergency Management Agency, and U.S. Department of Transportation, December, 1987, p. 17.
4. Title III of CERCLA.
5. R. V. Kolloru, 1994, *Environmental Strategies Handbook: A Guide to Effective Policies & Practice*, McGraw-Hill, Inc., p. A.1-6.

8 THE RESOURCE CONSERVATION AND RECOVERY ACT

INTRODUCTION

This chapter discusses the four programs of the act responsible for the management of the huge volume of municipal and industrial solid waste that is generated by our society. Specifically, it will provide the reader with a discussion on managing non-hazardous waste; on controlling, managing and tracking hazardous waste from the time it is generated until its ultimate disposal; on regulating underground storage tanks; and on managing, controlling, and tracking waste generated and used in the medical community.

In addition, this chapter discusses the process for obtaining a permit for a facility regulated by RCRA; the means by which RCRA is enforced; the impact RCRA has on the industrial community; how RCRA impacts or interrelates with other environmental laws, regulations, or acts.

It became clear, in the mid 1970s to Congress and the nation alike, that action had to be taken to ensure that solid wastes are managed properly. This realization began the process of the Resource Conservation and Recovery Act (RCRA), an amendment to the Solid Waste Disposal Act.

RCRA was enacted in 1976 to address a huge problem of enormous magnitude--how to safely dispose of huge volumes of municipal and industrial solid waste generated nationwide. RCRA goals are to (1) protect human health and the environment; (2) reduce waste and conserve energy and natural resources; (3) reduce or eliminate the generation of hazardous waste as expeditiously as possible.

To achieve these goals, four district programs exist under RCRA. The first program, under Subtitle D of RCRA encourages states to develop comprehensive plans to manage primarily non-hazardous wastes (e.g., household waste). The second program under Subtitle C of RCRA establish a system for controlling hazardous waste from the time it is generated until its ultimate disposal. The third program, under Subtitle I of RCRA regulated certain underground storage tanks. It establishes performance standards for new tanks and require leak detection, prevention, and correction action at underground storage tank sites. The newest program to be established is the medical waste program under RCRA Subtitle J which establishes a demonstration program to track medical waste from generation to disposal.

RCRA's cradle to grave program regulates the management of hazardous waste. The program imposes comprehensive obligations and carries significant sanctions for non-compliance. The RCRA program identifies a broad universe of waste materials as hazardous and regulates the handling of these wastes by generators, transporters, and treatment, storage, and disposal facilities. Additionally RCRA imposes corrective action requirements and provides standards for cleanup under the Superfund statute.

While the Superfund statutes focuses on remedy past waste disposal at abandoned sites, RCRA addresses the ongoing management of hazardous wastes at manufacturing plants and other facilities thus affecting a large segment of business engaged in manufacturing.

The EPA published its first RCRA regulations as a result from increased public and congressional concern with hazardous waste beginning with the Love Canal episode, which led to the passage of the Superfund Law of 1980.

In 1984, Congress amended RCRA extensively in the Hazardous and Solid Waste Amendments of 1984 (HSWA). The HSWA authorized the regulations of underground tanks, the cleanup of contaminated areas of industrial sites not covered by the original law, and increased restriction on the disposal of wastes on land. The amendments contained detailed provisions and established strict deadlines. Today EPA and the states are involved primarily in implementing these amendments.

OVERALL THRUSTS OF RCRA

The Resource Conservation and Recovery Act (RCRA), an amendment to the Solid Waste Disposal Act, was enacted in 1976 to address a problem of enormous magnitude. That is how to safely dispose of the huge volumes of municipal and industrial solid waste generated nationwide. It is a problem with roots that goes back well before 1976. It is estimated that we generate approximately 275 million metric tons of hazardous waste nationwide annually. By the mid-1970s Congress and the nation alike took actions to ensure that our solid waste problems are managed properly. Thus, began the process that resulted in the passage of RCRA. Its primary goals were to protect human health and the environment, to reduce waste and conserve energy and natural resources, and to reduce or eliminate the generation of hazardous waste as expeditiously as possible.

To achieve the above goals, four distinct yet interrelated programs were established under RCRA. The first program, established under Subtitle D of RCRA, encourages states to develop comprehensive plans to manage primarily non-hazardous solid wastes, e.g., household waste. The second program, established under Subtitle C, developed a system for controlling hazardous waste from the time it is generated until its ultimate disposal, in effect, from "cradle to grave." The third program, established under Subtitle I of RCRA, regulates certain underground storage tanks. It establishes performance standards for new tanks and requires leak detection, prevention, and corrective action at underground storage tank sites. The newest program to be established is the medical waste program under RCRA Subtitle J. It establishes a demonstration program to track medical waste from generation to disposal.

The structure of the RCRA is straightforward. It is currently divided into ten subtitles, A through J. Subtitles A, B, E, F, G, and H outline general provisions; authorities of the Administrator; duties of the Secretary of Commerce; federal responsibilities; miscellaneous provisions; and research, development, demonstration, and information. Subtitles C, D, I, and J lay out the framework of the four programs, as outlined above, that make up RCRA. These four programs will be described in detail in the ensuing sections. The solid waste program (Subtitle D) is discussed before the hazardous waste program (Subtitle C), because the former defines solid waste, and this definition is required before the Subtitle C program can be discussed.

Subtitle D--Managing Solid Waste

The primary objectives of Subtitle D are to assist and encourage states and local governments in developing and implementing solid waste management plans to manage non-hazardous solid waste. These plans are intended to promote recycling of solid wastes, to require closing or upgrading of all environmentally unsound dumps, and to encourage utilization and maximization of valuable resources including energy and materials. To achieve the above EPA developed criteria for the classification of solid waste disposal facilities and practices and established a framework for state solid waste management plans. The criteria provide the basis for enforcing the prohibition on "open dumps" and provide the public with a legal basis for bring suits in federal district courts.

Non-hazardous solid waste as defined by Subtitle D are garbage, e.g., milk cartons and coffee grounds; refuse, e.g., metal scrap, wallboard, and empty containers; sludge from a waste treatment plant, water supply treatment plant, or pollution control facility, e.g., scrubber sludge; and other discarded material, including solid, semisolid, liquid, or contained gaseous material resulting from industrial, commercial, mining, agricultural, and community activities, e.g., boiler slag, fly ash. The term solid waste as defined by Subtitle D is very broad including not only the traditional non-hazardous solid wastes, but also some household hazardous wastes and hazardous wastes generated by small quantity generators.

Classifying Solid Waste Disposal Facilities/Practices

The first component of the Subtitle D program is the "Criteria for Classification of Solid Waste Disposal Facilities and Practices," commonly called the "Subtitle D Criteria." The criteria are used to determine which solid waste disposal facilities and practices pose an adverse effect on human health and the environment. Facilities failing to achieve this criteria will be considered an open dump for purposes of state solid waste management planning.

The Subtitle D Criteria include general environmental performance standards for addressing floodplains by specifying that facilities or practices in floodplains shall not interfere with the floodplain or result in washout of solid waste so as to pose a hazard to human health, wildlife, or land or water resources; endangered species by prohibiting solid waste disposal facilities and practices that cause or contribute to the taking of any endangered or threatened species or results in the destruction or adverse modification of the critical habitats or such species; surface water by specifying that disposal facilities shall not cause a discharge of pollutants or dredged or fill material to waters of the United States; groundwater by specifying that facilities or practices that will require a discharge to groundwater must comply with the Safe Drinking Water Act Maximum Contaminant Levels; land application by requiring that a facility or practice meet certain restrictions with respect to the concentrations of cadmium and PCBs contained in waste applied to land used for producing food crops; disease by specifying that waste disposal facilities and practices must institute appropriate disease vector controls, such as periodic application of cover material; air by prohibiting open burning of solid waste and specifies that the applicable requirements of the State Implementation Plans must comply with the Clean Air Act; and finally, safety by requiring control of explosive gases, fires, bird hazards to aircraft, and public access to the facility.

The above Subtitle D Criteria serve as a minimum technical standard for solid waste disposal facilities. Facilities must comply with these minimum criteria to ensure that ongoing operations are protective to human health and the environment. Failure to meet these criteria will classify a facility as an open dump and require its upgrade or closure immediately. Compliance with the Subtitle D Criteria and the ban on dumping can be enforced through citizen suits or by the state. In addition to the Subtitle D Criteria serving as minimum technical standards, they can be used by states as a benchmark to identify open dumps and to force them into compliance or closure.

Guidelines for Developing/Implementing State Plans

The second component of Subtitle D is to provide the states with guidelines for developing and implementing state solid waste plans. Such plans serve to ensure environmentally sound solid waste management and disposal, resource conservation, and maximum utilization of valuable resources. The guideline assists by outlining the minimum requirements for state plans and detailing how these plans are approved by EPA. The state plan must include as a minimum the following elements:

1. It must identify the responsibilities of state, local, and regional authorities in implementing the plan.

2. It must establish a solid waste disposal program that prohibits new open dumps, provides for the closing or upgrading of all existing open dumps, and establish a state regulatory power necessary to implement the plan.

3. It must develop a strategy for encouraging resource recovery and conservation activities.

4. It must ensure that adequate facility capacity exists to dispose of solid waste in an environmentally sound manner.

5. It must coordinate with other environmental programs.

6. It must ensure public participation in plan development, regulatory development, facility permitting, and open drum inventory.

Once the plan has been developed and adopted by the state, it is then submitted to EPA for approval. Approval is within 6 months from the date of submittal, and is often granted if the plan fulfills all of the minimum requirements outlined above.

Subtitle C--Managing Hazardous Waste

The Subtitle C of RCRA establishes a program to manage hazardous wastes from "cradle to grave." Its objective is to ensure that hazardous waste is handled in a manner that is protective to human health and the environment. The Subtitle C program does this by first identifying those solid wastes that are "hazardous" and then establish various administrative requirements for the three categories of hazardous waste handlers: generators, transporters, and owners or operators of treatment, storage, and disposal facilities (TSDFs). In addition, the Subtitle C regulations set technical standards for the design and safe operation of TSDFs. The standards are designed to minimize the release of hazardous waste into the environment. Furthermore, the regulations for TSDFs serve as the basis for developing and issuing the permits required for each facility. Issuing permits is essential to making the Subtitle C regulatory program work, since it is through the permitting process that we are able to apply the technical standards to facilities.

Definition and ID of Hazardous Waste

RCRA defined the term "hazardous waste" as a "solid waste or combination of solid wastes, which because of its quantity, concentration, or physical, chemical, or infectious characteristics may cause, or significantly contribute to an increase in mortality or an increase in serious irreversible, or incapacitating reversible, illness or; pose a substantial present or potential hazard to human health or the environment when improperly treated, stored, transported, or disposed of, or otherwise managed."

Therefore, we can see that the universe of potential hazardous wastes is large and diverse, consisting of chemical substance, mixtures, generic waste streams, and specific products. Subtitle C requires that all solid waste generators, from national manufacturers to the corner dry cleaners, to determine if their solid waste is hazardous and thus subjected to regulation under Subtitle C. If your solid waste stream is ignitable, corrosive, reactive, fails an extraction procedure (EP) toxicity test, is listed as a hazardous waste, is mixed with a hazardous waste, or is derived from the treatment, storage, or disposal of a listed waste then that waste stream is hazardous pursuant to RCRA Subtitle C.

A solid waste stream is ignitable if its liquid steam has a flash point less than 60°C; if its nonliquid waste stream, under normal conditions, is capable of spontaneous and sustained combustion; if its waste stream is an ignitable compressed gas per DOT regulations; or if its waste stream is an oxidizer per DOT regulations. A waste stream is corrosive if its aqueous waste stream has a pH less than or equal to 2 or greater than or equal to 12.5; or if its liquid waste stream can corrode steel at a rate greater than 1/4 inch per year at a temperature of 55°C. A waste stream is considered reactive if under normal conditions it reacts violently without detonating; if it reacts violently with water; if its explosive when mixed with water; if it generates toxic gases, vapors, or fumes when mixed with water; if it contains cyanide or sulfide and generates toxic gases, vapors, or fumes at a pH of between 2 and 12.5; if its capable of detonating when heated in confinement or subjected to strong imitating sources; if its capable of detonation at standard temperature and pressure; and if its listed by DOT as Class A or B explosives. If a solid waste stream leaches hazardous concentrations of particular toxic constituents into the ground water as a result of improper management, as exhibited by an EP Toxicity Test, then it is considered hazardous.

As noted above the responsibility for determining if a particular solid waste is hazardous falls on the generators. If a solid waste is neither excluded nor listed, then it must be assessed to determine if it exhibits any of the above characteristics. If it does then it is a hazardous waste and must be handled accordingly.

Generators of Hazardous Waste

The Subtitle C regulations broadly define the term "generator" to include any facility owner or operator or person who first creates a hazardous waste; or a person who imports a hazardous waste, initiates a shipment of a hazardous waste from TSDF, or mixes hazardous wastes of different DOT shipping descriptions by placing them into a single container.

Under RCRA, there are three categories of generators. The first is known as a large quantity generator and is defined as those facilities that generate over 1000 kg/mo of hazardous waste or over 1 kg/mo of acutely hazardous waste. The second is known as the small quantity generator and include those generators who produced greater than 100 but less than 1000 kg/mo of hazardous waste at a site or less than 1 kg/mo of acutely hazardous waste. Those facilities who generate less than 100 kg/mo of hazardous waste and less than 1 kg/mo of acutely hazardous waste is known as the conditionally exempt small quantity generators. However, this kind of generator must identify the waste to determine whether it is a hazardous waste; not accumulate more than 1000 kg of hazardous waste at any time; and treat or dispose of the waste on site or ensure that the waste is sent to a permitted TSDF or permitted municipal or industrial solid waste facility or recycling facility.

Regulatory Requirements for Generators

Generators of hazardous waste are the first link in the cradle-to-grave chain of hazardous waste management under RCRA. Subtitle C requires generators to ensure and fully document that the hazardous waste they produce is properly identified and transported to a RCRA TSDF. Thus, large and small quantity generators are required by regulations to have an EPA ID number, to properly handle hazardous waste before transportation, to manifest hazardous waste, and to recordkeep and report on the hazardous waste they have generated.

Subtitle C requires that each generator be assigned a unique identification number to help EPA track and monitor the generation of hazardous waste. Without this number the generator is barred from treating, storing, disposing of, transporting, or offering for transportation any hazardous waste. Furthermore, the generator is forbidden from offering hazardous waste to any transporter, or treatment, storage, or disposal facility that does not also have an EPA ID number. If the waste is to be shipped off site, the generator must ensure that it is properly packed to prevent leakage, and properly labeled, marked, and placarded to identify the characteristics and danger associated with transporting the waste.

The Subtitle C regulations allows large quantity generators to accumulate hazardous waste on site for 90 days or less, prior to transportation, if the waste is properly stored in containers or tanks marked with the words "hazardous waste" and the date on which accumulation began; if they have a written contingency plan and procedures in place in the event of an emergency; and if they have trained facility personnel to properly handle hazardous waste. The 90-day period allows a generator to collect enough waste to make transportation more cost-effective. If a generator accumulates hazardous waste more than 90 days, he or she is considered an operator of a storage facility and must comply with the Subtitle C requirements for such a facility. A small quantity generator can store waste on site for up to 180 days providing the quantity does not exceed 6000 kg at any time, the facility has a basic safety information system in place, and that facility personnel are familiar with emergency procedures that must be followed during spills and accidents.

The manifest is part of a control tracking system. Through its use generators can track the movement of hazardous waste from the point of generation to the point of ultimate treatment, storage, or disposal. The manifest contains the name and EPA ID number of the generator, transporter, and the facility where the waste is to be treated, stored, or disposed; DOT description of the waste being transported; quantities of the waste being transported; and the address of the TSDF to which the generator is sending waste. Each time the waste is transferred the manifest must be signed to acknowledge receipt of the waste. A copy of the manifest is retained by each link in the transportation chain. Once the waste is delivered to the designated facility the owner or operator of the facility must send a copy of the manifest back to the generator. This system ensures that the generator has documentation that his or her hazardous waste has made it to its ultimate destination.

The recordkeeping and reporting requirements for generators provide EPA and the states with a method to track the quantities of waste generated and the movement of hazardous wastes. The three primary recordkeeping and reporting requirements are biennial reporting, which details the generator's activities during the previous calendar year; exception reporting, which requires reporting a missing or unsigned and dated manifest; and keeping a copy of the biennial report, exception report, and manifest for three years.

Transporters of Hazardous Waste

The transporters of hazardous waste are the critical link between the generator and the ultimate off-site hazardous waste TSDF. A transporter under Subtitle C is any person engaged in the off-site transportation of hazardous waste by air, rail, highway, or water within the United States.

Regulatory Requirements for Transporters

A transporter is required to obtain an EPA ID number, comply with the manifest system, and be able to handle hazardous waste discharges in the event of a serious accident or spill. Transporting hazardous waste can be dangerous, there is always the possibility that an accident will occur. Thus, the regulations require transporters to take immediate action to protect human health and the environment. It requires transporters to notify local authorities and/or dike off the discharge area, notify the National Response Center, the Center for Disease Control, and file an incident report to DOT within 15 days of a spill. In addition to meeting the federal requirements, transporters must also comply with the requirements of each state they travel through.

Treatment/Storage/Disposal Facilities of Hazardous Waste

Treatment, storage, and disposal facilities (TSDFs) are the last link in the cradle-to-grave hazardous waste management system. Subtitle C requires all TSDFs that handle hazardous waste to obtain an operating permit and abide by the treatment, storage, and disposal (TDS) regulations. These regulations establish design and operating criteria as well as performance standards that owners and operators must meet to protect human health and the environment.

A TSDF is comprised of three different systems. Treatment is the first and consists of any method, technique, or process, including neutralization, designed to change the physical, chemical, or biological character or composition of any hazardous waste so as to neutralize it, or render it non-hazardous or less hazardous, or to recover it, make it safer to transport, store or dispose of, or amenable for recovery, storage, or volume reduction. Storage is the second system and involves the holding of hazardous waste for a temporary period, at the end of which the hazardous waste is treated, disposed, or stored elsewhere. The third system is disposal and involves the discharge, deposit, injection, dumping, spilling, leaking, or placing of any solid waste into or on any land or water so that the waste or any constituent thereof may enter the environment or be emitted into the air or discharged into any waters, including groundwaters.

Regulatory Requirements for TSDFs

TSDFs falls into two categories called the interim status facilities and the permitted facilities. The former was established to allow owners and operators of facilities in existence, who meet certain conditions, to continue operating as if they have a permit until their permit application is issued or denied; the letter is made up of facilities that have permits. The interim status standards are primarily "good housekeeping practices" that owners and operators must follow to properly manage hazardous wastes during the interim status period. The permit standards are a mix of performance standards and "design and operating" criteria that permit writers include in facility-specific permits.

Both interim status and permit standards consist of administrative and nontechnical requirements and technical and unit-specific requirements. The former is identical for both TSDFs categories, but the latter are significantly different for the interim status facilities and permitted facilities. This chapter will limit its focus on permitted facilities.

Administrative/Nontechnical Requirements for Permitted TSDFs

The purpose of the administrative and nontechnical requirements is to ensure that owners and operators of TSDFs establish the necessary procedures and plans to run a facility properly and to handle any emergencies or accidents. It requires all owners or operations of facilities treating, storing, or disposing of hazardous waste to meet the appropriate TSD regulations; it requires that every facility obtain an EPA ID number, ensuring that their waste are properly identified and handled (conduct waste analyses prior to treatment, storage, and disposal), that facilities are secure and operating properly, and that personnel working at facilities are trained in hazardous waste management; it requires an owner or operator to have a contingency plan/emergency procedures in place and be prepared and ready to minimize the possibility and effects of a release, fire or explosion; and finally to have a manifest, recordkeeping, and reporting system established to track and monitor the waste being treated, stored, or disposed.

Technical Requirements for Permitted TSDFs

The permitted standards are very extensive, and compel the owners and operators of the different waste management methods to design their management units to prevent the release of hazardous waste. The technical standards are broken down into general and specific standards. With the general standards covering the requirements for groundwater monitoring, the requirements for closure/post-closure, and the proper financial assurance needed to maintain or close a facility. The specific standards on the other hand are tailored for containers, tanks, surface impoundments, waste piles, land treatment units, landfills, incinerators, and miscellaneous units.

General Standards

The general standards requires a groundwater protection system in place at surface impoundments, waste piles, land treatment units, and landfills. The system consists of a detection monitoring program to determine whether hazardous wastes are leaking from a TSDF at levels great enough to warrant compliance monitoring, it includes background monitoring, and semiannual monitoring for indicator parameters monitoring is conducted at a compliance point located at the edge of the waste management area; a compliance monitoring program to evaluate the concentration of certain hazardous constituents in groundwater to determine whether groundwater contamination is occurring at a level requiring corrective action, protective standards can be either background, MCLs or site-specific standards; and a corrective action program that will require remediation of the groundwater or soil if cleanup is warranted.

For closure/post-closure an owner or operator must develop a plan for closing the facility and keep it on file at the facility until closure is completed and certified. The plan requires a description of the facility to be closed, an estimate of the maximum amount of waste the facility will be able to handle, a description of the steps needed to decontaminate equipment and remove soils and debris during closure, an estimate of the year of closure, and a schedule for closure. Post closure, which applies only to land-disposal facilities, is normally a 30-year period after closure during which owners or operators of disposal facilities conduct monitoring and maintenance activities to preserve the integrity of the disposal system.

Financial requirements were established to assure that funds are available to pay for closing a facility, for rendering post-closure care at disposal facilities, and to compensate third parties for bodily injury and property damage caused by sudden and nonsudden accidents related to the facility's operation.

Specific Standards

The specific technical standards for a container requires that its placed in a containment system that is capable of containing leaks and spills, when closing its required that all hazardous waste and residues be removed, and after closure all contaminated equipment or soil must be decontaminated or removed; for tanks an assessment is required to evaluate the tank system's structural integrity and compatibility with the wastes that it will hold, a secondary containment system must be designed, installed, and operated to prevent the migration of liquid out of the tank system, and to detect and collect any releases that do occur, an O&M system must be in place to respond to leaks or spills immediately, and closure/post-closure requires removal of all contaminated soils from the tanks storage areas and decontamination of tanks; for surface impoundments at least one linear is required and the impoundment should be located on an impermeable base, a leachate collection system between liners is required, groundwater monitoring is a must, and for closure/post-closure its required that all waste residue be removed and properly covered and maintained to prevent leaks from occurring.

For waste piles an impermeable base with a linear designed and constructed to prevent any migration of wastes out of the pile into adjacent soil or waters is required, leachate collection system immediately above the liner must be installed, and run-on and run-off systems must be constructed to prevent water from flowing onto the active portion of the waste pile, and during operation inspection is required weekly to ensure no deterioration of the linear; for land treatment the standards requires owner or operator to ensure that hazardous constituents placed in or on the treatment zone are degraded, transformed, or immobilized within the treatment zone; for landfills the standards requires the installation of two or more liners, two leachate collection systems (one above and one between liners), and groundwater monitoring; and for incinerators the owner or operator must determine the operating methods that will result in 99.99 percent of each principal organic hazardous constituent that must be destroyed or removed, must minimize the hydrogen chloride emission, and limit particulate emissions.

Subtitle I--Managing Underground Storage Tanks

This subtitle was enacted to control and prevent leaks from underground storage tanks (USTs). Specifically, it regulates underground tanks storing regulated substances. An underground storage tank is any tank with at least 10 percent of its volume buried below ground. Subtitle I developed performance standards for new tanks that include design, construction, installation, release detection, and compatibility standards for new tanks and requirements applicable to all tank owners and operators concerning leak detection, recordkeeping, reporting, corrective action, and closure. USTs generally release contaminants into the environment as a result of tank corrosion, faulty installation, piping failure, and overfills.

Regulatory Requirements for USTs

Subtitle I requires that all new tank systems, piping, and cathodic protection systems be designed and installed carefully and in accordance with industry codes. Tank system installation must ensure that the substances to be stored are compatible with the tank system. Tanks must be properly installed following manufacturers specifications, and certified when installation is satisfactorily completed. They must also be fitted with equipment to prevent spills and overfills. The regulation requires that owners or operators of, both existing and new, tanks to report their age, size, type, location, and use to state agencies.

To prevent spills or overfills, tank owners and operators must ensure that the capacity of the tank is greater than the volume of product to be transferred, have someone present at all times during the transfer, and use equipment that can prevent or severely limit spills. Subtitle I requires release detection systems on all tanks. These systems must be capable of detecting a release from any portion of the UST, is installed and maintained in accordance with the manufacturer's instruction, and is capable of meeting performance standards that have been designed for the chosen release detection method.

The regulation requires that an owner or operator must report within 24 hours any suspected releases to the appropriate state agency. In addition, if release is suspected, the owner must investigate and confirm the release; if confirmed, then cleanup will be required. Any spill or overfill over 25 gallons (for petroleum) must be reported, and spills of regulated substances not exceeding this amount must be cleaned up immediately.

The closure requirement for UST systems depends on the amount of time that the tank system is out of service. Unless permanently closed, all systems containing regulated substances must continue to comply with all the normal regulatory requirements. Prior to closing the UST system, the owner or operator must assess the site to ensure that no releases have occurred. This is usually done by sampling nearby soil and groundwater. Any release that is discovered will be subject to corrective action. Closure procedures follow accepted industry codes, including emptying the tank and filling it with an inert material, or removing the tank from the ground.

Subtitle J--Managing Medical Waste

Subtitle J addresses the problems associated with mismanagement of medical wastes. This program applies to generations, transporters, and facilities who handle medical waste. Medical waste is defined as any solid waste which is generated in the diagnosis, treatment, or immunization of human being or animals, in related research, biological production, or testing. Regulated medical wastes are a subset of all medical wastes and include cultures and stocks of infectious agents, human pathological wastes, human blood and blood products, and sharps (such as hypodermic needles and syringes used in animal or human patient care), certain animal wastes, certain isolation wastes (such as wastes from patients with highly communicable diseases), and unused sharps.

Medical Waste Generator Requirements

Subtitle J requires all generators, including transporters who repackage shipments, to properly handle medical waste before shipping it off site. The waste must be segregated into sharps, fluids, and other wastes and then packaged in rigid, leak-resistance containers. Untreated medical waste must have a water resistant label on the outside of the packaging identifying it as "infectious waste" or "medical waste" or displaying the biohazard symbol. Regulated medical waste shipments must be marked with information identifying the generator and each individual transporter. Treated waste need not be labeled. A tracking form is required under Subtitle J, similar to the hazardous waste manifest, to trace medical waste from generation through disposal. Generators must initiate a tracking form for most off-site shipments of waste. Copies of the form must be retained by the generator, watch transporter, and each facility handling the waste.

Medical Waste Transporter Requirements

Transporters who ship regulated medical waste must obtain a medical waste identification number from EPA to be used on all tracking forms and on all transporter reports. Subtitle J requires transporters only to accept medical waste shipments that are properly packaged, labeled, marked, and are accompanied by a tracking form. Upon delivery of the shipment to a subsequent transporter or to the facility receiving the waste, a copy of the form must be signed and retained by the transporter. In addition, transporters must submit semiannual reports to EPA detailing information about their medical waste shipments.

**Medical Waste Treatment/Destruction/Disposal
Facility Requirements**

Medical Waste Treatments, Destruction, and Disposal Facilities (TDDs) must sign and return tracking forms to the generator. Facilities that receive regulated medical waste shipments that differ in certain respects from the waste indicated on the tracking form must attempt to resolve the discrepancy. If the discrepancy cannot be resolved within 15 days, the TDDs is required to report the issue to EPA. Owners and operators of incinerators that burn medical wastes generated on site must submit reports to EPA covering the first and third six-month periods of their process. These reports should describe, among other things, the amounts of medical waste received and burned on site, and the date and length of each incineration cycle.

PERMITTING

Permits detail the administrative and technical performance standards that TSDFs must adhere to, and thus are the key to implementing Subtitle C regulations. Therefore, owners and operators of facilities that treat, store, or dispose of hazardous waste must obtain an operating permit under Subtitle C. There are several categories of permits issued under Subtitle C, however, this chapter will limit its discussion to permits required for operating TSDFs and for post-closure of land disposal facilities. The permitting process involves submitting a permit application to the appropriate state agency or EPA, reviewing the permit application, preparation of the draft permit, releasing it to the public for their input, and finalizing the permit.

Submitting a Permit Application

Owners and operators of facilities that fall under the permitting regulations are required to submit a comprehensive permit application covering all aspects of the design, operation, and maintenance of the facility. This submission is necessary to provide the permit reviewer with sufficient information to determine if the facility is in compliance with Subtitle C regulations and to assist in the development of facility-specific permit requirements.

The permit application is divided into Part A and Part B. The former is a short, standard form that collects general information about a facility such as its name and a description of the activities that it conducts. The latter is much more extensive, and requires the owner or operator to supply detailed and highly technical information on the hazardous waste to be handled at the facility. Depending on the situation, Parts A and B may be submitted at different times.

Reviewing the Permit Application

Once the owner or operator of a facility has submitted an application (both Parts A and B), the first step is to determine if all the required information has been submitted. If the application is not complete, a notice of deficiency (NOD) letter is sent to the owner or operator describing the additional information that is required for a complete application. Once the owner or operator has submitted all of the required information, the application is considered complete. Once the owner or operator is informed that his application is complete, an in-depth evaluation of the permit application begins.

The purpose of the evaluation of the permit is to determine if the application satisfies the technical requirements of RCRA. After the permit application is evaluated, a tentative decision either to issue or deny the permit is made. If the decision is to deny, the owner is sent a notice of intent to deny, and if its to issue, a draft permit is prepared.

The evaluation of a permit application is a lengthy process, and can take from one to three years.

Preparing the Draft Permit

The draft permit incorporates applicable technical requirements and other conditions pertaining to the facility's operation. These other conditions consists of conditions that are applicable to all permits and those that are applicable on a case-by-case basis. The former is referred to as the general permit conditions and requires the owner or operator of a facility to comply with all conditions listed in the permit; to notify the appropriate regulatory agency of any planned alterations or additions to the facility; to certify annually that a program is in place to reduce the volume and toxicity of waste, and that the proposed method of treatment, storage, and disposal minimizes threats to human health and the environment; and to ensure to submit required reports, e.g., Unmanifested Waste Report, Biennial Report, and Manifest Discrepancy Report.

The case-by-case permit conditions may include a compliance schedule if the facility is in need of corrective action; a duration of permit where the permit is valid for up to ten years, for land disposal facility a permit must be reviewed every five years; and technical requirements as specific by the various type of solid waste management units at the facility.

Taking Public Comments

Once the draft permit is completed it is released with a fact sheet, explaining the permit process and the nature of the facility, to the public for 45 days of public comments. If information submitted during the initial comment period appears to raise substantial new questions concerning the permit, the appropriate regulatory agency must re-open or extend the comment period. In this situation the agency may also decide to revise the draft permit.

Finalizing the Permit

After the comment period closes, a response to all significant public comments is prepared and the permit is either made final or denied.

This decision may be appealed, and when appeals are exhausted, the petitioner may seek judicial review of the final permit decision.

ENFORCEMENT

For EPA's RCRA program to substantially work, all regulated groups must comply. To ensure compliance, program personnel establish an enforcement program. The two most important aspects of the enforcement program are: compliance monitoring and enforcement actions.

Compliance Monitoring

Close monitoring facility activities is required to verify their compliance with RCRA's regulatory requirements. This monitoring serves several purposes. It allows EPA and authorized states to find out which facilities are not in compliance. It also acts as a deterrent, encouraging compliance with the regulations by making noncompliers susceptible to enforcement actions.

One of the program's most important monitoring tools is site inspection. During an inspection, regulatory personnel, that is, EPA or an authorized state official or both, can enter any hazardous waste handling facility to review the facility's records, take waste samples, and assess the facility's operating methods. A number of different types of inspections conducted under the RCRA program are summarized below.

- **Compliance Evaluation Inspection (CEI)**--CEIs encompass a file review prior to the site visit, an on-site examination of generation, treatment, storage or disposal areas, a review of records, and an evaluation of the facility's compliance with the requirements of RCRA.
- **Case Development Inspection (CDI)**--CDIs are conducted when significant RCRA violations are known, suspected, or revealed. Activities conducted during a CDI are specific to the type of information required to document the violation (e.g., incinerator investigations, closure/post-closure investigations).
- **Comprehensive Groundwater Monitoring Evaluations (CME)**--CMEs include sampling and an analysis of the facility's

groundwater monitoring system and hydrogeological conditions. This ensures that groundwater monitoring systems are designed and function properly at RCRA land disposal facilities.

- **Operations and Maintenance Inspection (O&M)**--The purpose of O&M inspections is to ensure that groundwater monitoring and other systems continue to function properly after a land-disposal facility has closed. O&M inspections are usually conducted at facilities that have already received a thorough evaluation of the groundwater monitoring system under a CME inspection.
- **Laboratory Audits**--These are inspections of laboratories performing groundwater monitoring analyses. Audits ensure that these laboratories are using proper sample handling and analysis protocols.

All federal- or state-operated facilities must be inspected annually. Facilities may also be inspected at any time if EPA or the state has reason to suspect that a violation has occurred. Finally, facilities are chosen for an inspection when specific information is needed to support the development of RCRA regulations. Inspection can also be done through examination of the reports that handlers are required to submit. Reports may contain information about the wastes being handled, the method of handling, and the ultimate disposition of wastes. Reports are submitted as required in a permit or enforcement order (e.g., corrective action schedules of compliance) and by regulation (e.g., biennial report).

The most important rule of any inspection is the determination of whether the handler is in compliance with the regulations. If the handler is not complying with all of the appropriate state or federal requirements, enforcement action may be taken.

Enforcement Actions

The goal of enforcement actions is to bring hazardous waste facilities into compliance with applicable RCRA Subtitle C regulations and to ensure:

- Proper handling of hazardous waste.
- Compliance with RCRA's recordkeeping and reporting requirements.

- Monitoring and corrective action in response to past and present releases of hazardous and non-hazardous waste, and hazardous constituents.

Enforcement may include administrative actions, civil and criminal actions, fines and/or imprisonment to correct the violations.

Administrative Actions

These actions are taken by EPA or state under its own authority. They tend to be less complicated than a lawsuit and can often be quite effective in forcing a handler to comply with regulations or to remedy a potential threat to health or the environment. Where the violation is minor, such as record maintenance requirement, EPA or state official notifies the facility that it is not in compliance with some provisions of the regulations. This is an informal action which can be in the form of a letter or a phone call. If the owner or operator does not take steps to comply within a certain time period, a warning letter will be sent, setting out specific actions to be taken to move the handler into compliance. The warning letter also sets out the enforcement actions that will follow if the handler fails to take the required steps. Where the violation is more severe, RCRA authorities can impose enforceable legal duties through administrative orders. Orders can be used to force a facility to comply with specific regulations, to take corrective action, to perform monitoring, testing, and analysis, or to address a threat of harm to human health and the environment. Different types of administrative orders can be issued under RCRA:

- **Compliance orders**--A compliance order may require immediate compliance or may set out a timetable to be followed in moving toward compliance. The order can contain a penalty of up to $25,000 per day for each day of noncompliance and can suspend or revoke the facility's permit or interim status. The person to whom the order is issued can request a hearing on any factual provisions of the order. If no hearing is requested, the order will become final 30 days after it is issued.
- **Corrective action order**--Section 3008(h) allows EPA to issue an order requiring corrective action at an interim status

facility when there is evidence of a release of a hazardous waste or constituent into the environment. These orders can be issued to require corrective action activities ranging from investigations to repairing liners or pumping to treat a plume of contamination. Corrective action can be required regardless of when waste was placed at the facility. Thus, past problems at RCRA facilities may be cleaned up using this mechanism. In addition to requiring corrective action, these orders can suspend interim status and impose penalties of up to $25,000 for each day of noncompliance with the order.

- **Section 3013 orders**--A 3013 order is used to evaluate the nature and extent of the problem at a facility through monitoring, analysis, and testing. These orders can be issued either to the current owner of the facility or to a past owner or operator if the facility is not currently in operation, or if the present owner could not be expected to have actual knowledge of the potential release.

- **Section 7003 orders**--This order can be used against any party contributing to an "imminent and substantial endangerment to health or environment" at a facility. This includes past or present of the facility. Violation can result in penalties of up to $5000 per day.

Civil Actions

A civil action is defined as a formal lawsuit, filed in court, against a person who has either failed to comply with some statutory or regulatory requirement or administrative order or has contributed to a release of hazardous wastes or constituents. Civil actions are generally employed in situations that present repeated or significant violations or where there are serious environmental concerns. The courts may impose penalties in order to force the handler to comply. Where a long-term solution to a problem is desired, a civil action may be helpful to endure proper supervision of the handler's actions. Civil actions may be used to stop conduct that is too dangerous to risk noncompliance with an administrative order, and they also may set a stronger example to other facility operators in order to deter their noncompliance. Different types of civil actions under RCRA include:

- **Compliance action**--Under Section 3008(a) the federal government can file suit to force a person to comply with any applicable RCRA regulations. In federal actions the court can also impose a penalty of up to $25,000 per day violation for noncompliance.
- **Corrective action**--In a situation where there has been a release of hazardous waste from a facility, the federal government can sue to have the court order the facility to correct the problem and take any necessary response measures under Section 3008(h). The court can also suspend or revoke a facility's interim status as a part of its order.
- **Monitoring and analysis**--If EPA has issued a monitoring and analysis order under Section 3013 or RCRA and the person to whom the order was issued fails to comply, the federal government can sue to get a court to require compliance with the order. In this type of case, the court can assess a penalty of up to $5000 for each day of noncompliance with the order.
- **Imminent hazard**--As with a Section 7003 administrative order, when any person contributed or is contributing to an imminent hazard to human health and the environment, the federal government can sue the person to require action to remove the hazard or remedy any problem. If the agency had first issued an administrative order, the court can also impose a penalty of up to $5000 for each day of noncompliance with the order.

Criminal Actions

A criminal action initiated by the federal government or a state can result in the imposition of fines or imprisonment. The penalties range from a fine of $50,000 per day or a prison sentence of up to five years, to a total fine of $1,000,000. Criminal actions are usually reserved for only the most serious violations. Different criminal actions are as described below:

- Knowing transportation of waste to a nonpermitted facility.

- Treating, storing, or disposing of waste without a permit or in violation of a material condition of a permit or interim status standard.
- Omitting important information from, or making a false statement in a label, manifest, report, permit, or compliance document.
- Generating, storing, treating, or disposing of waste without complying with RCRA's recordkeeping and reporting requirements.
- Transporting waste without a manifest.
- Exporting a waste without the consent of the receiving country.

Knowing transportation, treatment, storage, or disposal, or export of any hazardous waste in such a way that another person is placed in imminent danger or death or serious bodily injury can result in a penalty of up to $250,000 or 15 years in prison for an individual or a $1,000,000 fine for a corporation.

Citizen Actions

RCRA and its 1984 amendments provide ongoing opportunities for public participation in all facets of implementation, enable citizens to take legal action against any individual or group involved in the mismanagement of hazardous waste, and require that citizens be notified about hazardous waste issues. The RCRA program requires that:

- Citizens have access to information obtained by EPA or the states during a facility inspection.
- Citizens are allowed to participate in the permitting process from the beginning.
- Citizens may bring suits against anyone whose hazardous waste management activities may constitute an imminent hazard or substantial endangerment.
- Citizens may bring suits against anyone who may be violating a RCRA permit, standard, or requirement.

- EPA or the state must notify local officials and post a sign at sites that pose an imminent and substantial threat to human health and the environment.

Federal Facilities Enforcement

Enforcing compliance under RCRA is different at federal facilities. EPA may only issue Section 3008(h) corrective action orders at federal facilities; no other orders may be used. When waste management activities at federal facilities is managed by a private contractor, as it often is, EPA has full authority to take enforcement activities against the contractor for violations of RCRA. States, however, may utilize the full range of their enforcement authorities at federal facilities.

When a federal facility is out of compliance with the RCRA regulations, EPA issues a notice of noncompliance, outlining violations at the facility and containing a compliance schedule, and a timetable for regaining compliance with RCRA. After the notice of noncompliance has been issued, EPA and the federal facility will negotiate an agreement outlining the steps to bring the facility back into compliance.

In cases where corrective action is required at a federal facility, EPA may issue either a Section 3008(h) corrective action order or a permit schedule of compliance to achieve compliance with the corrective action requirements.

RCRA IMPACT ON INDUSTRY

RCRA has significant impacts on industry and industrial trail manufacturing practices. Any person violating any of the RCRA requirements could be subject to at least a civil penalty as explained in the previous section.

Since 1986, EPA has promoted self-initiated compliance programs by promising not to file criminal charges for promptly identified and fully corrected violations. However, the agency has never offered to give up its civil remedies.

When companies carry out environmental audits and uncover past violations, the law compels them to notify state and federal regulators. Nonetheless, enforcement officials always try to establish a paper trail that links responsible company officials with knowledge of the problems.

WMX Technology, for instance, audited one of its municipal solid waste landfills in Pennsylvania and found that certain operating personnel and managers ignored permit limits on incoming waste. Taking the situation in hand, WMX fired several employees who were responsible for the violation and notified the state. The state of Pennsylvania fined the company $4 million--at least 20 times the amount that WMX might have benefited from taking in extra loads. This cost will unavoidably be passed unto customers. In short, WMX was penalized for having a system that uncovered misdeeds, punished culpable employees, corrected compliance problems, and, ultimately, for being candid with state regulators.

Complying with RCRA Subtitle D--A Case Study

South Valley Disposal and Recycling (SVDR) is a mid-sized solid waste and recycling company in California, serving residents and commercial customers in Gilroy, Morgan Hill, and South Santa Clare County. The services of the company include:

- Residential garbage collection, curbside recycling, hauling and disposal for 18,000 households in an 80 square-mile service area.
- Operation and maintenance of the 136-acre Pacheco Pass landfill.
- Commercial waste handling and disposal for about 1000 businesses.
- Special programs including phone book, office paper, and Christmas tree recycling.
- An annual household hazardous waste drop-off day.
- Collection and sorting of recyclables at the San Martin transfer station.

Materials collected as part of the curbside recycling program include glass, aluminum, newsprint, used motor oil, plastics, and tin. In addition, the company's transfer station accepts white goods, cardboard, steel, demolition debris, wood, tires, and mattresses from the public.

SVDR, with annual revenues of $9.2 million, has 56 full-time employees and a fleet of 28 waste trucks and three recycling rigs on 20 routes that manage approximately 75,000 tons of waste and recycle 10,000 tons annually. Aside from the potentially damaging impact on the environment or human health, the financial consequences of noncompliance with Subtitle D regulations are unthinkable. Landfill violations with a high potential for causing health problems are penalized with fines of $25,000 per day.

To ensure compliance with Subtitle D regulations, SVDR plans to install a methane gas collection and migration monitoring system, a composite liner and leachate collection system, and groundwater monitoring wells. Construction of a methane gas collection and migration monitoring system consisting of 18 vertical gas extraction wells with a blower and flare complex. The project's estimated cost is $723,000 with an annual anticipated operating cost of $18,000 through the remaining active life of the landfill. The composite liner and leachate collection system estimated cost will be over $500,000. The full system includes:

- A compacted clay base with a maximum permeability of 1 x 10^{-7} cm/sec using high-quality clay located on site.
- A minimum 60-mil double-sided textured HDPE liner.
- A layer of Amoco 4512-12 ounce nonwoven geotextile fabric.
- HDPE perforated drain pipe in a gravel drain layer for the leachate control and recovery system.
- A layer of Amoco 4508-8 ounce nonwoven geotextile fabric.
- A final soil layer.
- A special double-walled leachate storage tank.

The estimated cost to install additional groundwater monitoring wells to an already extensive well network was $48,000 for the new wells, plus $98,000 annually for monitoring. SVDR also increased contributions to an existing financial assurance trust for closure and post-closure maintenance. The estimated total cost of the project will exceed $8.6 million.

SVDR experts want to spend approximately 35 percent of its annual budget on environmental compliance. But this is not the only factor that is costly to the company. The California Solid Waste Management Act of 1989, broadly known as Assembly Bill 939 (AB939), was enacted to reduce the amount of waste going to landfills; improve the amount of materials recycled; encourage the processing of recyclable materials diverted from landfills into new products; and strengthen regulation of existing and new sanitary landfills.

The most widely known provision of the bill requires municipalities to reduce the amount of solid waste they send to landfills by 25 percent by 1995 and by 50 percent before 2000. Much of the actual recordkeeping falls to the companies that collect recyclable and trash curbside and run the local landfill.

Perhaps resulting from the fact that strapped municipalities are looking for new sources of revenue. SVDR's fees have climbed to among the highest in the two-county surrounding area. The company predicts that it will pay $1.6 million on revenues of $9.2 million to the cities of Gilroy, Morgan Hill and unincorporated Santa Clare County this year.

Finally, the company's profitability picture, which has been dismal over the past five years, must be improved if the company is to survive. SVDR did not build any profit into its rate structure from 1991 until now, in an attempt to keep rates lower.

In order for the company to maintain continuing operation, it developed a thorough study plan. This study computed where rate-payers' money was going and where it would need to go in the future. It also compared SVDR's operating expenses and wages to similar jurisdictions in the area. The major cost categories were defined as labor, equipment, repairs, landfill costs, franchise fees, profit, and other costs.

Based on this information, a detailed rate increase application was prepared. The study recommended the company request a 30 to 47 percent rate increase from the relevant city councils.

Finally, the company initiated a public information campaign to explain the necessity of the rate increase to customers and the city council. If the new rate increase is implemented, customers will pay between $19 and $21 a month under the new plan, up from $13.44 to $15.68.

As it can be seen, meeting new environmental regulations means higher operating cost for companies. EPA estimates that Subtitle D compliance alone will cost $330 million a year. The costs of environmental compliance are typically passed from companies to those that generate waste in the first place. Solid waste regulatory costs are no exception.

RCRA INTERACTION WITH OTHER REGULATIONS

RCRA does not operate alone, but in conjunction with other environmental acts. This linkage includes Superfund sites having to abide by RCRA requirements; the Clean Water Act regarding water discharge; air emission covered by the Clean Air Act; and waste disposal covered by the Solid Waste Disposal Act.

Additionally there is overlap between RCRA and some of the lesser environmental regulatory statutes including the Safe Drinking Water Act; the Toxic Substance Control Act; and the Hazardous Material Transportation Act.

This section will provide a brief synopsis of RCRA's interaction with other regulations.

RCRA and the Clean Water Act

The discharge of pollutants into surface waters and publicly-owned treatment works (POTWs) is regulated by the Clean Water Act (CWA). Industrial facilities that discharge pollutants directly into surface waters must obtain and comply with a permit issued and pretreatment standards established under the Clean Water Act.

RCRA intersects with the CWA in at least three ways: First, certain wastewater streams that are subject to regulations under the CWA are excluded from regulations under RCRA. For example, "domestic sewage" and "industrial discharge" which are subject to permits issued under the CWA are excluded from the definition of "solid waste" for the purpose of RCRA.

Second, the intersection involves the status under RCRA of certain wastewater treatment systems that are already regulated under the CWA. Most industrial facilities are required to treat their wastewater before discharging to surface water. The permit and other related requirements that apply to hazardous waste treatment facilities might be expected to apply under RCRA also.

And third, the intersection between RCRA and CWA involves the RCRA land disposal restriction program (LDR). This program prohibits the land disposal of hazardous waste.

RCRA and the Clean Air Act

The Clean Air Act (CAA) is designed to limit the emission of pollutants into the air in order to protect human health and environment from the effects of airborne pollution. The CAA seeks to control the emission of hazardous air pollutants (HAP). The EPA is required to ensure that the requirements of Subtitle C of RCRA is consistent and practicable with the provisions of Section 112 of the CAA.

The major interactions between RCRA and CAA include (1) air emissions from incinerators and other types of TSDFs regulated under RCRA which must comply with applicable ambient standards and/or emission limitation of the CAA and (2) extraction of pollutants from air emissions under CAA controls can create hazardous waste or sludges containing such waste. Disposal of these materials must comply with RCRA.

The practical impact that the CAA and RCRA has on each other is one important areas where the CAA jurisdiction and the RCRA jurisdiction overlap. The practical impact is that air pollution control devices installed to meet Clean Air Act requirements produce dusts and sludges that may be subject to regulations as hazardous waste under RCRA. Several types of waste from air pollution control devices such as K061 emission control dust/sludge from the primary production of steel in electric furnaces have been listed by EPA as hazardous. RCRA also regulates waste from air pollution devices that exhibit hazard characteristics.

The EPA intends to make sure that the resulting HAP rules are consistent with the existing RCRA regulations in the future.

RCRA and the Safe Drinking Water Act

The Safe Drinking Water Act (SDWA) intended purpose is to protect the quality of drinking water in two ways. First, the SDWA directs the EPA to set national drinking water regulations for contaminants in public drinking water systems. The maximum contaminant levels (MCLs) for the various chemicals are set as close to the health based maximum contaminant level goals (MCLGs) as feasible.

Second, the SDWA prohibits the underground injection of pollutants except in compliance with an underground injection control (UIC) permit.

The relationship between RCRA and both aspects of the SDWA regulatory program is that MCLs are used as criteria for setting action levels in the water under the proposed rule to establish comprehensive procedures and technical standards for conducting corrective action under RCRA. MCLs also are used as the health basis for setting the toxicity characteristics levels (TCLs) that cause a waste to be identified as a hazardous waste. MCLs are applied as the health-based value in evaluating delisting petition under RCRA. They also serve as a basis for calculating generic exclusion levels for certain high temperature metal recovery residues under the RCRA derived-from rule.

RCRA also relates to the Underground Injection Control Program. The term "land disposal" includes the placement of hazardous waste in an injection well also subject to RCRA regulations as well as SDWA regulation. RCRA prohibits the disposal of hazardous waste by underground injection into or above a formation that contains an underground source of drinking water within one quarter mile of the relevant well. Therefore, EPA has had to integrate the two regulatory schemes.

RCRA and the Toxic Substance Control Act

Toxic Substance and Control Act (TSCA) authorizes the EPA to impose a variety of reporting requirements and control on the manufacture, processing, distribution, and disposal of chemical substances.

Hazardous waste generally fall within the TSCA's definition of "chemical substances" and therefore potentially are subject to the full range of TSCA regulation. However, the EPA has exempted most hazardous waste from key TSCA requirements.

TSCA does direct the EPA to regulate polychlorinated biphenyls (PCBs). The EPA has adopted rules which generally prohibits the manufacture, processing, and distribution in commerce of PCBs, prohibits their use except as authorized by EPA and impose strict requirements on their disposal. PCBs are not regulated as closely under RCRA because they are so comprehensively regulated under TSCA.

RCRA and the Hazardous Material Transportation Act

Materials that are hazardous waste under RCRA are also likely to be hazardous material for purpose of the Hazardous Material Transportation Act. Generators and transporters of hazardous waste must comply with the Department of Transportation (DOT) regulations and requirements as well as separate requirements imposed by EPA under RCRA.

The EPA has expressly incorporated applicable DOT requirements such as packaging, labeling, marking, placarding, and discharging reporting into RCRA regulation.

RCRA's Relation to the Superfund Cleanup Statute

CERCLA is most commonly known as the Superfund Statute. CERCLA and RCRA deals primarily with waste sites. RCRA and CERCLA are unique because their primary purpose is to protect human health and the environment from the danger of hazardous waste. CERCLA imposes cleanup responsibility, whereas RCRA imposes management standards for the handling of hazardous waste.

CERCLA is considered the more comprehensive statute. CERCLA hazardous substance encompass RCRA hazardous wastes. All RCRA hazardous wastes may trigger CERCLA response action when released into the environment, however, RCRA may not necessarily trigger CERCLA response action unless as pollutants or contaminants they present an imminent and substantial danger.

Once hazardous wastes are transported from a site, they are considered as having been generated under RCRA. Therefore, all generation, transportation, and TSD requirements under RCRA must be followed. This means off-site shipments must be accompanied by a manifest.

For off-site land disposal of wastes, CERCLA contains two additional requirements. First, the unit in which the wastes are to be disposed must be releasing hazardous wastes or constituents into groundwater, surface water, or soil. Second, any releases from other units of the facility must be under an approved RCRA corrective action program.

The RCRA and Superfund program uses different labels, but follow roughly parallel procedures in responding to releases. However, the two programs overlap. RCRA's corrective action and CERCLA's remedial action utilize parallel, but not identical, procedures, and both statutes authorizes EPA to act in the event of an imminent hazard.

SUGGESTED READING

1. B. Shanoff, "EPA Struggles to Balance Self-Audited Programs," *World Waste*, 37:9 (1994).
2. D. M. Steinway, A. K. Olick & Oshinsky, "RCRA Title IV: Managing Solid Waste," *Pollution Engineering*, October, 1991.
3. D. Wright, "Coping With the High Cost of Subtitle D Regs," *World Waste*, 37:1 (1994).
4. U.S.E.P.A., "RCRA Orientation Manual," U.S.E.P.A. Office of Solid Waste, Washington, D.C., 1990.
5. U.S.E.P.A., "Solving the Hazardous Waste Problem--EPA's RCRA Program," EPA/530-SW-86-037, November 1986.
6. S. M. Briggum, "Hazardous Waste Regulation Handbook--A Practical Guide to RCRA & Superfund," Executive Enterprise Publications, Inc., New York, N.Y., 1985.

9 MANAGING FACILITIES AND FACILITY TRANSFERS

REGULATORY OVERVIEW

Principal Federal Regulations

In 1986, the Superfund Amendments Reauthorization Act (SARA) was signed into law to provide important corollaries to CERCLA. SARA significantly broadened the definition of parties potentially responsible for a property's cleanup. For example, SARA § 107 provides for financial liability for environmental cleanup regardless of "ownership." Under SARA, liability extends to owners, operators, and legal entities holding title to the property, regardless of whether such ownership was transferred through bankruptcy, foreclosure, abandonment, or payment of delinquent taxes.

Together, CERCLA (Comprehensive Environmental Response and Cleanup Liability Act) and SARA define "strict, joint, several, and retroactive liability" for hazardous waste cleanups. **Strict liability** indicates that contributory negligence is not a prerequisite for determining responsibility under the statute. The purchaser, current owner, or operator on the property, may be liable for cleanup costs even if the property was contaminated prior to purchase. The original owner can be held liable for all or part of the cleanup costs despite compliance with all regulations in effect at the time of property transfer. **Joint and several liability** suggests that one or several parties may be responsible for cleanup costs. Furthermore, corporate lenders, creditors, and shareholders can be named potentially responsible parties (PRP's) and may have to assume some, or all, of a property's cleanup costs. Only when the financial resources of identified responsible parties have been exhausted do the federal Superfund monies (and/or the state "superfunds") become available.

Costs associated with environmental impairment of real property resulting from releases of hazardous substances are eligible for cost recovery under Superfund. The original intent of SARA (the 1986 amendments to CERCLA) was to provide potential defense against liability for "innocent" purchasers of property affected by listed hazardous substances. The environmental site assessment process, developed to respond to the need to perform due diligence under SARA, has expanded to include evaluations of environmental issues such as wetlands and degradation of property by petroleum product releases, asbestos, radon, and lead; for instance, all directly affecting the collateral value, though not necessarily a liability under Superfund, to potential owners and parties to the transaction process. How these related issues will be evaluated and the corresponding level of risk assessed should be carefully and completely discussed by all parties to the transaction prior to commencement of the environmental assessment process. The purchase price and ability of the purchaser to obtain financing are directly affected by actual cleanup costs and perceived risks associated with the presence of toxic and hazardous substances in site buildings, soil, and groundwater.

Following CERCLA, several states adopted hazardous waste liability laws. Some states, including Massachusetts, New Jersey, Connecticut, and New Hampshire, have enacted so-called **super lien** laws which provides states the authority to impose a lien on any property requiring cleanup that involves state expense. The super lien law takes precedence over all other encumbrances, including first mortgages.

While various states passed super lien legislation, New Jersey enacted the Environmental Cleanup Responsibility Act (ECRA). Under ECRA, the New Jersey government has become aggressively involved in regulating property transfers by requiring proof that commercial and industrial properties are "clean" prior to a change in ownership. In effect, New Jersey has the authority to void property transactions if an environmental site assessment and cleanup of hazardous materials present on the property have not been completed. New Jersey's ECRA has been amended and retitled ISRA (discussed later). It is important to understand both the ECRA and ISRA legislation, because many aspects of ECRA still apply and are under enforcement.

Other states require disclosure of known environmental impacts during transfer of residential property under civil codes. In California, sellers must disclose knowledge of the presence of substances which may be an environmental hazard such as asbestos, formaldehyde, radon gas, lead-based paint, fuel or chemical storage tanks, and contaminated soil or water on the property.

CERCLA, SARA, ECRA (or ISRA), and similar federal and state environmental acts have established the legal boundaries within which liability can be assigned. Buyers and lenders are now sensitized to the costs associated with encountering and resolving environmental problems. Indeed, hazardous waste cleanup costs can potentially exceed the value of the property itself.

It is not uncommon for regulatory agencies to impose fines up to $25,000 per day for environmental violations: for instance ten to hundreds of thousands of dollars in fines can be imposed for ongoing noncompliance with air quality regulations. Environmental issues have affected all aspects of property transactions. Because of the potential magnitude if financial liabilities, property transfers are now subject to unprecedented scrutiny by borrowers, lenders, and other potentially responsible parties (PRPs) financially involved in the transaction. Identifying and evaluating environmental liabilities and risks is essential in limiting liabilities to parties to the transaction.

Objectives of Property Transaction
Environmental Site Assessments

A standard for performing real property transaction environmental site assessments involve independent investigation of key issues or facts related to potential environmental liabilities associated with the property transaction. A complete site assessment includes independent verification of historical documents and facts about the property's use.

Objectives of the environmental site assessment include identification of:

- On-site liabilities associated with past or current practices involving the use, storage, treatment, or disposal of hazardous materials (hazmat) or substances.
- Off-site contingent liabilities involving past or current offsite hazmat storage or disposal practices.

Regulatory compliance and permit status of the site operations may also be evaluated, depending on specifics of the transaction.

The property value is typically a significant factor in establishing the extent and content of the site assessment. Real estate transactions on lower value properties generally require a lower level of effort than transactions involving high-risk industrial properties, higher value properties, or transactions with larger loan-to-value ratios. Some lenders impose assessment requirements that remain standard, regardless of the size of the transaction involved.

Laws Directly Affecting Property Transfers

In response to public outcry following the discovery of dangerous contamination at Love Canal, New York, and thousands of other sites around the country, the U.S. Congress and state legislatures enacted laws intended to identify and ensure the cleanup of contaminated sites. Some contamination has resulted from disposal site practices that were previously accepted as adequate by the responsible government agencies. Other contaminated sites are associated with leaking underground storage tanks (USTs), or with leaks or spills that occurred during chemical use at industrial sites or in transit. Additional contamination results from conscious illegal disposal: in remote areas, along roadsides, into sewers and ditches.

To fund cleanup of contaminated sites, Congress enacted the Comprehensive Environmental Response, Compensation, and Liability Act of 1980 (CERCLA--also known as "Superfund") and its subsequent amendments. New Jersey's Spill Compensation and Control Act (N.J. Stat. Ann. § 58:10-23.11) served as the model for the federal legislation. The contamination at Love Canal drew national public attention to toxic contamination and helped solidify action to pass Superfund. Love Canal was such a dramatic incident that it gained a reputation throughout the nation and galvanized efforts at the federal and state levels to pass cleanup legislation. It signaled the beginning of an era of heightened political and public awareness and recognition of the threat chemicals pose to land and water.

The Love Canal incident continues to draw attention. Many residents had to be relocated and their properties purchased; part of Superfund's purpose was to compensate victims for their losses.

Another high profile example is Times Beach, Missouri, where waste oil laced with dioxin was spread on the town's dirt roads to suppress the dust. There, too, residents had to be evacuated and millions have been spent to relocate them.

CERCLA, SARA, AND SUPERFUND IN PERSPECTIVE

Overview

The basic premise of Superfund is that the polluter pays. Like no other environmental legislation, however, Superfund has invoked extreme emotional criticism. Industry representatives argue that society shares the blame for contamination because modern lifestyles are dependent on the chemical industry, which simply responds to what society demands. Moreover, many caught in the Superfund net are troubled by the fact that the practices that led to contamination often conformed to standards accepted at that time. From this standpoint it seems highly unfair to apply present-day standards to the results of disposal methods that were practiced only a few decades ago. On the other hand, advocates of Superfund argue that although there may be some inequities in the system, it if fairer for the responsible parties to pay than for the taxpayer to bear the financial burden of environmental remediation. Moreover, Superfund acts as a deterrent to prevent irresponsible practices that might lead to contamination.

There are numerous other criticisms of the Superfund process. Many proponents claim that the process is ineffective, too slow, and hindered by litigation. Litigation is a particularly sore spot for many critics of Superfund since it is claimed that the only real winners are the lawyers who reap huge profits from the litigation process, thereby diverting money away from cleanups.

There is also the long-debated question of "how clean is clean," amidst allegations that the program's cleanup standards are too rigid or stringent given the relative risks, and that the risks associated with sites slated for cleanup are inflated. Lending institutions also assail the program as too far-reaching and perhaps too altruistic.

Some suggestions for changing the Superfund process include altering the strict liability standard of the law and restructuring the funding mechanism to force localities to bear some of the cleanup costs. The latter idea is based on the premise that, because communities are not faced with remediation costs themselves, they don't appreciate the cost of cleanup in proportion to the risks of a particular site. If communities had to pay directly for part of the cleanup, they would be less likely to demand that stringent standards be met when less comprehensive cleanup methods would suffice. Conversely, many argue that a purely economic analysis ignores certain aspects of fairness. Lower cleanup standards or cost-sharing may make some sense when limited federal funds are being used for cleanup and responsible parties cannot be identified or made liable for reimbursement. However, when responsible parties can be identified, it seems only equitable to force those parties to restore contaminated property to its original state since their actions were the direct cause of the contamination. Many see it as especially fair and necessary if large corporations with "deep pockets" had profited from activities that resulted in contamination.

State Superfund

Many states have created programs similar to Superfund. Generally, these state laws are intended to help finance the state's share for cleanup of sites under the federal Superfund program, and to finance cleanups at state sites that are not considered a priority or slated for cleanup under the federal program. While some contaminated sites are considered extremely important to a particular state, or are in fact a threat to public health and the environment, federal resources are spread thin, and cleanup under the federal program is unlikely unless a site poses a tremendous danger to public health and the environment. Consequently, only the most pervasively contaminated sites are addressed under the federal program.

**Comprehensive Environmental Response,
Compensation, and Liability Act**

CERCLA created national policy and procedures for containing and removing releases of hazardous substances, and for identifying and cleaning up sites contaminated with hazardous substances. It was amended and strengthened by the Superfund Amendments and Reauthorization Act of 1986 (SARA). SARA left the objectives and the basic structure of CERCLA intact, but substantially expanded the scope of hazardous waste cleanup and the size of the Superfund, and imposed tougher and more specific cleanup requirements.

Superfund creates a reporting scheme to assure adequate emergency response to contain and clean up unauthorized hazardous substance releases. The statutes most notable purpose is to provide standards and financial assistance for site cleanups and to impose liability on parties responsible for such contamination. In addition to correcting environmental damages, Superfund is also designed to ensure that victims of hazardous substance releases are compensated for their injuries. Responsible parties, however, are often unable to fund expensive cleanups, may be difficult to identify, and/or no longer exist. For this reason, Superfund provides governmental funding when necessary for remediation and removal projects.

CERCLA is implemented by the U.S. Environmental Protection Agency (EPA), but specific elements allow state agencies to lead site cleanups; there are also extensive provisions in the law for public participation. Local governments are not explicitly assigned any Superfund responsibilities, but are eligible for reimbursement of certain site mitigation expenditures, and are generally included in the provisions for "public participation" at Superfund sites.

Under Superfund, "hazardous substances" are defined to include:

- All toxic pollutants and hazardous substances listed under the federal Clean Water Act.
- Hazardous wastes regulated under RCRA.
- Any hazardous air pollutant under the federal Clean Air Act.
- Chemicals designated as "imminently hazardous" under the Toxic Substances Control Act (TSCA).

CERCLA excludes crude oil, petroleum products, and natural gas products [although the National Oil and Hazardous Substances Pollution Contingency Plan (commonly called the National Contingency Plan, or simply NCP) does address oil spills pursuant to CWA]. CERCLA allows EPA to designate additional substances, if they present a substantial danger to the public health or welfare or the environment when released. By early 1989, EPA had established reportable quantities (RQs) for 719 hazardous materials and wastes. On May 24, 1989, RQs were added for approximately 1500 radionuclides; RQs were set based on the levels of radiation emitted from the individual materials.

Notification Requirements

The initial step in the Superfund process involves identification of sites that may be contaminated with hazardous substances. Two general requirements imposed on owners and operators of facilities and vessels are intended to identify contaminated sites: release reporting requirements for facilities and vessels, and notification of the existence of hazardous waste disposal sites by owners and operators of these facilities.

What Happens If There Is a Release?

Superfund requires owners and operators of facilities or vessels who know of a release of hazardous substances to immediately report to the National Response Center all such releases which equal or exceed specified RQs established by EPA. This reporting requirement, as well as the designation of hazardous substances and their associated RQs, is part of the NCP, and closely parallels provisions of CWA which originally required the development of the NCP. CERCLA expands the scope of the NCP and reporting requirements to include additional substances. Moreover, CERCLA requirements apply to all spills and releases into the environment, rather than just actual or threatened spills into waterways. If notifying the National Response Center is not applicable, notification may be made to the Coast Guard, EPA, or the On-Scene Coordinator (OSC) designated for the geographic area where the discharge has occurred. The OSC is designated by EPA or the Coast Guard to coordinate and direct federal cleanup efforts.

Failure to notify the National Response Center in the event of a release, or knowing submission of false or misleading information, is punishable by a fine or term of imprisonment of not more than three years, or five years for a second or subsequent conviction. Notification of a release may not be used in a criminal case against the person reporting the information, except in prosecutions for perjury or giving a false statement. Therefore, even if cleanup costs are charged or incurred, the consequences of not reporting may be more severe than if reporting is satisfied.

After a spill or release is reported, EPA (or the Coast Guard if the release is into navigable waterways) then notifies other appropriate agencies and begins any necessary emergency response or cleanup actions. The lead agency is authorized to undertake removal or remedial action in the event of a release or substantial threat of a release into the environment that may present an imminent and substantial danger. Response actions must conform with the NCP. Responsible parties are liable for costs associated with removal and abatement. Sites that have been severely contaminated by releases may subsequently be evaluated for listing as a "Superfund site" on the National Priorities List (NPL; also known as the Superfund list).

All owners and operators (including former owners and operators) of hazardous substance TSD facilities were required to report the existence of these facilities to EPA by June 11, 1981. This notification was to include the location of the site, the amount and type of material, and any known or suspected releases. These reports are intended to identify sites where wastes were disposed of routinely, as opposed to the reporting of accidental or unauthorized releases.

The hazardous waste disposal site reporting requirements were designed to locate facilities that were not already regulated by EPA as TSD facilities under RCRA. In fact, there was no duty to report a hazardous waste facility operating with a RCRA permit. Since 1981, many additional facilities and hazardous waste dump sites have been identified by state and local governments as well as by the public. EPA has incorporated information on approximately 30,000 sites into its CERCLA Information System, i.e., "CERCLIS" data base.

What About Cleanup?

The ultimate goal of Superfund is the cleanup of contaminated sites. The program therefore includes extensive provisions for site investigations, selection of methods to be used for cleanup, and levels of eventual cleanup to be achieved. Cleanup operations are generally directed by EPA. EPA also has authority to approve response actions by responsible parties after the agency determines that the person carrying out these actions will investigate and respond promptly and properly to site conditions. States may also be granted responsibility to conduct cleanup operations and enforce CERCLA.

States are also required to enter into "cooperative agreements" with EPA as a condition for any remedial action under Superfund. These agreements reflect a variety of procedural and financial commitments. Procedurally, states must comply with EPA requirements, and assure the availability of licensed hazardous waste disposal facilities. Financially, states pay a 10 percent share of remedial action costs not forthcoming from responsible parties, including all future maintenance, at sites where the federal Superfund pays for cleanup; states pay 50 percent or more of such costs if the facility in question was operated by the state, either directly or through a contractual relationship, at the time of disposal.

What Are Removal and Remedial Actions?

There are two types of response actions for cleanup. Removal actions are short-term actions of limited scope and are carried out by the EPA or the Coast Guard when there is a reported release of a hazardous substance. Other cleanups are categorized as remedial actions.

When a release occurs the lead agency may remove or arrange for removal of the contamination. Under SARA, removal actions are generally limited to those which take no more than one year and cost no more than $2 million. However, there are exceptions that allow the lead agency to continue removal actions or roll removal actions into ongoing site remediation. Also, when EPA or the Coast Guard determines that an actual or threatened release may present "imminent and substantial endangerment" to the public health and welfare or the environment, EPA or the Coast Guard may request that the Attorney General secure an abatement order in federal district court to force the property owner to stop the release and/or prevent future releases. The courts have considered various factors in determining whether there has been imminent and substantial endangerment, including evidence of amounts of, and hazards associated with, the substances released, as well as the potential for exposure.

SARA also establishes a mechanism for reimbursement by the Superfund of costs incurred by a person who receives and complies with an abatement order. To obtain reimbursement, however, a party must show that it is not liable for response costs, and that the reimbursable costs are reasonable as measured by the terms of the EPA order.

What Is Remedial Action?

Superfund establishes priorities for cleanup of sites severely contaminated through releases and past hazardous waste disposal practices based on a Hazard Ranking System (HRS). A part of NCP, EPA has established the

NPL, a list of contaminated sites ranked most hazardous by the HRS to guide the expenditure of cleanup funds. The NPL includes abandoned and uncontrolled hazardous waste sites, which EPA updates periodically. The NCP excludes sites already subject to EPA's jurisdiction under RCRA, where facility operators are required, under their hazardous waste permits, to prevent and clean up contamination.

EPA lists sites on the NPL based on the quantitative HRS. The HRS consists of several analytical methodologies for estimating the potential health risks through any of five potential pathways of exposure:

- Groundwater.
- Surface water.
- Air.
- Direct contact with materials.
- Fire and explosion.

The HRS employs a weighting process to assure that a high risk, via any one or more of the pathways described above, will tend to produce a high ranking, and so a high priority for cleanup. Sites which receive the highest ranking under HRS are placed on the NPL and thus become eligible to have cleanup activities financed by the Superfund. The NPL includes abandoned and uncontrolled hazardous waste sites.

What Do Site Evaluation, Remedial Action Selection, and Cleanup Standards Mean?

The site evaluation and cleanup selection (or Remedial Investigation/Feasibility Study) process is referred to as the "RI/FS" process. Remedial investigation covers site assessment activities, under which lead agencies evaluate the nature and extent of site contamination and general site conditions, and begin to identify possible cleanup methods. The remedial action selected must attain a specified degree of cleanup and control of further releases which, at a minimum, assure protection of human health and the environment. EPA establishes the cleanup standards to impose, taking into account the risk posed to human health and the environment, as well as "applicable or relevant and appropriate requirements" (ARARs) for environmental quality found in other federal, state, and local environmental and health laws. This includes selection of a remedial action that enables attainment of maximum contaminant level (MCL) goals established under the federal Safe Drinking Water Act (SDWA) and water quality criteria established under CWA.

In the feasibility study process, comprehensive cleanup options are developed and evaluated to select alternatives. SARA specifies a list of seven minimum factors which EPA must consider in assessing alternative remedial actions. However, in 1990 EPA listed nine criteria to be considered when evaluating and selecting alternatives:

- Overall protection of human health and the environment.
- Compliance with ARARs.
- Long-term effectiveness and permanence.
- Reduction of toxicity, mobility, or volume through treatment.
- Short-term effectiveness.
- Ability to implement.
- Cost.
- State acceptance.
- Community acceptance.

SARA states that cleanup methods in which treatment "permanently and significantly reduces the volume, toxicity or mobility of ... hazardous substances ... are to be preferred over remedial actions not involving such treatment." Consequently, permanent solutions to hazardous waste problems are preferred in site cleanups, as opposed to mere containment or redisposal of contaminated materials (in potentially leaky landfills, for example). Consistent with the emphasis on treatment technologies, SARA does not favor the transport and disposal offsite of hazardous substances.

EPA approves cleanup plans, including cleanup standards, in a formal document called the Record of Decision (ROD). Final cleanups should reduce contamination to levels that meet CWA and SDWA standards, as well as potentially more stringent ARARs standards. Provisions are made, however, for cost-based exceptions to these requirements.

CERCLA provides that Superfund response action contractors (RACs) are not liable to any person for injuries, costs, damages, expenses, or other liability resulting from an actual or threatened release not caused by RACs' negligence or intentional misconduct. In 1990, this was amended to clarify that issuers of surety bonds for cleanups have the same protection from liability. The amendment applies only to sureties that provide bid, performance, or payment bonds to RACs.

CERCLA also gives EPA discretionary authority to indemnify RACs for releases of hazardous substances or pollutants, or for contamination arising out of negligence in conducting response activities at sites on the NPL and in removal actions.

To be eligible for indemnification by EPA, a RAC must have made diligent efforts to obtain insurance coverage from non-federal sources. The goal of the guidelines is to ensure that an adequate pool of qualified RACs is willing to work at Superfund sites. However, EPA does not intend to offer indemnification if it receives a sufficient number of qualified bids or proposals but only to offer it if lack of response can be linked to the absence of indemnification. This is disappointing to contractors since the policy will favor those contractors that carry their own insurance. Moreover, many are concerned that the liability coverage of $50 million ($75 million for long-term contracts of five years or more) is insufficient, given the high risk of liability to which they are exposed. The term of the coverage offered by EPA is for 10 years.

Where Does the Term "Superfund" Come From?

The purpose of CERCLA was to create a substantial fund (hence, the name "Superfund") to finance cleanup at sites where no financially viable responsible parties could be identified, and to cover costs of the extensive RI/FS evaluation process. The Superfund was set at $1.6 billion for its first five years; SARA expanded the fund to $8.5 billion for the following five years. The Superfund was originally financed by a tax on domestic crude oil, imported petroleum products, and sales of certain feedstock chemicals. SARA raised the tax on petroleum and added a broad-based tax on business income to finance the Superfund's expansion. Both imported and domestic oil are charged a tax of 9.7 cents per barrel.

When no financially viable responsible parties can be located or identified, Superfund's federal money is available for 90 percent of the full range of cleanup activities in states that contribute the remaining 10 percent. At state-owned sites on the NPL, the cost division between federal and state is 50:50. States are not required to contribute matching funds to the cleanup of federal facilities.

Who Are Responsible Parties and What Are Their Liabilities?

Superfund includes extensive provisions for the identification of parties responsible for site contaminations. EPA and state agencies seek to identify "potentially responsible parties" (PRPs) and ultimately "responsible parties" who can be required to finance cleanup activities, either directly or through reimbursement of expenditures from the federal Superfund.

Owners and operators of vessels or facilities from which releases occur are considered PRPs. These owners and operators are usually discovered through the release reporting requirements discussed above. However, PRPs may also be identified through the hazardous waste disposal site notification requirements.

A PRP may be any person who:

- Currently owns or operates a facility where hazardous substances have been or are being released.
- Owned or operated a facility when the disposal of hazardous materials occurred.
- Arranged for the treatment, disposal, or transportation of a hazardous substance to the facility from which the release has occurred or may occur.
- Transported a hazardous waste to a facility from which a release or threatened release occurs.

Responsible parties are strictly liable under CERCLA. Thus, CERCLA requires only a past or present release or threatened release from a facility to impose liability. This means that negligence or other wrongdoing is not required. Parties identified may be held liable for cleanup costs even if procedures followed at the time of disposal were reasonable and met then-current regulatory requirements. It is because of the strict liability nature of CERCLA that site assessments have become routine practice in the transfer of any commercial property. Purchasers that ignore this practice may be subjecting themselves to potential liabilities.

The courts have also agreed that CERCLA authorizes the imposition of joint and several liability. Whether or not joint and several liability applies in a given case depends on whether the harm caused is "divisible" or "indivisible." If the harm is indivisible, any single responsible party may be held liable for the entire harm. Courts will not impose joint and several liability, however, when the harm is divisible and a reasonable basis exists for apportioning the harm. Superfund's liability provisions are so broad that even state governments may be held liable for response costs. The U.S. Supreme Court held that SARA's broad liability provisions strip state governments of their traditional immunities against lawsuit, so that states may now be named as responsible parties and charged with cleanup costs.

What Are the Liabilities?

Under Superfund, responsible parties are ultimately liable for:

- All costs of a removal or remedial action incurred by the federal or state government not inconsistent with the NCP.
- Any other necessary costs incurred by any other persons consistent with the NCP.
- Damages for injury, destruction, or loss of natural resources and the cost of possessing such damages.

SARA also establishes responsibility for interest on the cost of response activities. However, Superfund establishes dollar limits on liability based on the type of "facility" involved. These limits are as follows:

- Vessels--the greater of $300 per gross ton or $5 million.
- Motor vehicles (including aircraft)--$5 million.
- Pipelines--$50 million.
- All other facilities, including incineration vessels--all response costs plus $50 million for any damages.

Failure to give notice of an unauthorized release waives these limitations. Also, failure to comply with an applicable federal standard through willful misconduct or willful negligence resulting in the release of a hazardous material also vitiates these limitations.

Lender Liability and the Security Interest Exemption

The term "owner or operator" is defined to specifically exclude any "person, who, without participating in the management of a ... facility, holds indicia of ownership primarily to protect his security interest in the ... facility." This provision, known as the security interest exemption, may be invoked to shield secured creditors from liability as "owners or operators" under CERCLA. A great deal of interest has therefore arisen regarding the exact meaning of the exemption, and in particular about what constitutes (1) "participating in the management" of a facility, and (2) holding "indicia of ownership primarily to protect" a security interest.

The federal courts have been asked to distinguish between activities that a secured creditor may engage in that are consistent with the security interest exemption and activities that expose the creditor to CERCLA liability. After the Eleventh Circuit Court of Appeals advanced a particularly controversial interpretation of the security interest exemption in 1990, EPA formulated a rule purporting to establish, with precision as well as finality, the precise contours of the exemption.

The Eleventh Circuit's decision in the so-called *Fleet Factors* case disturbed many in the lending community, particularly those who read the opinion to suggest that the mere "capacity" to affect hazardous waste treatment or disposal activities could subject a creditor to CERCLA liability. The ensuing debate over the meaning and implications of *Fleet Factors* was interrupted by the Ninth Circuit's decision in *Bergsoe Metal Corp. vs. The East Asiatic Co.* Although the court formally refused to adopt a rule delineating the degree of control a secured creditor may exert before it incurs liability under CERCLA, it was careful to emphasize that some actual management of the facility must be involved. Since the conduct of the secured party in the case did not amount to actual management, the court found it unnecessary to define the precise parameters of "participation in management." The court did assert that the mere holding or reservation of a right to engage in activities at the secured property, in the absence of the actual exercise of that right, did not constitute participation in management for purposes of the security interest exemption.

The Lender Liability Rule

Fleet Factors increased the risk that lenders would be subject to CERCLA liability when attempting to protect their interests, and it resulted in a mass outcry from lenders and financial institutions for reform. Otherwise incompatible decisions, as exemplified by the divergent Ninth and Eleventh Circuit philosophies on the issue, provided a source of consternation for the financial and lending community. Many commentaries on the subject also exaggerated the implications of *Fleet Factors* and caused added confusion and turmoil. EPA and Congress were then heavily lobbied to ameliorate the possible damaging results of *Fleet Factors*. Consequently, in 1992, EPA published a final rule clarifying the scope of CERCLA's security interest exemption and specifying a range of activities that a secured creditor might engage in without losing the protection of the exemption. The rule provides relief or certainty to lenders in the wake of *Fleet Factors* and related case law.

One reason for EPA's diligence in promulgating this rule was the predicament of the Resolution Trust Corporation (RTC) and the Federal Deposit Insurance Corporation (FDIC). RTC and the FDIC were created by Congress to handle failed banking institutions and are now conservators and receivers of many real property holdings--which include contaminated parcels--in the aftermath of the savings and loan debacle. A clear rule on the issue dispels any anxiety that these institutions might have had.

The key provisions of the rule are those defining the phrase "participation in management." The term is limited to actual participation in the management or operation of a facility, and excludes "the mere capacity to influence, or ability to influence, or the unexercised right to control facility operations." When the debtor is in possession of the facility, the secured party is considered to be participating in management only if at least one of the following two circumstances applies:

- The secured party exercises decision-making control over the debtor's environmental compliance, such that the secured party has undertaken responsibility for the debtor's hazardous substance handling or disposal practices.

- The secured party exercises control at a level comparable to that of a manager of the debtor's enterprise, such that the secured party has assumed responsibility for the overall management of the enterprise encompassing the day-to-day decision-making of the enterprise with respect to either (1) environmental compliance, or (2) all or substantially all of the operational aspects of the enterprise other than environmental compliance.

The term "operational aspects" refers to functions handled by a facility or operations manager, chief operating officer, or chief executive officer. Operational aspects do not include "financial or administrative aspects," which encompass functions similar to those of a credit, accounts, or personnel manager; controller; or chief financial officer. The rule further specifies activities of secured parties that do not constitute management participation for purposes of the security interest exemption. These include conducting or requiring an environmental inspection of a prospective debtor's facility. Included are "policing" or "work out" activities performed prior to foreclosure, provided that the secured party does not by such actions participate in the management of the facility. "Policing" activities include requiring the debtor to clean up the facility or to comply with applicable environmental and other laws, and monitoring or inspecting the facility or the debtor's business or financial condition. "Work out" activities are those undertaken by the secured party to prevent, cure, or mitigate a default by the debtor or to preserve or prevent the diminution of the security's value. Restructuring or renegotiating the terms of a security interest and providing specific or general financial or other advice or suggestions are examples of work out activities.

The rule also addresses post-foreclosure activities. "Indicia of ownership" includes legal or equitable title acquired via foreclosure. These indicia are deemed to be held after foreclosure primarily to protect a security interest if both of the following are true:

- The holder undertakes to divest itself of the property "in a reasonably expeditious manner, using whatever commercially reasonable means are relevant or appropriate."
- The holder did not participate in management prior to foreclosure.

A holder affirmatively establishes that ownership indicia continue to be held primarily to protect a security interest when it does either of the following within 12 months following foreclosure (or acquisition of marketable title):

- Lists the facility with a broker, dealer, or agent who deals with the type of property in question.
- Advertises the facility at least monthly in a publication or newspaper specified in the rule.

A holder that did not participate in management prior to foreclosure and that otherwise complies with the above rules regarding post-foreclosure may conduct any of the following activities without voiding the security interest exemption:

- Sell or release property held pursuant to a lease financing transaction.
- Maintain business activities.
- Liquidate or wind up operations.
- Undertake a response action under CERCLA.
- Take measures to preserve, protect, or prepare the secured asset prior to sale or other disposition.

Such a holder will incur CERCLA liability with respect to a facility it possesses after foreclosure only if it does either of the following:

- Arranges for disposal or treatment of a hazardous substance, as provided by CERCLA.
- Accepts for transportation and disposes of hazardous substances at a facility selected by the holder, as provided by CERCLA.

A holder does not incur liability by virtue of taking any response action under CERCLA.

It is important to note that if a plaintiff brings suit under CERCLA, he has the burden of establishing that the defendant is liable as an owner or operator.

States are also developing lender liability rules under their state programs. For instance, the Oregon Environmental Quality Commission adopted rules exempting those lenders and trust companies that act as fiduciaries from liability for contaminated properties if certain procedures and rules are followed; government entities are also exempt from liability. Oregon rule also exempts trust companies.

Finally, the most dramatic break for lending institutions has come from the state with the most prolific cleanup programs: New Jersey. New legislation limits the liability of banks and other lenders under the state's superfund law, the Spill Compensation and Control Act (N.J. Stat. Ann. § 58:10-23.11). The lender exemption provided by this law is based on the same principles as EPA's lender liability rule (e.g., lenders will be exempt as long as they do not actively participate in the management of the facility prior to foreclosure). Lenders could still be held liable under the law for hazardous substance releases that continue after foreclosure. However, they can only be held liable for such contamination if they are found "negligent." For instance, if a bank was aware of a release from drums of hazardous wastes after foreclosure, the bank could be held liable under a negligence standard if proper containment precautions were not taken to prevent the spread of contamination. This is a striking departure from the usual strict liability standard imposed under federal and state Superfund laws and is quite a coup for lending institutions.

What Are Defenses Against Liabilities?

Superfund does not impose liability when a release is caused solely by an "act of God" or an act of war. There is also no liability when the sole cause of a release is the act of a third party (other than an employee, agent, or independent contractor of the defendant). A key point though is that the defendant must prove that due care was exercised and precautions taken against foreseeable acts. These defenses are not available to persons who fail to report releases.

SARA added an important defense for property owners who acquire land and subsequently discover that hazardous substances were disposed of on the property without their knowledge. This is known as the "innocent landowner defense." This defense is available only if a person acquired property after the disposal or placement of the hazardous substances on the property, exercised due care with respect to the substances, and took reasonable precautions against foreseeable acts or omissions of third parties. The property owner must also establish at least one of the following:

- He or she did not know and "had no reason to know" of the presence of hazardous substances on the property when it was acquired.
- The property owner is a governmental entity that acquired the property involuntarily or by eminent domain.
- The property was acquired by inheritance or bequest.

The courts generally consider a variety of factors to assess the property owner's level of knowledge or innocence, including any specialized knowledge or experience and the ability to detect contamination by an appropriate inspection. As awareness of the likelihood of site contamination spreads, the viability of this defense narrows.

STATE SUPERFUND PROGRAMS AND PROPERTY TRANSFER LAWS

Introduction

State superfund programs are designed to provide for the state to share in projects funded under CERCLA and to provide added resources for remediation of sites not slated for cleanup under the federal program. These programs generally parallel the federal cleanup program, with some exceptions: for instance, CERCLA excludes petroleum, but relevant state cleanup programs do not.

The New Jersey Spill Compensation and Control Act

The New Jersey Spill Compensation and Control Act (the Spill Act--N.J. Stat. Ann. § 58:10-23.11) focuses on discharge prevention and standards for facilities storing hazardous substances and petroleum. The program adds site cleanup provisions to many of the mechanisms found in other federal laws, such as the Clean Water Act. Like CERCLA, the Spill Act also has provisions for notification [which is given to the state Department of Environmental Protection and Energy (DEPE)], response, and removal of unauthorized or accidental discharges. Liability under the Spill Act is strict, joint, and several--just as it is under CERCLA. All removal and cleanup under the Spill Act must, to the greatest extent possible, be conducted in accordance with the NCP for removal of oil and hazardous substances.

Like CERCLA's Superfund, the Spill Act created the New Jersey Spill Compensation Fund (the Fund) to support cleanup and removal costs incurred by DEPE and third parties, and to pay direct and indirect damages to innocent persons who sustained losses due to hazardous substance discharges. The Fund derives its money from a state tax on barrels of hazardous substances transferred, and by costs and damages recovered from dischargers. Like EPA's authority under CERCLA, the administrator of the Fund may settle disputes with responsible parties over monies disbursed by the Fund. The Spill Act directs the administrator to promote and arrange for settlements between claimants and responsible parties--where identifiable--to avoid recourse against the Fund. If responsible parties cannot be identified, the administrator is directed to seek settlement of claims against the Fund.

As under CERCLA, any person who has discharged a hazardous substance or is in any way responsible for any hazardous substance is held strictly liable, jointly and severally, without regard to fault, for all cleanup and removal costs, no matter who incurred them. In contrast, CERCLA does not explicitly set forth the standard of liability to be imposed. Strict liability under the Spill Act and CERCLA is applied retroactively to discharges that occurred before the enactment of the Spill Act.

Under the Spill Act, liability for cleanup and removal costs can reach up to $50 million for each major facility and $150 per gross ton for each vessel. These limitations do not apply in cases of gross negligence, willful misconduct, or gross or willful violations of safety, construction, or operating standards.

The Spill Act provides for more extreme penalties than CERCLA does for certain types of violations. Also the Spill Act imposes "punitive" measures for severe discharges to the land and/or waters of the state: "[A]ny person whose intentional or unintentional act or omission proximately results in an unauthorized releasing, spilling, pumping, pouring, emitting, emptying, or dumping of 100,000 gallons or more of a hazardous substance, or combination of hazardous substances, into the waters or onto the lands of the state, or entering the lands or waters of the state from a discharge occurring outside the jurisdiction of the state, is liable to a civil administrative penalty or civil penalty of not more than **$10,000,000** ... In assessing a penalty pursuant to this section, [DEPE] shall take into account the circumstances of the discharge, the conduct and culpability of the discharger, or both, prior to, during, and after the discharge, and the extent of the harm resulting from the discharge to persons, property, wildlife, or natural resources." N.J. Stat. Ann. § 58:10-23.11.1.

New York State Toxic Cleanup Law

In 1978, the state legislature passed a measure directing the New York State Department of Health (DOH) to conduct a study to evaluate the effects on public health associated with "exposure to toxic substances emanating from certain landfills." N.Y. Pub. Health Law § 1386. This study was the direct result of Love Canal. Subsequent to this measure, the state legislature passed the New York Inactive Hazardous Waste Sites Law (the Inactive Sites Law). Although the Inactive Sites Law was enacted prior to the federal Superfund program, the Inactive Sites Law's provisions for public financing of contaminated sites were not born until after the passage of CERCLA.

Under the *New York State Hazardous Waste Site Remedial Plan*, which has been mandated by the Inactive Sites Law, the New York State Department of Environmental Conservation (DEC) has established an aggressive cleanup schedule. DEC is attempting to begin remediation at 500 of the state's identified sites by the year 2000, a clear sign of intense commitment. The total number of sites that will require remediation under the program is expected to reach over 700.

Although DEC appears to be on track toward meeting its self-imposed deadline of the year 2000 for beginning remedial actions, final cleanup of these sites will take many more years. DEC estimates that the average time to complete remediation efforts at contaminated sites is five years. However, many sites are sure to take much longer to fully remediate. Nevertheless, DEC's attempt to begin cleanup at 500 sites is a sign of its strong commitment.

The scope of the New York program is more narrow than that of CERCLA. The Inactive Sites Law provides for the identification, listing, and remediation of "inactive hazardous waste disposal sites," which it defines as "any area or structure used for the long term storage or final placement of hazardous waste including, but not limited to, dumps, landfills, lagoons, and artificial treatment ponds, as to which area or structure no permit or authorization issued by [DEC] or a federal agency for the disposal of hazardous waste was in effect after the effective date of this [law]." N.Y. Envtl. Conserv. Law § 27-1301(2). This definition is much narrower than that employed by the federal Superfund program, which does not exclude hazardous waste sites that are permitted after 1979, the effective date of the Inactive Sites Law. Moreover, the Inactive Sites Law merely applies to "hazardous waste sites" and does not include the broader category of all sites contaminated with "hazardous substances" covered under CERCLA. Because of this limited coverage, remediation of New York sites that pose a threat to public health and the environment may require the assistance of the federal Superfund program. However, the use of state funds for the state matching share under the federal Superfund program is permitted. Therefore, if a site is being addressed under the federal program, the state's "superfund" may be used for the state matching share.

The New York State DEC (under the Inactive Waste Sites Law) has developed a comprehensive registry of inactive hazardous waste sites (the registry) in the state. The registry lists inactive sites, defines the scope of cleanup problems, sets priorities, and tracks progress at individual sites. The registry is reviewed continuously and updated annually on March 31. In maintaining the registry, DEC annually reassesses, in cooperation with DOH, the relative need for action at each site. DEC classifies each site similarly to the way EPA does under HRS. The ultimate purpose of the Inactive Sites Law is to provide for cleanup of contaminated sites. In pursuing this goal, the most severely contaminated sites are usually addressed first since they generally pose the greatest threat to public health and the environment. In this way the expenditure of funds and time are approached with reference to the relative need for action at the sites. By ranking and prioritizing the sites, DEC determines which enter the remedial process first and schedules enforcement efforts in pursuing responsible parties.

The Inactive Sites Law's preferred source of funding cleanups is responsible parties. DEC is directed to identify private parties responsible for contamination, and to enforce payment of cleanup costs. DEC attempts to negotiate consent orders to secure voluntary cleanup by responsible parties. Where no financially solvent responsible party can be located, DEC may develop and implement any remedial program. If responsible parties are later identified, DEC can recover from them costs, penalties, and monetary damages, or may require such parties to continue the development and implementation of a remedial program.

The Hazardous Waste Remedial Fund is the state superfund for funding emergency abatement measures, remedial activities that responsible parties are unwilling to perform, remedial activities when responsible parties cannot be identified, and the state share of cleanup costs under the federal Superfund program. This state superfund is financed through assessments on the generation and disposal of hazardous wastes and petroleum surcharge fees, fines, and penalties, and it also receives appropriations from the state's general fund. Where possible, DEC attempts to secure funding for site remediation through the federal Superfund program. In 1986, the state Legislature responded by passing the Environmental Quality Bond Act of 1986 (Bond Act). The Bond Act added a considerable financial commitment to the state superfund effort, providing $1.45 billion for a variety of environmental programs, with $1.2 billion of that targeted for hazardous waste remediation projects.

Under the Inactive Sites Law, owners and operators of sites on the registry must notify DEC and DOH before substantially changing the use of their sites. Written notice must be provided at least 60 days prior to a change in use or physical alteration of land or construction. Substantial changes include erection of buildings, paving of roadways and parking lots, or the creation of a park or recreation facility. A substantial change in use requires notice only, and not DEC and DOH approval, unless DOH declared "a condition dangerous to life or health resulting from an inactive hazardous waste disposal site." In these cases, initiation of changes to the site may not begin prior to written approval being issued by both DEC and DOH. The agencies cannot approve the changes if the new use would interfere with a remedial program or increase risk to the environment or human health.

The "Super Lien" Laws

Some states have gone beyond CERCLA and SARA by enacting a priority lien or "super lien" provision as part of their Superfund laws. A priority lien allows the state to impose the lien with priority over all other claims. New Jersey has led the way in allowing liens for cleanup costs.

In 1980, the New Jersey's Spill Act was amended to include a super lien provision designed to prevent responsible parties from escaping liability by claiming bankruptcy. This predated SARA, which included a much weaker federal lien provision.

Any expenditure made for cleanup and removal is a debt of the discharger to the New Jersey Spill Compensation Fund:

> The debt shall constitute a lien on all property owned by the discharger when a notice of lien, incorporating a description of the property of the discharger subject to the cleanup and removal and an identification of the amount of cleanup, removal and related costs expended from the fund is duly filed with the clerk of the Superior Court ... Upon entry by the clerk, the lien, to the amount committed by the administrator for cleanup and removal, shall attach to the revenues and all real and personal property of the discharger, whether or not the discharger is insolvent. N.J. Stat. Ann. § 58:10-23.11f(f).

The lien constitutes a priority lien--meaning it creates a lien with priority over all past and future claims or liens filed--on the property which is the subject of the cleanup and removal costs. A typical lien may apply to all other property that the discharger owns:

> The notice of lien ... which affects any property of a discharger other than the property subject to the cleanup and removal, shall have priority from the day of the filing of the notice of the lien over all other claims and liens filed against the property, but shall not affect any valid lien, right, or interest in the property filed in accordance with established procedure prior to the filing of a notice of lien ... N.J. Stat. Ann. § 58:10-23.11f(f).

The priority lien or super lien "does not come into existence and is not recorded until expenditures are made out of the Spill Compensation Fund. Therefore, the state cannot simply assert the lien on property in anticipation of, or prior to, cleanup; it can only assert the lien once it has spent money on cleanup efforts.

As originally enacted, the priority lien provision extended to all assets of the responsible party. However, mass criticism of the statute's scope forced later amendments. In 1985, the priority lien provision became limited to "dirty assets"--those associated with un-authorized discharges--although a typical lien is available for other assets.

With the advent of liabilities stemming from the so-called super lien laws, it has become standard practice for purchasers to perform site assessments prior to real property transfers. Severe contamination can not only result in excessive liability for responsible parties and landowners, but also be a "deal breaker": if a site is severely contaminated, a potential buyer may walk away from purchasing the property. Lenders are also particularly leery of such properties. The site could still be useful as a commercial property, however, even if there are leaking USTs or the land has been contaminated through other commercial activities, e.g., a bus yard where years of leaking fuel and oil have contaminated the land (if such problems are not remediated, they can, of course, lead to further troubles at a later date). In many of these circumstances, a buyer may still be willing to purchase a contaminated property if remediation of the property is a condition of the sale.

Many contractual options are available. Agreements between sellers and purchasers can be structured so that the seller either performs cleanup or reimburses the buyer for the cleanup costs. As with many types of contractual arrangements, there are pitfalls. The type of agreement chosen will depend on the buyer's and seller's respective needs and their willingness to negotiate. For instance, if the seller takes on the burden of cleaning the property, there may be a dispute over "how clean is clean." Conversely, if the buyer agrees to remediate the property conditional upon reimbursement from the seller, a dispute may arise over remediation costs with the seller refusing to pay above a certain level. A cap on remediation costs in the sales agreement will prevent the buyer from forcing the seller to restore the property to a pristine condition if it is unnecessary under the circumstances.

Despite the now-routine site assessments conducted when commercial properties are transferred, there are additional state requirements that mandate either the performance of these assessments or the notification of buyers that contamination exists. These types of state laws provide added protection for buyers and place the burden on sellers to perform site assessments and to be candid about the site's history. In states without such statutes, the burden is on the buyer to ensure that the property is clean before it is purchased. In either type of state, however, both buyers and sellers may wish to get their own consultants to ensure the accuracy and honesty of the assessments; the buyer wants to avoid liabilities and the seller does not want to be saddled with unnecessary or inflated cleanup costs.

The types of pretransfer statutes may vary from those that merely require sellers to notify buyers of contamination (for example, the Illinois

Responsible Property Transfer Act of 1988 and the Indiana Responsible Property Transfer Law) to those that require an actual pretransfer cleanup as a condition of the sale (such as the Connecticut Property Transfer Act). New Jersey, however, has been the benchmark by which all these statutes are judged as it has developed one of the most innovative programs in the nation.

New Jersey's Industrial Site Recovery Act (ISRA) is a pretransfer cleanup law developed to promote cleanup of toxic contamination. ISRA requires industrial establishments to disclose and remove contamination located on their properties prior to transfer of the establishment or the contaminated property, or when operations at these sites cease. The owners and operators (sellers in the case of a transfer) of the properties or businesses are responsible for the costs and implementation of cleanup. In essence, ISRA imposes a precondition on the transfer or closure of an industrial site or establishment and forces the use of private funds--rather than public funds like Superfund--to clean up contaminated industrial sites.

ISRA serves as a model to other states. Among the national and state legislation dealing with toxic contamination, ISRA is unique; only a small number of states (e.g., Connecticut) have enacted similar laws that require actual cleanup prior to transfer. Even among those states with similar programs, ISRA is recognized as the most powerful law of its kind.

Although some states, such as California and Massachusetts, discussed the possibility of enacting similar pretransfer statutes, the vast majority avoided such a program, possibly for fear of disenfranchising industry and creating turmoil in the commercial real estate market. Many also viewed it as unnecessary since fear of Superfund liability had made site assessments routine practice as part of commercial property transfers. ISRA, however, also mandates site assessments for an expanded variety of activities. For instance, site assessments are required every time operations at a site are discontinued or drastically changed. In these instances, it is not even necessary to transfer property to trigger cleanup requirements. New Jersey does not want site owners merely to cease operations, let their sites deteriorate, and later become a burden for taxpayers by needing cleanup under the state Superfund program. Moreover, even though cleanup may occur as a practical matter in states that do not have pretransfer statutes, in New Jersey such cleanups have been conducted with oversight by the state.

The (New Jersey) Industrial Site Recovery Act

New Jersey has the distinction of being the state with the highest number of NPL hazardous waste sites, i.e., sites requiring cleanup under CERCLA. The residents of the most densely populated state in the nation have to contend not only with Superfund sites, but also numerous chemical and pharmaceutical facilities. Stringent environmental initiatives, considered some of the most aggressive in the country, have been created as a result of these circumstances.

ISRA requires industrial establishments to disclose and remove contamination located on their properties prior to the transfer of these establishments or contaminated properties, or when the operations at such sites cease. The owners and operators (sellers in the case of a transfer) of the properties (or businesses) are responsible for the cleanup costs and implementation. In essence, ISRA imposes a precondition on the transfer or closure of an industrial site or establishment and forces the use of private funds--rather than public funds like CERCLA--to clean up contaminated industrial sites. Thus, ISRA has made it mandatory to perform site assessments prior to the transfer of property.

Until a major legislative amendment in 1993, ISRA was called the Environmental Cleanup Responsibility Act (ECRA). The original law became effective in 1983.

While ECRA supporters hailed the program as a huge success, critics claimed that it strained New Jersey's economy (and local economies) by causing both delays in real estate transactions and additional financial and legal burdens to be placed on business--ultimately forcing business to leave the state in search of more favorable treatment. ECRA was viewed as a continuing cause of the relocation of New Jersey's manufacturing base and a major reason for chilling the movement of new business into the state. Even ECRA supporters recognized that ECRA has been problematic for urban redevelopment efforts. Both the state Department of Environmental Protection and Energy (DEPE) and a state legislator who sponsored the original legislation have been sensitive to these problems and have implemented initiatives or proposed legislation to address these shortcomings.

Despite the fact that the law was very successful in forcing an astonishing number of site cleanups, the political landscape had entirely changed and complete reform of the landmark law became inevitable. Pressure by industry lobby groups together with a deep recession and a dramatic shift from a Democrat-controlled legislature to a Republican-controlled one, helped create the momentum for an overhaul of ECRA. Many Republicans criticized the law as a symbol of government overregulation.

On June 16, 1993, the Governor signed the long-awaited revision to ECRA. When signing the 66-page bill (S 1070), Governor Florio commented that the bill restored a proper balance between environmental protection and economic development. S 1070 was the result of months of discussions with environmentalists, business interests, and DEPE. The intent of S 1070's sponsors was to improve New Jersey's business climate by reducing regulatory burdens and spurring redevelopment of New Jersey's vast industrial and commercial lands. ECRA was renamed ISRA; the stigma attached to the old law was apparently enough to warrant the change of name. Not surprisingly, neither environmentalists nor industry viewed the amendments as going far enough in their respective directions. It was, indeed, a carefully prescribed compromise.

While ECRA was a law characterized by its inflexibility in application, ISRA is expressly written to provide for waivers and deferrals that can apply depending upon the circumstances of a given facility or site history. Specifically, ISRA made the following changes in ECRA:

- Site owners no longer have to provide separate financial assurance--such as a bond--for cleanup while using other financial resources to undertake the cleanup. Although a site owner must still establish how cleanup will be funded, so as to ensure that actual cleanup proceeds, money can now be drawn from that source to pay for cleanup.
- Environmental reviews at sites that have undergone state-approved cleanups in the past will be expedited.
- Most soil cleanups will be allowed to proceed without state oversight, however, oversight of ground water and surface waters will be increased.
- Property owners will be permitted to transfer ownership of up to one third--or larger in some cases--of the value of the site without triggering mandatory cleanup.
- The utilization of caps, fences, restrictions on site use, and other practices will be permitted to a greater extent as alternatives to permanent remediation.

- Government entities that acquire property involuntarily (e.g., from tax delinquency, bankruptcy) will be exempt.

ISRA also permits the use of differential cleanup standards depending on whether property will be used for residential or non-residential purposes. The premise is that non-residential properties need not be as clean as residential properties. Nevertheless, the law does impose an across-the-board risk level no matter what cleanup standard is used. That is, ISRA standards prevent exposure to any pollutant that would result in an additional cancer in one-in-one million persons during a lifetime of exposure.

Finally, ISRA establishes a $50-million Hazardous Discharge Site Remediation Fund that will provide grants and loans to small businesses and municipalities to aid in cleanup. This program will be funded by a one-percent annual surcharge on cleanup funding sources.

ECRA was created with two purposes in mind. The program provides insurance against the creation of future Superfund sites in New Jersey--a state with more than its fair share when you consider its size--and provides for a unique "buyer protection plan" by requiring that all contamination be disclosed and cleanup completed prior to sale or transfer. These basic principles still operate under ISRA, although in a less rigid form.

It is important to note that ISRA pertains only to industrial establishments engaged in activities falling into the major Standard Industrial Classification (SIC) code groups 22-39 (manufacturing), 46-49 (transportation; communications; electric, gas, and sanitary services), 51 (wholesale trade, nondurable goods), and 76 (miscellaneous repair services). Refer to Table 1. Additionally, these businesses must be engaged in the generation, manufacture, refining, transportation, treatment, storage, handling, or disposal of hazardous substances and/or wastes. ISRA exempts facilities subject to certain state laws. DEPE also has exempted certain operations and transactions, and certain subgroups or classes within these SIC categories (e.g., sewage systems) from the ISRA program. New Jersey courts have thus far deferred to DEPE's interpretation of ISRA and its applicability. Given the fact that ISRA slightly narrows the scope of the state's cleanup law, courts may continue to defer to DEPE's interpretation of the law.

TABLE 1	
INDUSTRIES COVERED UNDER ISRA	
SIC Code	**Industry Description**
	Manufacturing Groups
22	Textile Mill Products
23	Apparel and Other Finished Products Made from Fabrics and Other Similar Materials
24	Lumber and Wood Products, except Furniture
25	Furniture and Fixtures
26	Paper and Allied Products
27	Printing, Publishing, and Allied Products
28	Chemicals and Allied Products
29	Petroleum Refining and Related Industries
30	Rubber and Miscellaneous Plastics Products
31	Leather and Leather Products
32	Stone, Clay, Glass, and Concrete Products
33	Primary Metals Industries
34	Fabricated Metal Products, except Machinery and Transportation Equipment
35	Machinery, except Electrical
36	Electrical and Electronic Machinery, Equipment, and Supplies
37	Transportation Equipment
38	Measuring, Analyzing, and Controlling Instruments; Photographic, Medical and Optical Goods; Watches and Clocks
39	Miscellaneous Manufacturing Industries
	Transportation, Communications, Electric, Gas, and Sanitary Services Groups
46	Pipe Lines, except Natural Gas
47	Transportation Services
48	Communication
49	Electric, Gas, and Sanitary Services
	Wholesale Trade Groups
51	Wholesale Trade, Nondurable Goods
	Services Group
76	Miscellaneous Repair Services

ISRA compliance is necessary in the following two instances:

- The transfer of ownership of a property or a business.
- The closure of a business (cessation of operations).

The statute and DEPE regulations list a number of specific circumstances that constitute transfer. As mentioned above, DEPE amended its regulations regarding applicability. Despite the fact that ISRA had not yet become law, DEPE proceeded with amendments to ECRA rules to comply with a court-imposed deadline and ruling. In *In re adoption of N.J.A.C. 7:26B*, the New Jersey Superior Court upheld DEPE's rules promulgated under ECRA, but remanded certain provisions regarding which transactions trigger an ECRA review (i.e., applicability) to DEPE for further rulemaking. DEPE proposed amendments to its rules on March 30, 1992, to conform to the court's ruling, but, on July 23, 1992, ISRA was introduced in the state legislature. DEPE still proceeded with its rulemaking despite the possibility that ISRA would affect the applicability provisions of ECRA.

DEPE finalized the proposed rules on March 1, 1993; portions of the new rules were not mandated by the court's decision. The rules clarified which business transactions trigger ECRA. DEPE has said that the new rule was consistent with the then-pending ISRA. DEPE's new regulations became effective when Governor Florio signed ISRA. Further rulemaking on the applicability issue may be forthcoming.

If requested by a site owner or operator, DEPE will perform an applicability determination for a fee. Applicability determinations enable establishments to be certain of whether they need to comply with ISRA.

Owners and operators of industrial establishments are responsible for compliance with ISRA. Once ISRA is triggered, the owner or operator must submit to DEPE a Pre-transaction Notice [corresponding to the General Information Submission (GIS) that was formerly required] in conformity with N.J. Stat. Ann. § 13:1K-9(4). Unless a waiver or deferral should apply, after the Notice has been submitted, the ISRA-triggering party must remediate the property "in accordance with criteria, procedures, and time schedules established by the department." N.J. Stat. Ann. § 13:1K-9. Approvals for an ISRA-regulated transaction include either an approved negative declaration, an approved remedial action workplan, a no further action letter, or a remediation agreement approval.

ISRA has addressed several sensitive issues involving landlord/tenant relations when the cleanup law has been triggered. Pursuant to ISRA, the tenant must supply the landlord with the information the landlord needs to comply with the law and vice versa. Additionally, when a lease makes it clear who (either landlord or tenant) is to comply with ISRA in the case of a trigger, the other party may petition DEPE to compel the responsible party's compliance.

With respect to cleanup criteria, ISRA, for the first time, requires DEPE to establish minimum soil remediation standards that differentiate between residential and non-residential uses. There is additional flexibility built into ISRA: alternative cleanup criteria may be adopted by DEPE for a given site, and engineering controls (such as capping) and institutional controls (such as deed restrictions) may be enlisted with the permission of DEPE. Furthermore, remediation beyond natural background levels of a given contaminant will not be required and remediation of contamination originating from offsite sources will not be demanded of an innocent party.

ISRA softens the impact of ECRA liability by introducing new provisions allowing for exemptions or deferrals. Some of these requirements, such as the deferrals, codify existing DEPE regulations. Under ISRA, for example, certain sites will not require pre-transaction cleanups or will be entitled to deferrals allowing the sale of a business or property prior to a cleanup. Additionally, financial security requirements have been significantly relaxed so that bonds will no longer have to be posted in the case of transactions that proceed before cleanups are undertaken.

One very important ISRA exemption is the so-called de minimis exemption for facilities whose usage of hazardous materials is comparatively small. The state must still be notified of an ISRA trigger, in the same manner as under ECRA, by way of a pre-transaction Notice filed with DEPE.

Another important exemption applies if the only environmental problems are related to one or more USTs. The USTs must still be remediated under the State Bureau of Underground Storage Tanks (BUST) program; however, a transaction that would have been covered by ECRA is no longer regulated under ISRA if the pollution is only tank-related. Once again, the state must be notified of the situation through the filing of a pre-transaction notice.

Four of the other options available under ISRA include:

- Deferral of a site cleanup when the transferee will continue the use of the property.
- Expedited review of sites already remediated under CERCLA, RCRA, or other hazardous waste law.
- Area of Concern waiver for any section of a site that has already been remediated.
- Waiver for a cleanup in progress.

A party required to perform an ISRA cleanup must establish and maintain a "remediation funding source" in the amount necessary to pay the estimated cost of the required remediation. Unlike past practice regarding "financial assurances" under ECRA, however, money from the remediation funding source may be used to pay for the actual cost of the cleanup and no further financial assurances can be required by DEPE.

In order to assist in financing ISRA-required remediation efforts, a new revolving fund known as the "Hazardous Discharge Site Remediation Fund" has been established. Loans from the Fund may be obtained by an owner or operator that cannot otherwise establish a remediation funding source. Grants are also available under certain circumstances where the ISRA party did not cause or have reason to know about the environmental problem.

ISRA specifically states that "[n]o obligations imposed by this act shall constitute a lien or claim which may be limited or discharged in a bankruptcy proceeding. All obligations imposed by this act shall constitute continuing regulatory obligations imposed by the state." N.J. Stat. Ann. § 13:1K-12.

Under CERCLA, the owners of contaminated property are strictly liable for the contamination regardless of actual responsibility. Owners must then seek recovery of cleanup costs from the site's previous owners or responsible parties. As a consequence, buyers normally conduct environmental assessments of sites prior to their purchase to avoid future liability. If proper environmental assessments are not conducted before purchase, buyers and lenders are taking unnecessary risks because of the imposition of strict liability. Although a purchaser could later seek indemnity for contamination, the cost of legal fees and the possible difficulties in obtaining money from prior owners or responsible parties make this option unattractive.

ISRA provides more than mere incentive for buyers to perform site assessments of properties prior to purchase; it provides the buyer with unique protections and shifts the burden of performing site assessments to transferrers from buyers or transferees. Transferrers must perform environmental assessments under the scrutiny of DEPE and clean the site, if it is contaminated, as a condition of a transaction. ISRA allows purchasers to void transfers of an industrial establishment or real property if the transferor does not disclose all contamination and perform the required cleanup, and if the transferor fails to comply with any ISRA provisions. The transferee is also entitled to recover damages resulting from the failure to implement a cleanup plan as well as all cleanup and removal costs.

DEPE performs inspections of sites at different stages of the ISRA process and oversees actual cleanup operations, an added comfort to purchasers of industrial property.

Of course, there are instances under ISRA--as under CERCLA--where parties required to comply with the statute may not have been responsible for the contamination. Many properties transferred prior to the passage of ECRA were contaminated. Cleanups were not performed in conjunction with these transactions and purchasers were often unaware of the extent or existence of contamination. ISRA, like CERCLA, provides for strict liability, without regard to fault, for all cleanup and removal costs. Therefore, parties who acquired contaminated real estate prior to ECRA/ISRA and who then attempt to sell this property may be held responsible for cleanup costs. Parties who find themselves in these seemingly unfair positions must then seek indemnity for cleanup costs from prior owners or responsible parties.

ISRA applies under the following guidelines only:

- There must be a legally defined pending transaction.
- The facility's SIC code number must be specified in ISRA.
- Hazardous substances or wastes as defined by the regulations must be present on the site.

SUMMARY OF FEDERAL REGULATIONS

Introduction

Property transfers are potentially affected by a broad range of federal legislation that deal with toxic and hazardous materials. For example, due diligence requires knowledge of the Toxic Substances Control Act (TSCA) and the Clean Air Act (CAA), which has been amended; as well as the Comprehensive Environmental Response, Compensation, and Liability Act (CERCLA), the Resource Conservation and Recovery Act (RCRA), and the Clean Water Act (CWA).

Certain legislation and regulations impact property transfers by limiting property uses. Examples include the Rivers and Harbors Act of 1899, the Endangered Species Act (1973), and the Historic Preservation Act. Violation of these regulations may result in criminal or civil penalties and removal of the offending activity. This chapter provides an overview of the federal regulations that may impact on a property transaction. Table 2 provides a summary of these regulations.

SARA Title III

The Emergency Planning and Community Right-to-Know Act was enacted as Title III of SARA in October 1986 and is intended to increase community awareness of the quantity and types of hazardous chemicals used by, and discharged from, local industries. SARA Title III requires emergency response plans to be developed for use in the event of releases of hazardous chemicals.

Under this act, the governor of each state must appoint a State Emergency Response Commission (SERC) which shall, in turn, appoint, supervise and coordinate the activities of Local Emergency Planning Committees (LEPCs). LEPCs are to consist of state and local officials, representatives of law enforcement, civil defense, fire departments, first aid and health personnel, and owners and operators of facilities subject to emergency planning and notification requirements. LEPCs develop plans for responding to hazardous chemical discharges and information requests from the public.

TABLE 2

SUMMARY OF MAJOR FEDERAL LEGISLATION

Legislation	Implementing Agency	Property Use Limitations
CAA (42 U.S.C. § 7401 et seq.)	U.S. Environmental Protection Agency (EPA)	Pollution control equipment or other offsets may be required to meet emission standards. Regulates construction and destruction activities involving asbestos-containing materials (ACMs).
CWA (33 U.S.C. § 1251 et seq.)	EPA, U.S. Army Corps of Engineers	Prohibits discharge of pollutants into U.S. waters and dredging or filling waters, including wetlands, without a permit.
Coastal Zone Management Act (CZMA--16 U.S.C. § 1451 et seq.)	National Oceanic and Atmospheric Administration	State Coastal Zone Management Commission may control coastal land uses. [16 U.S.C. § 1456(C)]
Endangered Species Act (16 U.S.C. § 1531 et seq.)	U.S. Department of the Interior, Fish & Wildlife Service	Prohibits federal actions that threaten species or affect critical habitat.
Fish & Wildlife Coordination Act (16 U.S.C. § 661 et seq.)	U.S. Department of the Interior, Fish & Wildlife Service	Interior may recommend modifying projects involving impoundment, diversion, or other control of a water body in order to reduce impacts on fish and wildlife.
Floodplain Management Order (Executive Order 11988)	Executive Office of the President, Council on Environmental Quality (CEQ)	Prohibits approval of construction permits for projects within the 100-year floodplain.
Marine Protection, Research and Sanctuaries Act of 1972 (16 U.S.C. § 1431 et seq., 33 U.S.C. § 1401 et seq.)	National Oceanic and Atmospheric Administration, Office of Coastal Zone Management	Protects marine sanctuaries.

Statute	Agency	Description
National Environmental Policy Act (42 U.S.C. § 4321 *et seq.*)	Executive Office of the President, CEQ	An Environmental Impact Statement (EIS) is required on all federal projects involving environmental impact.
National Historic Preservation Act (16 U.S.C. § 470 *et seq.*)	Advisory Council on Historic Preservation	Whenever a proposed action may adversely affect properties listed on the National Register of Historic Places, mitigation is required.
Prime and Unique Farmland Policy (CEQ Memorandum of 30 August 1976)	U.S. Department of Agriculture	Discourages conversion of highly productive farmland unless national interests are at stake.
RCRA (42 U.S.C. § 6901 *et seq.*)	EPA	Regulates and defines the generation, treatment, storage, and disposal of hazardous wastes. RCRA regulates underground storage tank (UST) projects.
Safe Drinking Water Act (SDWA) (42 U.S.C. § 300 *et seq.*)	EPA	Prohibits construction activities that may contaminate a municipality's watershed if used as the primary drinking water supply.
Surface Mining Control and Reclamation Act of 1977 (30 U.S.C. § 1201 *et seq.*)	U.S. Department of the Interior, Office of Surface Mining	Regulates mining and associated impacts.
Wetlands Protection Order (Executive Order 11990)	Office of the President, CEQ	Discourages approval of permits required for construction in a wetland.
Wild and Scenic Rivers Act (16 U.S.C. § 1271 *et seq.*)	U.S. Department of the Interior	Prohibits any project that may adversely impact a national "wild and scenic river."

Emergency planning and notification requirements apply to facilities containing one or more extremely hazardous substance (EHS) equal to or in excess of the threshold planning quantity (TPQ). EPA has established a complex set of six different thresholds for 360 EHS, ranging from 1 to 10,000 pounds. EPA regulations also require thresholds for any mixture containing EHS to be set individually, based on the percentage of EHS in the mixture (if above one percent for most EHS). The owner or operator of a facility subject to these requirements must notify the SERC within 60 days of becoming subject to the requirements. The owner or operator must also designate a representative to participate in the local emergency planning process as a facility emergency response coordinator; within 30 days of establishment of an LEPC, the owner or operator must notify the LEPC of the existence of the facility. The facility owner or operator must provide information necessary for developing and implementing the emergency plan upon request from the LEPC.

The Resource Conservation Recovery Act

The Resource Conservation and Recovery Act (RCRA), 42 U.S.C. §§ 6901-6992k, provides the basic framework for federal regulation of hazardous waste. RCRA controls the generation, transportation, treatment, storage and disposal of hazardous waste through a comprehensive "cradle to grave" system of hazardous waste management techniques and requirements.

RCRA [Pub. L. No. 94-580, 90 Stat. 2795 (1976)] was adopted in 1976 as a revision and expansion of the Solid Waste Disposal Act (SWDA) of 1965 which, until then, had focused on disposal of municipal solid wastes. RCRA introduced a detailed nationwide program for management of hazardous wastes. Subsequent amendments, most notably the 1980 Solid Waste Disposal Act Amendments [Pub. L. No. 96-463, 90 Stat. 1982 (1976)] and the Hazardous and Solid Waste Amendments of 1984 (HSWA), have refined this regulatory framework and introduced new substantive requirements. RCRA is administered nationally by the United States Environmental Protection Agency (EPA), with major components of the law delegated to the states for ongoing implementation.

RCRA contains the official definition of hazardous waste; certain solid wastes are exempted under 40 C.F.R. § 261.4 and include the following:

- Domestic sewage.
- Household wastes.
- Industrial wastewater (point source) discharges subject to regulation under § 402 of the Clean Water Act, i.e., 33 U.S.C. § 1342.
- Agriculturally derived solid wastes.
- Mining overburden returned to the mine site.
- Solid waste generated from the extraction and processing of ores and minerals.
- Drilling fluids and other wastes associated with the exploration, development or production of crude oil, natural gas, or geothermal energy.
- Cement kiln dust waste.
- Discarded wood products treated with arsenic.

The following are exempted wastes under 40 C.F.R. § 261.6:

- Spent lead-acid batteries to be sent offsite for reclamation.
- Used oil not mixed with hazardous waste.
- Dry cleaning solvents routinely reclaimed onsite without being stored.

Hazardous waste generators and transporters, and owners and operators of hazardous waste treatment, storage, or disposal (TSD) facilities must comply with the applicable regulations. Regulatory compliance includes manifesting and record keeping, maintaining facility standards, groundwater protection standards, preparing and submitting contingency and emergency preparedness plans, closure and post-closure standards, and contingent financial responsibility measures. In addition, owners and operators of hazardous waste treatment, storage (for greater than 90 days) or disposal facilities must obtain a RCRA permit from EPA or an authorized state agency.

A Comparison of RCRA and CERCLA

Although both RCRA and CERCLA were developed to protect human
health and the environment, substantial differences do exist. CERCLA
is a goal-oriented program giving EPA the authority to perform cleanups,
or to compel potentially responsible parties to remediate NPL sites.
RCRA is a process-oriented law which compels owners to manage their
facilities in a specified manner. Also, RCRA is a relatively inflexible
program, whereas CERCLA is flexible and practical.

Given these distinctions, regulatory entities must apply innovative
technical and policy interpretations when applying RCRA regulations and
policies to CERCLA actions as when RCRA properties are transferred
to CERCLA sites. EPA has ruled that any of the following four
conditions must apply before a RCRA facility can be considered for
transfer to the Superfund program for cleanup:

- The owner or operator of a RCRA facility declares bankruptcy,
  and the courts protect the facility's assets.
- A RCRA facility loses its authorization to operate. For example,
  EPA may deny a facility the permit required to operate, or EPA
  may revoke interim status.
- A RCRA facility is negligent in submitting or executing an
  acceptable closure plan.
- A RCRA facility violates other RCRA directives.

RCRA corrective action enforcement is currently the sole
responsibility of EPA under the Office of Waste Programs Enforcement.
Refer to Table 3.

Underground Storage Tanks

HSWA also included provisions for regulating **underground storage
tanks** (USTs) containing any substance defined as hazardous under
CERCLA and petroleum. 40 C.F.R. § 280.12 defines an UST as a tank
that stores regulated substances and has at least 10 percent of its volume,
including the contents of connected pipes, underground.

TABLE 3	
RCRA AND CERCLA COMPARISON	
RCRA	**CERCLA**
Purpose:	**Purpose:**
To regulate all applicable hazardous waste management activities.	To perform remedial action on NPL sites.
To protect human health and the environment.	To protect human health and the environment.
Enacted to regulate hazardous waste generators, transporters, and TSD facility operators.	Aimed at hazardous waste generators, transporters, and TSD facility operators.
Only specified TSD components can be regulated. These are: • Containers. • Incinerators. • Landfills. • Land treatment units. • Surface impoundments. • Tanks. • Waste piles. HSWA also regulates solid waste units on TSD facilities.	Aimed at any threat to human health and the environment due to release of hazardous substances.
Incinerator operation is subject to minimum acceptable performance standards.	Standards are interpretative, health based, and set on a case-by-case basis.
Cost effectiveness is not a consideration under this program.	According to 42 U.S.C. § 9604(c) (CERCLA), remedial actions must be cost effective.

TABLE 3 (Continued)	
RCRA AND CERCLA COMPARISON	
RCRA	**CERCLA**
Purpose:	**Purpose:**
Regulated material includes hazardous waste, and all listed and designated wastes per 40 C.F.R. part 261.	Regulated material includes substances designated in the following sections: Federal Water Pollution Control Act 33 U.S.C. §1321(b)(2)(A) 40 C.F.R. part 261 33 U.S.C. § 1317(a) (FWPCA) 42 U.S.C. § 7412 (CAA) 15 U.S.C. § 2606 (TSCA) 42 U.S.C. § 9602 (CERCLA), which allows EPA to designate any element, compound, mixture, solution or substance as a hazardous substance.

HSWA subtitle I grants EPA the authority to regulate USTs, including registration, and establishing technical performance standards. EPA implemented the UST registration program and enjoined anyone from installing unprotected USTs in 1984, under 42 U.S.C. § 6991a (HSWA). However, the program was not enforced until 1986. EPA proposed technical performance standards for USTs in April 1987. 52 Fed. Reg. 12662. Interim technical performance standards dictate design, construction, installation, and release detection; EPA issued final technical performance standards in September 1988. 53 Fed. Reg. 37082. Notification became mandatory as of October 1988. Anyone selling an UST on or after October, 1988, must notify the purchaser. 40 C.F.R. § 280.22.

USTs containing radioactive wastes and materials are regulated by 40 C.F.R. Part 280 Subpart A, and the corrective action provisions of 40 C.F.R. Part 280 Subpart F, only.

Field-constructed tanks, including underground bulk storage tanks, must comply only with 40 C.F.R. Part 280 Subparts A and F. **Field-constructed tanks** are vertical cylinders with a capacity of greater than 50,000 gallons.

USTs larger than 110 gallons storing oil used for emergency power generators are subject to all UST regulations except for release detection requirements.

Liability and Enforcement Actions Under RCRA

Because RCRA provides "cradle-to-grave" regulation of hazardous wastes covering generation, transportation, storage, treatment, and disposal, the hazardous wastes generator is faced with a nearly limitless period of liability. A generator who has properly managed and disposed of wastes at a licensed offsite disposal facility may still be required to contribute funds to clean up the disposal facility in the future. The federal government can order such payment by authority of CERCLA or RCRA. In effect, implementing proper practices at a RCRA permitted facility is no guarantee against incurring financial liability for past practices. Cost recovery provisions covering leaking USTs also exist. The authority for these decisions was granted through SARA, via 42 U.S.C. § 6991(d).

The enforcement provisions of 42 U.S.C. § 6928 authorize the imposition of civil penalties at a maximum rate of $25,000 per day per violation. Knowingly treating, storing, transporting to an unpermitted disposal facility, or disposing of hazardous wastes without a RCRA permit can result in criminal penalties. Criminal fines can be up to $50,000 per day per violation and can include a five-year prison sentence. If the party responsible for the illegal activity knowingly places another person in imminent danger of death or serious bodily injury, criminal penalties can be expanded to a maximum total of $250,000 for an individual, and $1 million for a corporation. Individuals may also face up to 15 years of imprisonment. Finally, EPA enforcement actions can result in a facility's closure through the suspension of the RCRA operating permit.

Section 6973 of 42 U.S.C. grants additional authority to EPA to handle any imminent hazard that endangers human health or the environment due to past or present handling, storage, treatment, transportation, or disposal of any solid or hazardous waste. EPA can bring suit against generators, transporters, or past or present owners or operators of a treatment, storage or disposal facility at which an imminent hazard has been identified. This provision affects past and present facility owners. Enforcement action under this provision includes the authority to issue an abatement order requiring a facility to take any action necessary to cease any action responsible for posing an imminent hazard. Failure to comply may result in a fine of $5000 per day per violation.

Clean Water Act (Federal Water Pollution Control Act)

In 1972, Congress enacted Pub. L. No. 92-500, 86 Stat. 816 (1972), entitled the Federal Water Pollution Control Act. This legislation was referred to as the Clean Water Act (CWA) after the addition of the 1977 amendments; it is the government's principal statute for regulating water pollution. Public Law No. 95-217, 91 Stat. 1566 (1977) addresses the problem of toxic water pollutants and Pub. L. No. 100-4, 100 Stat. 7 (1986) refines enforcement priorities and increases EPA's enforcement authority. EPA was granted authority to implement CWA, but states can administer certain tenets of the National Pollutant Discharge Elimination System (NPDES) program.

The objective of CWA is to "restore and maintain the chemical, physical, and biological integrity of the Nation's waters." "The CWA can be divided into five policy areas:

1. National water quality standards.

2. Industry specific minimum national effluent standards.

3. A permit program to regulate point source discharges, and to otherwise enforce water quality standards.

4. Special problems including toxic chemical releases and oil spills.

5. Grants for construction of publicly owned treatment works (POTWs)."

Each state is required to divide water bodies into segments for CWA planning and implementation purposes. CWA requires states to submit plans to EPA defining water quality standards to be achieved for each segment identified. 33 U.S.C. § 1313. Water quality standards measure the attributes of a given body of water and address all discharges into it.

Water quality standards serve a dual role. They establish goals for the quality of water in a specific water body; and, they serve as the regulatory basis for defining and enforcing treatment controls and strategies beyond the national standards based on technology (discussed *infra*).

All dischargers must apply a minimum level of water pollution control technology, regardless of which water body receives their effluent discharge. These are termed "**technology-based limits**." Dischargers in selected locations must go further, applying additional pollution controls to ensure that their discharges do not cause violations of the water quality standards set for that receiving body. These are termed "**water quality-limited requirements**."

States designate uses for all water body segments (i.e., public water supplies, agricultural and industrial uses, protection and propagation of shellfish, fish and wildlife, and recreation), and then set criteria necessary to protect these uses. 33 U.S.C. § 1312(a). Consequently, the water quality standards developed for particular water segments are based on their designated use and vary depending on such use (e.g., recreational waters are subject to more stringent standards than industrial waters).

In addition, each state identifies areas failing to meet water quality standards, and then establishes maximum daily pollutant loads that will achieve the applicable standards. 33 U.S.C. § 1313(d). The states are also responsible for periodic review and modification of water quality standards. All water quality standards proposed by a state must be approved by EPA. 33 U.S.C. §1313(a)(l). Certain states have set water quality standards that are more stringent than the federal guidelines.

CWA § 402; *see* 33 U.S.C. § 1342 (1972) empowers the Director of EPA to "issue a permit for the discharge of any pollutant, or combination of pollutants ... as the Administrator determines are necessary to carry out the provisions of this act." The discharge of any pollutants directly into waters of the United States from a new or existing point source is prohibited unless the point source has an NPDES permit. 33 U.S.C. § 1342(a)(l).

Pollutants that industries discharge indirectly into U.S. waters through POTWs constitute **indirect point source discharges**, and do not require NPDES permits; however, indirect sources are regulated under separate state or local programs that involve compliance with general pretreatment standards. Certain industries, whether they contribute through direct or indirect sources, must also comply with specific industrial toxic pollutant standards which are directed to control conventional, nonconventional, and toxic pollutants from specific industries. Table 4 lists industries for which these categorical limits have been granted. By definition, "point source" excludes surface water runoff, though such sources are covered under separate provisions of the NPDES program. This term does not include agricultural storm water discharges and return flows from irrigated agriculture.

An NPDES permit is required before point source pollutants may be discharged directly into U.S. waters. 33 U.S.C. § 1342. EPA has granted most states permitting authority under the NPDES program.

Permit applications must be submitted at lest 180 days prior to the proposed discharge date, or at the expiration of the existing permit. NPDES permits must be renewed every five years. 40 C.F.R. § 122.46(a). NPDES permits set levels of performance for each discharger while EPA sets national permit limits, based on EPA effluent guidelines. Generally, effluent limitations must follow EPA guidelines, and may be further regulated by stricter receiving water quality standards.

EPA, authorized by the 1987 amendments to CWA, may grant variances from national effluent guidelines to certain industries, if those industries differ significantly from the industries considered when effluent guidelines were established. These variances are called the "Fundamentally Different Factors Variances."

NPDES permits generally include requirements for periodic monitoring and reporting. Such reports, called the Discharge Monitoring Reports (DMRs), must be submitted by the discharger to the appropriate regulatory agency. DMRs present the results of the industrial waste discharger's effluent sampling program.

NPDES Permit for Storm Water Discharges

A section, 33 U.S.C. § 1342(p), of the 1987 Water Quality Act (WQA), specifically addresses storm water discharges to be regulated under the NPDES program. The regulated discharges all constitute point source pollution. Uncontaminated storm water runoff that is considered a nonpoint source is regulated by EPA or the state by authority of 33 U.S.C. § 1329, titled "Non-point Source Management Programs."

TABLE 4

EFFLUENT GUIDELINES AND INDUSTRIAL CATEGORIES
(as of July 1, 1990)

40 C.F.R. Part	Industrial Category	40 C.F.R. Part	Industrial Category
467	Aluminum Forming	432	Meat Products
427	Asbestos Mfg.	433	Metal Finishing
461	Battery Mfg.	464	Metal Molding & Casting
431	Builders' Paper & Board Mills	436	Mineral Mining & Processing
407	Canned and Preserved Fruits & Vegetables Processing	421	Nonferrous Metals Manufacturing
408	Canned and Preserved Seafood Processing	471	Nonferrous Metals Forming & Metal Powders
458	Carbon Black	435	Oil & Gas Extraction
411	Cement Mfg.	440	Ore Mining and Dressing
434	Coal Mining	414	Organic Chemicals
465	Coil Coating	446	Paint Formulation
468	Copper Forming	443	Paving and Roofing Materials
405	Dairy Products	455	Pesticide Chemicals
469	Electrical and Electronic Components	419	Petroleum Refining
413	Electroplating	439	Pharmaceutical Mfg.
457	Explosives Mfg.	422	Phosphate Mfg.
412	Feedlots	459	Photographic
418	Fertilizer Mfg.	463	Plastics Molding and Forming
424	Ferroalloy Mfg.	466	Porcelain Enameling
426	Glass Mfg.	430	Pulp, Paper and Paperboard
406	Grain Mills	428	Rubber Mfg.
454	Gum and Wood Chemicals	417	Soap and Detergent Mfg.
460	Hospital	423	Steam Electric Power Generating
447	Ink Formulation	409	Textile Mills
415	Inorganic Chemicals	429	Timber
420	Iron and Steel Mfg.		
425	Leather Tanning and Finishing		

Storm water discharges must obtain permits prior to October 1, 1992, if:

- An NPDES permit was issued prior to February 4, 1987.
- The discharge is due to industrial activity.
- The discharge is from a municipal separate storm sewer system serving a population of 250,000 or more.
- The discharge is from a municipal separate storm sewer system serving a population of 100,000 or more but less than 250,000.
- The EPA Administrator or the state considers it violates a federal or state water quality standard, or it is a "significant contributor" of pollutants to U.S. waters.

Industrial Storm Water Dischargers

Industries that have current NPDES permits for the discharge of storm water from their properties are regulated by the current permit. Industries without a current NPDES permit for uncontaminated point source storm water discharges must obtain in NPDES permit. EPA was empowered to establish permit application requirements for such discharges by February 1989. Permit applications must have been filed by February 1990; by February 1991, EPA or the appropriate state regulatory agency must have acted on each permit application. Large municipal storm water dischargers (those serving populations in excess of 250,000) must adhere to the above schedule. Small municipal storm water dischargers (those serving populations between 100,000 and 250,000) are required to await EPA's permitting requirements have been developed by February 1991. Following the promulgation of permitting requirements, these permit applications must be filed no later than February 1993. All NPDES permit applicants must comply with permit provisions within three years of permit issuance. Finally, NPDES storm water permits issued to municipalities must contain a prohibition against discharging anything but storm water into the storm sewers.

Industry-Specific Minimum National Effluent Standards

The majority of industry's hazardous wastes are in liquid form. The treatment of industrial effluent requires dewatering, and frequently secondary wastewater treatment, before the treated effluent can be discharged to sanitary sewers, storm drains, surface impoundments, and waterways. Regardless of pretreatment method, industrial effluent typically retains some pollutants. Minimum National Effluent Standards are specified for each industry to control the types and quantities of pollutants entering sewers and receiving waters.

Publicly-Owned Treatment Works (POTWs)

NPDES Permits for POTWs: Like other direct dischargers, POTWs are required to apply for NPDES permits for their discharges to waters (*see* discussion *supra* for permit requirements). However, the technology-based effluent limitations for POTWs differ substantially from those required of all other point source discharges. These differences reflect the dominant role of POTWs in managing domestic pollutants and municipal/household wastes, and the dominant role of the federal government in providing funds to upgrade the pollution control capabilities of these public sewerage agencies. POTWs' unique role in managing industry's indirect discharges through their implementation of pretreatment requirements constitutes another important distinction.

The 1972 Amendments made all discharges from POTWs subject to **secondary treatment** as of July 1, 1977. As in the case for all other point sources, EPA determines what constitutes secondary treatment and more stringent requirements may be placed on POTWs if necessary to meet water quality standards for the receiving waters.

Requirements for Indirect Discharges (National Pretreatment Standards for Industrial Users of POTWs): In order to protect the operation of POTWs and to prevent the discharge from POTWs of pollutants which have not received adequate treatment, CWA requires EPA to adopt and amend, as necessary, national pretreatment standards for discharges **into** POTWs. Discharges into POTWs are often referred to as "indirect discharges" because they are not directly discharged into receiving waters, but are sent through POTWs to the receiving waters.

Industrial users of POTWs for such "indirect discharges" are not required to obtain NPDES permits. Rather, POTWs impose restrictions or "pretreatment standards" on these industrial users in order to ensure compliance with their own NPDES permit and its discharge limitations. POTWs regulate industrial discharges into their system to meet three objectives:

1. Prevent introduction of pollutants into POTWs which would interfere with equipment or operations, or endanger personnel.

2. Prevent introduction of pollutants that would pass through (i.e., would not be treated adequately before discharge) or be incompatible with the POTW.

3. Improve opportunities to recycle and reclaim municipal and industrial wastes and sludges.

POTW pretreatment programs must enforce national pretreatment standards. Many also establish and enforce additional local requirements that are more stringent and more comprehensive than the national standards. These local requirements are often imposed in response to unique concentrations of point or non-point discharges into receiving waters, or to provide additional protection to these waters.

National pretreatment standards developed by EPA take two forms: prohibitions on discharges to POTWs, and categorical standards.

Asbestos Regulations

The term "asbestos" is applied to a group of naturally occurring fibrous, inorganic hydrated mineral silicates. The group includes actinolite, amosite, anthophyllite, chrysotile, and crocidolite. From about 1946 until EPA banned its use, asbestos-containing materials (ACMs) were widely used for fireproofing, insulation, and soundproofing. EPA defines any material containing more than one percent asbestos as an ACM. EPA reported that ACM was used to simulate snow in movies such as the "Wizard of Oz" and "White Christmas."

Applications of ACM generally fall into one of the following categories:

- Sprayed onto surface material.
- Used as insulation around pipes, ducts, boilers, and tanks.
- Construction applications such as ceiling and floor tiles, wall insulation.
- Manufacturing applications such as cloth, cord, wicks, tape, twine rope, etc.

In a 1984 survey EPA determined that approximately 733,000 public and commercial buildings in this country contain friable asbestos. This number represents about 20 percent of some 3.6 million public and commercial buildings. Of this number, 28 percent are residential apartment buildings, 70 percent are private nonresidential buildings, and 2 percent are federal government buildings. EPA also estimated that approximately 30 percent of all school buildings, approximately 35,000 contain friable asbestos. **"Friable asbestos material"** is defined as any material that contains more than 1 percent asbestos by weight, and can be crumbled, pulverized, or reduced to powder by hand pressure. Table 5 provides a summary of ACM commonly found on sites.

With the increased use of ACM, the medical profession has become concerned about potential consequences of asbestos exposure. Aspirated fibers cause damage to the lungs that sometimes takes 20 years to manifest. The most common of these is asbestosis, a respiratory disease that scars the lungs causing respiratory difficulties. Exposure to asbestos fibers is also linked to mesothelioma, a rare cancer involving the thin membrane lining of the chest and abdomen that can develop following a single exposure to asbestos. Evidence suggests that smokers are particularly susceptible to this disease. The government first began to ban certain uses of asbestos in 1973. As more information became available on the health effects of asbestos, other forms of ACM also were banned. A chronology of the various forms of ACM banned from use is as follows:

- 1973: all spray-on applications of asbestos coating banned for fireproofing and insulation.
- 1975: installation of wet-applied and pre-formed asbestos pipe insulation banned; asbestos block insulation used on boilers, hot water tanks, and heat exchanger banned.
- 1978: all spray-applied asbestos coatings intended for decorative purposes banned; use of asbestos as an ingredient in spackle and joint compounds banned.

During the late 1970s, numerous lawsuits were filed against asbestos manufacturers. These legal actions sought billions of dollars in damages for injury and death resulting from worker exposure to asbestos. At least one manufacturer, Johns-Manville, sought protection under the federal bankruptcy laws due to the volume of actions against it. Today,

TABLE 5

TYPICAL ASBESTOS-CONTAINING MATERIALS (ACM)

Subdivision	Generic Name	Asbestos (%)	Dates of Use	Binder/Sizing
Surfacing material	Sprayed- or troweled-on	1-95	1935 - 1970	Sodium silicate, portland cement, organic binders
Preformed thermal insulating products	Batts, blocks and pipe covering		1926 - 1949	Magnesium carbonate
	85% magnesia	15		Calcium silicate
	Calcium Silicate	6-8		
Textiles	Cloth[a]			
	Blankets (fire)[a]	100	1910 - present	None
	Felts:	90-95	1920 - present	Cotton/wool
	Blue stripe	80	1920 - present	Cotton
	Red stripe	90	1920 - present	Cotton
	Green stripe	95	1920 - present	Cotton
	Sheets	50-95	1920 - present	Cotton/wool
	Cord/rope/yarn[a]	80-100	1920 - present	Cotton/wool
	Tubing	80-95	1920 - present	Cotton/wool
	Tape/strip	90	1920 - present	Cotton/wool
	Curtains[a] (theater, welding)	60-65	1945 - present	Cotton

Cementitious (concrete-like products)	Extrusion panels:			
	Corrugated	8	1965 - 1977	Portland cement
	Flat	20-45	1930 - present	Portland cement
	Flexible	40-50	1930 - present	Portland cement
	Flexible perforated	30-50	1930 - present	Portland cement
	Laminated (out surface)	35-50	1930 - present	Portland cement
	Roof tiles	20-30	1930 - present	Portland cement
	Clapboard & shingles:			
	Clapboard	12-15	1944 - present	Portland cement
	Siding shingles	12-14	Unknown - present	Portland cement
	Roofing shingles	20-32	Unknown - present	Portland cement
	Pipe			
Paper products	Corrugated:			
	High temperature	90	1935 - present	Sodium silicate
	Moderate temperature	35-70	1910 - present	Starch
	Indented	98	1935 - present	Cotton & organic binder
	Millboard	80-85	1925 - present	Starch, lime, clay
Roofing felts	Smooth surface	10-15	1910 - present	Asphalt
	Mineral surface	10-15	1910 - present	Asphalt
	Shingles	1	1971 - 1974	Asphalt
	Pipeline	10	1920 - present	Asphalt

TABLE 5 (Continued)

TYPICAL ASBESTOS-CONTAINING MATERIALS (ACM)

Subdivision	Generic Name	Asbestos (%)	Dates of Use	Binder/Sizing
Asbestos-containing compounds	Caulking putties	30	1930 - present	Linseed Oil
	Adhesive (cold applied)	5-25	1945 - present	Asphalt
	Joint Compound		1945 - 1975	Asphalt
	Roofing Asphalt	5	Unknown - present	Asphalt
	Mastics	5-25	1920 - present	Asphalt
	Asphalt tile cement	13-25	1959 - present	Asphalt
	Roof putty	10-25	Unknown - present	Asphalt
	Plaster/stucco	2-10	Unknown - present	Portland cement
	Spackles	3-5	1930 - 1975	Starch, casein, synthetic resins
	Sealants fire/water	50-55	1935 - present	Castor oil, polyisobutylene
	Cement, insulation	20-100	1900 - 1973	Clay
	Cement, finishing	55	1920 - 1973	Clay
	Cement, magnesia	15	1926 - 1950	Magnesium carbonate
Asbestos ebony products		50	1930 - present	Portland cement
Flooring tile and sheet goods	Vinyl/asbestos tile	21	1950 - present	Poly(vinyl)chloride
	Asphalt/asbestos tile	26-33	1920 - present	Asphalt
	Sheet goods/resilient	30	1950 - present	Dry goods
Wallcovering	Vinyl wallpaper	6-8	Unknown - present	
Paints and coatings	Roof coating	4-7	1900 - present	Asphalt
	Air tight	15	1940 - present	Asphalt

ªLaboratory aprons, gloves, cord, rope, fire blankets, and curtains may be common in schools.
Source: U.S. EPA. *Guidance for Controlling Asbestos-Containing Materials in Buildings*. Rockville, MD: Government Institutes, 1985.

multimillion dollar awards are common. With the passage of the Asbestos School Hazard Detection and Control Act, whereby Congress authorized funding for asbestos inspection and abatement in schools, a new wave of claims against asbestos manufacturers began. These new claims sought compensation for inspection and removal costs. Numerous class-action suits seeking property damage have been filed since 1980. These involve schools, hospitals, and governmental units. There even have been claims made by private parties, including commercial building owners.

At present, there are no federal regulations requiring the abatement of ACMs in commercial buildings based solely on the presence of ACM. However, there are two key federal regulations that involve control of asbestos; each is summarized below.

Federal Regulations Controlling Asbestos (Non-School Setting)

OSHA's 1986 Health Standard (29 C.F.R. §§ 1910.1001, 1926.58, effective July 20, 1986) adopted two standards for asbestos, one for general industry (§ 1910.1001), the other for the construction industry (§ 1926.58).

For general industry (all private sector workers in occupations other than construction) OSHA adopted a **permissible airborne exposure level** (PEL) of 0.2 fibers per cubic centimeter of air (f/cc), averaged over an 8-hour day. The standard also establishes an action level of 0.1 f/cc which triggers a need for employer compliance with air monitoring, employee training and medical surveillance.

For the construction industry, OSHA established a similar PEL; additionally, the construction standard includes requirements for proper respiratory protection, protective clothing, hygiene facilities and practices, and nonmandatory guidelines on the proper practices and engineering controls for major asbestos removal, renovation, or demolition operations.

Under EPA's NESHAPs (40 C.F.R. Part 61), asbestos has been designated a hazardous air pollutant. As such, the NESHAPs regulations prohibit visible asbestos emissions from mills and manufacturing plants, establish notification requirements and procedures for both the demolition and renovation of all buildings containing friable asbestos, and delineate procedures to be followed in the disposal of asbestos-containing waste material.

Of particular interest to owners of buildings with ACMs are the following NESHAPs provisions:

- When a building is demolished, or when 260 linear feet of asbestos pipe insulation or 160 square feet of asbestos surfacing material are removed during renovation, advance notice must be filed with EPA regional office and/or state, giving:

    -- Name and address of the building owner or manager.
    -- Description and location of the building.
    -- Scheduled start and completion date of ACM removal.
    -- Description of the planned removal methods.
    -- Name, address, and location of disposal site.

ACMs can be removed only with wet removal techniques. Dry removal is allowed only under special conditions and only with written EPA approval.

- No visible emissions of dust are allowed during removal, transportation, or disposal of ACM (the wet removal techniques are designed to satisfy this requirement).

None of the federal regulations require the removal of asbestos from commercial or industrial buildings, even if friable (crumbling). Additionally, at the present time, ACMs are not considered a hazardous waste and are not regulated under either RCRA or CERCLA. However, certain states or local governments may regulate asbestos and have stringent requirements in this regard. California, for example, has designated asbestos as a hazardous waste under 22 Cal. Code of Regs. § 66680. Recently enacted (January 1, 1988) U.S. Department of Transportation (DOT) regulations found that 49 C.F.R. Parts 171 and 172 do require national and international hazardous waste markings on all containers used to transport asbestos wastes, including asbestos debris that is removed from buildings by asbestos abatement contractors.

AHERA requires EPA to regulate response actions addressing friable asbestos in schools. AHERA provides for regulatory guidance from EPA on the issues of asbestos removal, a uniform program for accrediting persons involved in asbestos removal, and EPA guidance for adopting abatement alternatives, such as asbestos management.

Polychlorinated Biphenyls (PCBs)

Polychlorinated biphenyls (PCBs) constitute a group of 209 chemicals that are based on the biphenyl molecule. PCBs were produced in the United States between 1929 and 1976 for use as nonflammable cooling oils in electrical transformers, hydraulic equipment, capacitors, and other electrical equipment. Because PCBs are **uniquely stable and highly heat resistant**, they have found widespread use throughout manufacturing, power distribution, and in transportation industries. PCBs have numerous other uses such as hydraulic fluids, sealants and caulks. By some estimates, over one billion pounds of PCB have been manufactured; nearly all PCBs are still in the environment due to their extremely stable nature. In 1976, the Toxic Substances Control Act was passed to ban the manufacture of PCBs in order to limit their distribution and control their disposal. In 1979 the "Final Rule Ban" (44 Fed. Reg. 31514) regulated all PCBs to 50 ppm. This legislation bans the manufacture of new PCBs; distribution, unless in a totally enclosed manner (as in an electrical transformer), is also banned unless authorized.

EPA's PCB Regulations

EPA has devised a method for controlling the use, storage, and disposal of PCBs. EPA's method of PCB classification is based on establishing three concentration ranges: 0-49 ppm, 50-499 ppm, and concentrations greater than or equal to 500 ppm. The PCB definitions are given in Table 6.

Radon

Radon, a chemical element formed by the disintegration of radium, is a heavy, colorless, odorless, radioactive gas. Radon occurs naturally in geologic formations containing uranium, granite, shale, phosphate, or pitchblende and was used commercially in luminescent products. Where radon is found, its daughters are also present. Radon daughter products are a lung cancer risk and may cause genetic damage. Exposures to

TABLE 6	
EPA PCB DEFINITIONS **(40 C.F.R. part 761)**	
PCB	Any chemical substance or combination of substances that contains 50 ppm, or greater, of PCB.
PCB item	Any PCB article, PCB container, or equipment that contains a concentration of 50 ppm or more.
PCB article	Any manufactured item, other than PCB containers, that contain PCBs.
PCB unit	Any PCB transformer or PCB-contaminated transformer in use or stored for reuse.
PCB transformer	Any transformer containing 500 ppm, or greater, PCB.
PCB-contaminated transformer	Any transformer containing 50-499 ppm PCB.
Non-PCB transformer	Any transformer containing less than 50 ppm PCB as determined by manufacturer certification or laboratory analysis.
Large capacitor	Any capacitors, either high or low voltage, that contain three pounds or more of PCBs.
Small capacitor	Any capacitor containing less than three pounds of PCBs.
PCB container	A device (drum, barrel, etc.) used to contain PCBs or PCB article.
Leak	Any substance in which a PCB unit has any PCBs on any portion of its external surface.

radon gas typically occur in confined areas such as in public, commercial, or residential buildings.

At present, there are no federal regulations concerning naturally occurring radon, with the exception of the regulation of toxic air emissions from uranium mines. EPA, however, has set maximum action levels for radon. At present, the action level is 4 picocuries per liter (PCi/l) of air. EPA recommends the following remediation methods for radon:

- Barrier remediation to prevent radon from seeping into the enclosure.
- Dilution ventilation which increases the frequency of air exchange in the enclosure.

EPA's Office of Radiation is currently researching radon gas, authorized by the Radon Gas and Indoor Air Quality Research Act of 1986, Title IV of SARA and 42 U.S.C. § 7403. Although naturally occurring radon gas is not currently regulated, it is a recognized carcinogen. There is an increasing concern among lenders regarding the potential presence of radon gas in structures which could affect property values. Lenders also worry about radon-related toxic tort liabilities that could affect property owners.

Toxic Substances Control Act

Passed in 1976, TSCA regulates chemicals that may cause adverse health effects or may negatively impact the environment. 15 U.S.C. §§ 2601-2671. TSCA requires:

- Rigorous testing of new chemicals prior to commercial distribution.
- Reporting of any chemical that presents a substantial risk to human health or the environment.
- Maintenance of records by manufacturers that process or commercially distribute chemicals (records must document any possible adverse health reactions to the chemicals).
- The study of radon in schools (15 U.S.C. § 2667), creation of regional radon training centers (authorized by 15 U.S.C. §

2668), and the study of radon occurrence in federal buildings (15 U.S.C. § 2669).

Federal Insecticide, Fungicide, and Rodenticide Act

The Federal Insecticide, Fungicide, and Rodenticide Act (FIFRA) of 1972 mandates the registering of all pesticides intended for sale in the U.S. (15 U.S.C. §§ 136-136Y). **"Pesticide"** means any substance or mixture of substances intended to prevent, destroy, repel or mitigate pests. Pesticides are registered for five-year periods and are classified for either general or restricted use. Restricted-use pesticides must be applied under the supervision of a certified applicator. Under FIFRA, the registration of a pesticide may be withdrawn by EPA if it suspects that the substance poses an **"imminent hazard."** FIFRA regulations also authorize states to set standards and establish certification procedures for pesticide applications. Some FIFRA-regulated pesticides are also considered toxic pollutants under SDWA primary drinking water standards and 33 U.S.C. § 1317(a).

Safe Drinking Water Act

Enacted in 1974, SDWA was established to assure safe drinking water in public water systems. 42 U.S.C. § 300(f) *et seq.* SDWA establishes "primary drinking water standards to protect human health;" the secondary (non-health related) drinking water standards are intended to protect public welfare. To safeguard underground drinking water sources, another objective of the regulations, SDWA authorizes states to regulate deep well waste injection. Injection wells fall into one of the five following categories:

- Class I wells in which hazardous wastes are injected (regulated under RCRA).
- Class II wells in which oil and gas products are injected.
- Class III wells in which mining wastes are injected.
- Class IV wells, regulated under RCRA, in which generators of hazardous or radioactive wastes dispose of these wastes; existing Class IV wells must be abandoned within six months after an underground injection control (UIC) program is issued; new

Class IV wells are prohibited in formations located within one-quarter mile of an underground drinking water source.
- Class V wells for those that do not fall within the above classifications.

The 1986 amendments to SDWA accomplish the following:

- Require a schedule for the promulgation of primary public drinking water systems.
- Provide civil and criminal penalties for tampering with public water systems.
- Require stricter enforcement of drinking water standards.

Federal Clean Air Act

The Clean Air Act (CAA) created the national framework for protecting and enhancing the nation's air quality. As a mechanism for attaining air quality levels that will protect the public health and environment, CAA directs EPA to set air quality standards and emission limitations. CAA provides for enforcement of these standards and limitations by both federal and state agencies, and also has special provisions pertaining to hazardous air pollutants (HAPs).

CAA, enacted in 1970, received major amendments in 1977. Finally, more than a decade after the act was last amended, the Clean Air Act Amendments of 1990 (the 1990 Amendments) were passed by Congress and signed in October of that year. These amendments substantially revised the existing framework and included provisions for stricter tail pipe emission standards, as well as emissions linked to acid rain and air toxics.

National Ambient Air Quality Standards

National Ambient Air Quality Standards (NAAQS) are the guidelines used to measure the air quality in regions or basins. NAAQS set minimum standards for concentrations of specific pollutants (i.e., ceilings or attainment levels which may not be exceeded). EPA is required to set NAAQS according to established criteria which are to be reviewed at least every five years by an independent scientific committee (42 U.S.C. § 7409(a)-(d)). These standards are set on the basis of scientific data and analyses, notwithstanding cost or technical feasibility (42 U.S.C. § 7408(a)); *American Petroleum Institute v. Costle, see also Lead Industries Ass'n v. EPA*, where the court determined that EPA is not required or allowed to consider economic or technological feasibility in setting air quality standards.

There are two types of standards:

- **Primary**--those specifying a level of air quality necessary to protect the public health while allowing for an adequate margin of safety.
- **Secondary**--those specifying a level of air quality necessary to protect the public welfare from known or anticipated adverse effects, including the effects on economic values and personal comfort (e.g., protect against environmental damage such as damage to soils, crops, wildlife, weather, climate, and personal comfort).

42 U.S.C. §§ 7409(a), (b), 7602(h). In setting margins of safety when promulgating primary and secondary standards, EPA is not limited to considering known dangers to health, but may err on the side of overprotection, provided the conclusions EPA has arrived at are not the product of mere guesswork. *American Petroleum Institute v. Costle.* NAAQS have been set for the following "criteria pollutants."

- Carbon monoxide (CO).
- Lead.
- Nitrogen dioxide (NO_2).
- Ozone.
- Particulates.
- Sulfur dioxide (SO_2).

Some states also develop their own air quality standards, which may be more stringent than NAAQS or cover more pollutants. California, for example, has adopted more stringent standards. And New York's standards cover all the pollutants covered by NAAQS except for lead, but also add hydrocarbons (which were rescinded by EPA as mentioned above), fluorides, beryllium, and hydrogen sulfide (H_2S). In addition, New York does not list ozone as a criteria pollutant, but uses the broader designation "photochemical oxidants," which, under the state regulations, include ozone, peroxyacyl nitrates, and organic peroxides.

DUE DILIGENCE AUDITS

1. <u>Why Due Diligence</u>

   -- Liability and risk management.
   -- Superfund and RCRA Corrective Action liability for cleanup of prior releases. Note Superfund imposes strict, joint, and severe liability.
   -- Inadvertent exposure of publics, including customers, employees, vendors, contractors, customers, family, etc.
   -- Lack of Government set of standards for such audits.
   -- Due Diligence Objective:

      To ensure informed and prudent decision-making in environmental risk management by developing a due diligence tool designed to identify, quantify, and address environmental contamination.

   -- The major categories of risk and responsibility in purchase and sale transaction include the following:

      ● Contamination of site and adjacent impact.
      ● Off-site waste treatment, storage, and disposal.
      ● Facility equipment containing asbestos, radon and lead paint, solvent, etc.
      ● Current compliance costs.
      ● Future compliance costs due to new laws, new regulations, earthquakes, or other acts of God, etc.
      ● Future compliance costs due to prior contamination, releases, and/or exposures to humans or the environment.

2. <u>When and Who?</u>

   A. <u>When?</u>

-- Purchase and sale of corporation ownership and/or corporate assets, such as real estate.

-- Decision to develop or change or begin utilization of assets.

"Innocent Purchaser" defense in the purchase of contaminated site or facility under CERCLA Sec. 101(35)A providing lack of prior owner knowledge suggests there is an underlying "reason to know" but no more than that, except to allow "innocent purchaser defense," only if the purchaser made "all *appropriate* inquiry into the previous ownership and use of property consistent with good commercial or customary care." CERCLA does clarify that "specialized knowledge" of the purchaser, or lack thereof be taken into account. Moreover, CERCLA takes into account relationship of value or price of property to "reasonably ascertainable" information. HR 1643 introduced in 102d Congress provides that aerial photography, chain of title search and review, history of violation review and onsite inspection be conducted to allow "innocent purchaser" defense.

-- The American Land Title Association Forms Committee has standardized chain-of-title disclosure forms for use at closures.

-- Likewise the American Water Well Association is devising an equivalent form (which would apply to four types of real property: vacant, agricultural, commercial, and industrial) at closure.

-- The Association of Engineers Practicing the Geosciences published "Pre-acquisition Site Assessments: Recommended Management Procedures for Consulting Engineering Firms: as a check list for pre-closure review.

-- ASTM has devised a draft set of protocols for an initial transaction screening process (TSP) to trigger further inquiry *prior to transaction*. In this course, we will review this draft protocol in detail.

-- Some states have set guidance for site assessment prior to purchase and use of real property (e.g., Connecticut and New Jersey - see draft to follow)

B. Who?

-- Government, so far, plays no role in defining due diligence, setting standards for audits contractual shields against liability, except in the courts.
-- Buyers and users of corporate assets such as equipment, facilities, and real property need to address liability created under previous use, before beginning changed or new use.
-- Sellers of corporate shares or assets who wish to protect themselves from future liability from retroactive environmental release claims likewise need to address liability created under previous use (this would especially apply to banks selling foreclosed property).
-- Lenders who face retroactive liability for property purchases funded or foreclosed by them and for proactive liability for property that might be foreclosed.
-- The secured creditor exemption provides some defense to liability where the party without participating in the management of a facility, holds "indicia" of ownership primarily to protect its security interest in the facility. Lenders have insulation under the secured creditor exemption [CERCLA Section 101(20)(a)]. This defense is of limited utility because in order to realize its collateral the secured lender must be able to foreclose and dispose of real property with full disclosure of any environmental risk or liability. To the extent a lender's collateral is burdened by environmental risk from concentration, it will be unable to convey such collateral to third parties who cannot succeed to the protection of the secured creditor exemption. In this sense, its secured creditor rights are illusory.

3. What: Due Diligence vs. Contractual Cure

A. Due Diligence Option

-- **Uncontrolled liability** is driving force.
-- Level of concern is inversely proportional to size of transaction.

-- Little opportunity or justification for due diligence audit when transaction is limited to a tender offer or purchase amount is small (relative to risk level).
-- Small businesses pose more mismanagement risk because of entrepreneur unawareness of risks, problems, proper management, or regulations.
-- For very large transactions environmental risk may not be "material" (i.e., GE acquisition of RCA, NBC).
-- For medium-sized deals, contract provisions such as indemnification clause attempt to cover purchasers.
-- Audits try to cover the following risks:

  ● Onsite contamination (prior or present releases).
  ● Offsite disposal.
  ● Equipment or structures, newly regulated (asbestos, PCB's).
  ● Current compliance requirements.
  ● Future compliance requirements.
  ● Future permits.
  ● Anticipated future changes in regulations or laws affecting compliance, permits, or retroactive conditions.

B. Contractual Cures Option

1. Factors in decision making

   -- Buyer or user or seller naivety.
   -- Seller anxiety to sell, and reason.
   -- Buyer anxiety to buy and use.

2. Disclosures and Agreements

   -- Buyer should avoid all retroactive risk.
   -- Seller should agree to execute all needed corrective action at buyer convenience and cover legal expense for retroactive risks.
   -- Seller should agree with buyer regarding setting of corrective actions and standards to be met.

-- Buyers wishing to minimize risk need to buy stock ownership below corporate control levels (rather than total corporate ownership) or preferably assets. Sometimes corporate control results in piercing the corporate veil by the courts. If the acquired assets involve an active business, not even indemnification can prevent piercing transaction veil if liabilities of previous business relationships follow assets through the purchase transaction.

-- Warranties and seller representations (such as no prior releases) offer some relief from concern, but they must be properly written. Total disclosure by the seller will protect both sides of the transaction. Seller may agree to: retention of newly discovered contaminated property, leasing until site cleanup and delayed closing, and payment for subsequent environmental discovery.

-- Disclosures

- Regulated material managed on-site.
- All wastes managed on-site.
- All waste shipments off-site.
- All recipient facilities for above shipments.
- All PCB, dioxin, asbestos, and lead wastes managed.
- All tanks (ever).
- Air emissions.
- Water discharges.
- Permits, notifications, registrations, etc.
- Penalties, citations, notices of violation, etc.
- Complaints, claims, etc.
- Convictions, consent decrees, etc.
- Any of the above pending.
- Full compliance statement.
- Pending facility alterations to comply with any regulatory requirements.

3. Warranties

Warranties and representations in a purchase and sale agreement will be the basis for future claims, but are also important to provide information about the operation before closing. These disclosures should cover:

- All hazardous or other regulated materials and wastes relating to current and prior uses.
- All hazardous or other regulated materials and wastes relating to the operation, including information on their treatment, storage, and disposal.
- All off-site waste handling facilities and transportation used by the operation.
- All PCBs, asbestos, or lead currently or historically used by the operation.
- All aboveground and underground storage tanks ever used.
- Characterization of all air and water releases by the operation and associated permits.
- All notifications, registrations, applications, etc., filed by the operation, and all inspections, notices, citations, penalties, etc., received by the operation.
- Descriptions of all spills, leaks, or other uncontrolled releases of any hazardous substance to air, ground, or surface water and land.
- Any pending or threatened claims or complaints with respect to operations at the facility, or any reasonable basis for claims or complaints.
- A statement that the facility is in full compliance with all applicable state, federal, and local legal requirements, with any exceptions described in detail.
- A description of any pending or proposed changes in the law which may affect operations at the facility.

Sellers will seek to limit such disclosures and warranties to their knowledge and "materiality." Such limitations are not acceptable to the buyer or the seller because non-disclosure followed by liability inevitably leads to costly litigation.

4. Environmental Agreements

Sometimes the seller keeps responsibility for resolution for specific problems, such as on-site contamination. An environmental agreement, separate from the purchase and sale agreement, should define ground rules for such cooperation between the parties. A separate agreement is necessary because purchase and sale agreements may become moot after the transaction is completed. The separate agreement sets contractors and employees of buyer and seller who will be implementing the environmental provisions over the next several years after the closing. An environmental agreement also isolates the environmental issues from general business negotiations.

An environmental agreement defining seller's responsibilities should:

1. Describe the "environmental problems."

2. Allocate responsibility for defined "remedial measures."

3. Provide access to the property and allow reasonable needed interference with or interruption of operations.

4. Design communication with responsible authorities.

5. Provide needed indemnification.

6. Allocate buyer's responsibility for ongoing operations and liability from seller's prior operation.

7. Provide for claims and resolution of disputes.

8. Specify special circumstances and performance which will determine when the obligations of the parties have been satisfied.

5. Dealing with the Dynamic

The time lapse between negotiation, purchase, new use of a facility to occurrence of an environmental problem, and associated liability could be months or years. During this time period, the array regulatory requirements is truly a dynamic, certainly not static. The following circumstances could occur:

-- Bankruptcy by buyer/user and/or seller.
-- Resale.
-- Foreclosure.
-- Total use redesign.
-- Death of principals involved, placing assets and liability in limbo.
-- Further environmental tort or impairment that alters or totally overwhelms prevailing issues.
-- New federal or state laws or requirements, retroactive or not.
-- Jurisdictional changes, such as international changes in authority, treaties, interstate compacts, pre-emption of state rules by federal rules or vice versa, or statutory mandates that codify pre-emptive rules or standards.

Knowledge and comprehension of all the above potential dynamic is totally impossible even by the most sophisticated seller, buyer, or asset user.

Groundrules for disclosure and hence negotiation are likewise dynamic and diverse: Germany holds that seller disclosure does not remove liability. France holds the opposite position. However, EEC guidelines are approaching some middle ground. In the U.S. there are numerous supporting arguments on both sides.

-- Environmental agreements, referenced before, can freeze the dynamic by: (1) pre-empting regulatory framework changes by placing requirements under the venue of contract vs. environmental law, (2) taking into account that site conditions change relative to acquisition baseline, prevailing standards, and reasonable principles for subsequent settlement of disputes.

-- Environmental agreements should be products of *business*, not legal, not environmental, not political, *negotiations*, but with all these influencing factors taken into consideration and with all influences present at negotiation, in frank open argument.

6. Due Diligence Audits

A. Pre-Phase I Transaction Screening Assessment: concurs with or denies need for Phase I audit.

B. Phase I Assessment for all facilities with: hazardous substance/waste permits or adjacent or near to such facilities, and having RCRA or CWA or CAA or other federal or state hazardous material permits, and having any permit or performance violations alleged, or if it is on any CERCLA site list (i.e., CERCLIS).

-- Phase I Assessment includes: site classification based on use, use record, site review, negative reports by neighbors, local contamination potentially related, setting, hydrogeological and surficial geological usage, and site usage (spelled out in detail by ASTM).

C. Dealing with the Dynamics.

D. A framework for Due Diligence ASTM Standard.

7. Phase II Sampling/Analysis

A Phase II assessment is required where the Phase I assessment shows the presence or potential presence of hazardous substances above background levels and at locations not protective of public health and environment. The ASTM Subcommittee intends to complete work on the assessment process through the Phase II triggers before beginning work on the contents of a Phase II type assessment process through the Phase II triggers before beginning work on the contents of a Phase II type assessment.

Several existing standards for Phase II activities have been developed by ASTM, various associations and committees. ASTM Committee D-18 has soil and groundwater sampling and analysis standards including 35 under development with EPA support. EPA's RCRA Technical Enforcement Guidelines contain several methods of performing Phase II sampling and analysis. EPA has issued applicable test methods for evaluating solid waste, SW 846.

8. Phase III Corrective Action

A Phase III assessment or cleanup is required where Phase II data indicate the presence of hazardous substances constituting a threat or potential threat to public health and environment. The purpose of a Phase III assessment is to identify and plan the means of remediating identified hazards constituting a threat to public health and environment, and effecting such remedial or corrective action.

When a property is subject to Phase III, defenses to Superfund liability can be preserved if the hazard is removed or remediated before or during the acquisition, consistent with the National Contingency Plan (see 54FR34241 Aug. 18, 1989). Also CERCLA Section 107(b)(3) legislative history recognizes merits of due care in protecting human health and environment by remedial action may remove liability after "due care" to provide an adequate remedy, new regulations, earthquakes or other acts of God, etc. Future compliance costs due to prior contamination, releases and/or exposures to humans or the environment.

Note: This works in the absence of declared violation of "imminent and substantial endangerment to the public health or welfare or environment" per CERCLA Section 107(b)(3) or "abatement actions" at CERCLA Section 106 or "imminent danger to public health or welfare" per CERCLA Section 104 authorities.

9. EPA Has Not Responded to the ASTM Phase III Proposal

Don't hold your breath.

CONSULTANT ISSUES AND STAFFING CONSIDERATIONS

General Staffing Considerations

There are a variety of critical skills that the auditor should have. These include a strong background in the environmental regulations, preferably, although not mandatory, an engineering or geological academic background, and an inquisitive mind. The environmental audit process requires a significant degree of face-to-face questioning of all levels of personnel at a facility. Although no auditor should cause uneasiness, he or she must be able to pursue issues which a facility or plant manager may wish to leave uncovered.

Ideally, a two-staff team should conduct an audit. This is to insure that data collected onsite can be confirmed by each member of the team. Single-person audits are possible, but should only be performed if the staff member is well versed in the audit process.

There are two other reasons to use two-person teams to conduct an audit. Auditing an industrial facility requires taking written notes while walking through a facility. This is not an easy task. The chances are high that information given in this context will be misinterpreted. Even if both members of the audit team heard the same information, plant management may insist that that information is incorrect. It is not unusual in auditing for this situation to occur. By using two staff people in an audit, it allows them to confirm that a major difference exists between their report and the claims of plant management.

The second reason to use two-person audit teams is to reduce potential legal liabilities a firm may be subject to by conducting an audit. An audit may include facts and conclusions that may adversely affect the regulatory compliance status of the facility or lower the performance rating of plant personnel. Corporate legal counsel must have the assurance that *all* information included in the audit is based on observations or other factual information. This can best be accomplished by having each member of the audit team verify each other's information.

Many companies do not feel that they should use internal staff to audit their facilities because the auditor may feel inclined to give his or her colleagues the benefit of a doubt. Such fears can be assuaged if the auditing staff is drawn from another region or division and has never had any direct contact with the specific plant in question.

One staffing option in planning audits involves pairing consultants with internal staff. The two-person audit team consists of a senior auditor who generally is a corporate employee and is supported in the audit effort by a consultant. This option works well when a company suspects that a particular environmental concern, for example, potential groundwater contamination, exists at a site and the company retains a hydrogeologist consultant to go along on the audit. An in-house/consultant team also can avoid the problem of an in-house staff member being too gentle with his or her own employees, since the consultant should submit a copy of his or field notes as an attachment to the final audit report. The submittal of these notes generally keeps the audit process honest.

A final staffing option relies on the use of internal staff to conduct audits but uses consultants to reaudit facilities randomly. Having consultants perform the reaudit function provides a less expensive way of independently verifying internal audit reports. Reaudits should be completed one year from the initial audit.

The value of using an outside consultant is that in-house staff will not be pulled from their ongoing responsibility. Consultants should also be able to evaluate a plant independently, as no corporate or personal relations exist between the consultant and the facility.

An environmental audit of a 100,000-square-foot manufacturing plant should take two people two days on-site. Preliminary planning and report writing will take the senior-level person another two days. The junior-level person will also use one day to conduct an off-site telephone regulatory compliance assessment of the plant. Total staff time could run seven man-days, not including travel. It is advisable to conduct a sample matrix analysis to determine whether to use internal or consultant staff. Table 7 shows how to determine the value of staff time compared to a consultant's estimate.

The table assumes that the total in-house labor costs are a function of straight salaries multiplied by seven man-days plus a disruption factor. The disruption factor is a way of estimating the cost impact of pulling staff off their routine work assignments and then having them return to

---

TABLE 7

INTERNAL VS. CONSULTANTS AUDIT COST

1. In-house Senior Staff Member #1
    $ _____ (salary dollars per day) x 4 = _____

2. In-house Staff Member #2
    $ _____ (salary dollars per day) x 3 = _____

3. Disruption Factor Staff Member #1
    (.30 times item one) = _____

4. Disruption Factor Staff Member #2
    (.30 times item two) = _____

    TOTAL IN-HOUSE LABOR   _____

---

those jobs days or weeks later. The disruption factor is set conservatively at thirty percent of salaries. It often takes a day or two to re-enter the normal work flow after being away from the office for two or three days.

Travel and hotel costs will be fundamentally the same for in-house staff or consultants. Consultants, however, will often charge a ten percent fee on top of expenses.

Another factor to assess in deciding to use in-house staff or consultants is the number of audits, their locations, and when audit data are needed by upper management. The environmental audit process does not lend itself well to sending a team from one site to another for weeks on end. Under such circumstances, key audit data is often lost or impressions become blurred. No more than two audits per team per month should be implemented. Based on this limit, a manager should determine how many audits the team can accomplish in a month. If the number of audits to be completed in any period exceeds in-house staff capabilities, one either has to pull more in-house staff into the audit program or go outside and use consultants. There may also be circumstances where upper management needs audit data faster than the in-house staff can generate it. Under these conditions, consultants should also be used.

The seven-day requirement for conducting an audit and writing a report assumes complete cooperation from facility personnel being audited. This assumption may require careful scrutiny. Some facility managers feel that they must show a minimum level of cooperation to corporate staff, but will not actively cooperate with the audit. Since an audit relies heavily on documentation held by the plant, less than full cooperation can add another one to two days to an audit if documents have to be uncovered and copied by the audit team.

It is important to note that the above time estimate does not include a key labor demand element, which is the time needed to develop the onsite audit protocol. The protocol is a detailed series of questions the auditor is required to ask in order to ensure that certain issues are routinely identified, in the same level of detail, from one audit to another. The protocol serves as the analytical skeleton of the audit. Auditors should not limit inquiries only to those issues raised in the protocol. Audit protocols or checklists can be simple and run three or four pages or can be complex and comprise more than one hundred pages.

No audit program should be implemented until the audit protocol has been developed. At a minimum, protocol development could easily involve two or three staff members for at least a week. Upper management should approve the draft protocol, because data not included in the protocol is often left out of the audit. It is often improper or impossible to collect additional data after an audit has been completed, as the auditor will be forced to rely solely on information provided by plant personnel. Such data cannot be independently verified by the auditor.

After the protocol is written, it should take the lead auditor approximately one day to set up a date for the audit with the selected plant, arrange for travel and lodgings, and send letters out confirming the audit date. This eight-hour period should also be used to verify independently the plant's regulatory status, which was established during the regulatory compliance assessment.

The two-person, two-day on-site portion of the audit involves confirming information sent to the auditor by the plant prior to the audit, a detailed inspection of the plant, review of on-site records, and documents and a debriefing for senior plant management.

The typical plant audit report should take about eight hours to write. The report should be completed immediately upon return to the auditor's home office.

An audit of a relatively simple manufacturing plant should take two days on-site. Audits of large-scale manufacturers (i.e., steel mills, petrochemicals) can easily take three to four days on-site. Plan enough time on-site to inspect every process and support unit easily. An auditor cannot choose not to inspect something simply because there was not enough time on-site.

Audits do not involve the taking of any samples from plant emissions or discharges, nor do they involve asbestos sampling. The audit report should, however, point out where such sampling is necessary.

Aspects of Cost and Cost Control

The cost concerns are very much different if a facility decides to use its own staff versus using a consultant. Staff salaries are often not a line item in the budgets of corporate-run audit programs. In this case, budget costs may only include travel, hotels, secretarial support, and report reproduction. The inclusion of staff salaries implies that the internal staff has preplanned time actually to conduct audits. This is often not the case. Table 7 points out, audit labor costs should be calculated on a true labor utilization basis (salary times hours). Many companies do, however, include the cost of having their own staff manage or conduct audits.

All audits should have a detailed budget prepared prior to project implementation. Table 8 notes the basic elements of an audit budget. Labor costs are broken down into senior auditor, junior auditor, and secretarial. Key expenses will include travel, hotel, meals, and car rental. Other expenses noted, while minor in nature, can easily add thousands of dollars to an audit.

With consultants, there are three different types of contracts that directly affect cost control. A fixed-fee, lump-sum contract is one under which the consultant will conduct the audit (including expenses) for a specified dollar amount. Under these circumstances, the consultant may not be willing to disclose to the client how labor and expenses are budgeted. A time and materials (T&M) contract is one in which the consultant charges the client an hourly rate for the actual hours used on

---

TABLE 8

AUDIT BUDGET FORM

Direct Labor

Senior Auditor	_____ hrs at $ _____	/per hrs =	$ _____
Junior Auditor	_____ hrs at $ _____	/per hrs =	$ _____
Secretary	_____ hrs at $ _____	/per hrs =	$ _____

Total Labor $ _____

Expenses

Air Fare
Car Rental
Hotel
Meals ($ _____ per day)
Purchase of Reports/Maps
Photographs
Telephone Charges
Courier (overnight delivery)
Reproduction of draft/final report
Binding

Total Costs $ _____

---

the audit plus materials (i.e., expenses). Most T&M contracts also assess a ten percent fee on expenses. A variation on the T&M contract is the placement of an absolute labor cost limit by the client or the consultant. This limit allows the client to establish an outer cost barrier or cap, which the consultant cannot exceed. It is recommended that lump-sum or T&M (with labor and expenses not to exceed a dollar amount) contracts be used with consultants.

Corporate tracking of consultant labor costs and expenses on T&M contracts can be conducted by requiring the consultant to submit copies of receipts of major expenses (air fare, hotel, car rental) as well as copies of consultant time sheets that specify the number of hours actually expended on the audit by individual staff members. A company may wish to require contractually that the consultant submit such documentation at the time of invoicing.

Affect of Audit Types on Staffing Requirements

There are two basic types of audits: compliance and risk. The environmental compliance audit investigates whether a plant is in compliance with all environmental regulations as of a particular date. Compliance audits are key management tools used to verify that internal company compliance programs are running and to identify any gaps in compliance before regulators issue notices of violation. A well-run compliance audit program should alert management to the specific issues that must be addressed in order to remain in compliance with permits and what new steps may have to be taken to meet new permit conditions. Compliance audits can also be used to verify that ongoing remedial programs agreed to by the company as a result of consent decrees are being undertaken in a timely fashion.

A compliance audit requires staff that has an extensive knowledge of federal environmental laws such as the Clean Air Act, Clean Water Act, Resource Conservation Recovery Act (RCRA), Toxic Substances Control Act (TSCA), and Comprehensive Environmental Response, Compensation, and Liability Act (CERCLA). As a rule, if staff members have not been directly involved with regulatory analysis for at least five years, they should not conduct compliance audits.

The compliance auditor must act as if he or she were a regulatory official. An excellent reference document, which identifies the elements used by USEPA to investigate a site, is USEPA, RCRA *Facility Investigation* (RFI) *Guidance*, Volume 1, "Development of an RFI Work Plan and General Considerations for RCRA Facility Investigations," EPA 530/SW-87-001, Revised December 1987. This report is several hundred pages of detailed approaches used to identify air, water, and soil contamination.

The compliance audit will also require staff that has engineering knowledge of how both line equipment and pollution control devices work. Many environmental permits include conditions related to the efficiency of controls. The compliance audit team must have staff who can determine whether such efficiencies are being met.

Many companies feel that they should assign an attorney to the compliance audit team. That may not be necessary. Lawyers are well versed in exactly what the environmental law states, but often do not know how environmental regulations are practically applied by state or federal agencies. The compliance audit team needs someone who can differentiate between the types of violations an agency will deem serious versus those that will be treated as minor. This is not to imply that the auditor may disregard minor violations.

It is also important that the compliance audit determines whether any upcoming changes in regulations will pull the plant out of compliance.

In conclusion, the compliance audit is, in reality, an inspection married to a permit review. The greatest risk in staffing a compliance audit is allowing someone onsite who truly believes he or she understands regulations but in reality does not. Good regulatory affairs staff are hard to find. If a company does not have a regulatory specialist, it may be worthwhile to grant time to someone to learn the basics about the key federal and state regulations affecting that company's operations. Even if this person does not participate in audits, he or she will be essential by providing needed quality assurance review of a consultant's efforts. There are a wide number of commercially available books that summarize the key sections of the RCRA, TSCA, and Superfund laws. Summaries of state regulations can often be obtained from the state agencies themselves, the state association of manufacturers, or the state Chamber of Commerce.

Another alternative a company has to staffing a compliance audit internally is to use a consultant to prepare a regulatory review of a plant. Such a review is not a compliance audit, since it only identifies those state and federal laws and regulations the plant is subject to, summarizes the requirements of these laws and regulations, and reviews the specific environmental requirements placed on the plant in its permits. This exercise enables a company to gain some expertise regarding the regulations and may provide the company with enough regulatory data so that it can staff a compliance audit internally.

A third staffing option for compliance audits is obvious, that is, to hire a consultant. The company must be careful that the consultant has routinely conducted compliance audits for other members of the same industry. Otherwise, it is paying to educate a consultant about the industry.

The environmental risk audit incorporates some aspects of the compliance audit, but is much broader in scope. Environmental risks are *any* aspects of plant operations that have the probability of causing environmental contamination. Engineers often have difficulty grasping the concept of general risk; however, a well-designed, well-engineered facility may still impose considerable risk. For example, if plant exhaust roof fans remove solvent vapors, but those vapors blow off-site to a school, the fan system is doing its job, but an unacceptable risk exists.

The risk audit looks at the potential of contamination at three interconnected levels: the source, the route, and the receptor. The source is any potential emission or discharge of substances. The route is how emissions or discharge can leave the plant--via the air, water, or in solid form. Receptors are either plants, animals, or humans.

Risk audits should include at least one chemical engineer and one staff member with either a geology or chemistry background. There are many excellent risk auditors who have also been academically trained as biologists and regional planners.

There are no federal standards for what constitutes an acceptable environmental audit. Likewise, there are no national standards for what constitutes an approved environmental auditor. The entire field of environmental auditing is quite new and the buyer must beware.

A number of consulting firms that currently conduct environmental audits began their businesses by performing environmental impact statements (EIS) for state and federal agencies in the early 1970s. While EISs are not similar in nature or purpose to audits, staff members with EIS experience are often the best qualified to conduct audits. This is true because the EIS process requires knowledge of environmental regulations and a sensitivity to environmental risks.

The market for audits is so good that many firms that have entered the field may not have the complete range of skills needed to complete an audit successfully. Care should be given retaining firms whose principal business is limited to one aspect of environmental consulting, be it air modeling, geohydrology, or engineering. The audit process is multidisciplinary with a heavy emphasis on environmental engineering, geology, toxicology, hydrology, and air pollution modeling.

A company should seek out those firms that can show evidence that they routinely conduct risk audits. The consultant should be an environmental engineering consulting firm with a broad interdisciplinary staff so that the audit can be staffed with people particularly sensitive to the unique aspects of a plant. For example, if an audit is going to a plant that has multiple wastewater ponds, it probably is a good idea to assign a geologist to the team to assess if the ponds are leaking.

As in the purchase of any service, be sure the consultant has been in business at least ten years and can provide a long list of corporate references. The ten-year experience requirements is necessary, as a failed audit could have very serious legal ramifications for the company.

It is also important to find a consultant who is currently retained by or is on the approved consultant list of banks and insurance companies. Banks and insurance companies have instituted programs in which they prequalify consultants to conduct audits of properties prior to their financing or sale. Although these prequalification lists are not the sole criterion for selecting a consultant, they are useful as a cross-check in the selection process.

A final issue regarding consultant qualifications is whether they offer their services to state or federal environmental enforcement agencies. Many consulting firms have two sides to their businesses, one that works for private industries and the other that works for regulatory agencies. Firms that provide both services insist that their staffs do not affect one another. An audit of a company's plants, however, opens up all of its secrets. One should have grave doubts about working with a firm that on one hand promises to help a private company while at the same time is being paid by a regulatory agency. It is difficult to decide whether to hire one consultant or another, as such decisions often come down to instinct. However, consultants should be required to disclose whether they work for regulatory enforcement agencies and how such work might affect a private audit.

Contracting Issues

The hiring of an environmental consultant involves five distinct steps: request for statement of qualifications, prequalification of consultants, preparation of request for proposal, contractor selection, and negotiation of the contract. Each step in this process allows the company to fine tune its expectations of a consultant and to finalize all related costs.

The Request for Statement of Qualifications (SOQ) is a document under which a company asks consultants to describe their organizational structure, number and academic training of employees, services offered, brief descriptions of project experience, and references. The information provided by each consultant in SOQs obviously is chosen to make the consultant look best qualified. SOQs should therefore be used with a high degree of skepticism.

The Request for Statement of Qualifications should be as specific as possible. If a company is interested in conducting risk audits, its request should state that the SOQ must include descriptions of five similar audits conducted within the last twelve months. If the company is concerned that its operations are unique, then the request should indicate that the consultant must identify audits it has conducted within the same industry.

Obviously, the cost of travel is a significant factor in audits. Limiting the request to local firms could, however, be a major mistake. As stated many times before, an audit is not a place for a consultant to learn its craft. If local firms have experience, their bids should be given due consideration. However, firms that have offices quite a distance from the audit sites should not be discounted. Many of the top-notch audit consultants have grouped their staff in one or two offices and fly them out to audit sites. This set-up allows consultants to build up highly qualified auditors who conduct audits on a routine basis.

How can a company generate a list of environmental consultants to mail its Request for Statement of Qualifications? A good source of information about consultants is the state environmental regulatory agency. The agency will not make recommendations, but it will disclose with whom it holds contracts. Another good source of information is the advertisements in the back pages of trade magazines. A final source of information is outside legal counsel. Most law firms either retain or know of environmental consultants.

An SOQ should not be longer than twenty-five pages. A page limit is necessary, otherwise consultants will include superfluous marketing material.

A consultant should be given at least two weeks to prepare an SOQ. For planning purposes, a company should assume it will take two weeks to develop a mailing list of consultants, two weeks to draft, edit, and finalize the request, two weeks for the mail to get to the consultants, two weeks for the consultant to prepare the SOQ, and two weeks for the company to prequalify consultants. Total time elapsed for the SOQ process is (unfortunately) ten weeks.

The SOQ can, however, be used to prequalify consultants and therefore limit the number of consultants a company will send its request for proposal (RFP).

Companies should restrict a consultant from submitting SOQs, unless a company is absolutely sure that the consultant cannot perform the required tasks. Care should be given not to restrict SOQ submittals, because consultants could sue a company under unfair trade practice regulations if vague reasons for restricting access to the contract are given.

A two-man team within a company should be formed to review all SOQs. Each SOQ should be ranked on numerical scale (usually one to ten, with ten the highest score). Table 9 shows how a typical SOQ ranking sheet looks.

The SOQ ranking can be filled out either by using a weighing system or not. If a company is particularly sensitive to working with consultants located nearby, it can add a weight (a number) to the location criteria, which is multiplied by the score. For example, Consultant A's location score is 5 and the weight is 10 (total score for location = 50). Consultant B is located down the block from the company, and its located score would be 10 times 10 = 100. The use of weights and scores allows a company to develop preselection criteria that reflect company needs. This ranking technique also facilitates the review of consultants on a uniform basis.

The RFP is a formal contracting document that requests a consultant to submit a detailed description of how it will conduct the audit, a schedule noting when each task will be complete, identification of selected staff, and a detailed cost proposal.

The entire heart of the consultant contracting process is the RFP. This document is the means by which a company dictates exactly what type of audit it wants performed, the level of detail within each topic area, whether company-generated audit checklists must be used, and the format of final reports. The RFP should also specify the date by which the audit must be completed.

The RFP becomes the basis upon which the company judges the performance of the consultant as well as the basis the consultant uses to set the minimum requirements of the audit.

The RFP should also have a section covering cost and standard terms and conditions. The cost proposal will identify if the audit will be performed on a time-and-materials basis or a lump-sum basis. The cost section of a proposal should include all costs related to conducting and writing up the findings of an audit. Costs noted in the cost section should be verifiable, either backed up by hourly rate charts, travel agency notes, etc. Table 10 is an example of a typical cost proposal sheet. Notice that labor and expenses are clearly separated. It is unlikely that this level of detail will appear in a cost section if a company

TABLE 9			
CONSULTANT - SOQ RANKING SHEET			

1)	**Score**	**Weight**	**Score**
___	Location of office doing work		
___	Number and background of staff		
___	Amount of directly related experience		
___	Company reputation based on verification of references		
___	Company organization		
	TOTAL SCORE		_____

2) Other Relevant Factors About Consultant A

Prepared by:_____

Date: _____

decides to contract on a lump-sum basis, as the consultant guarantees delivery of an audit report for a fixed cost under a lump-sum contract.

Many companies are choosing to contract out audits where labor charges are lump sums, and expenses are set on a not-to-exceed basis. This allows the consultant to juggle expenses as they actually occur, while at the same time the company knows what the total cost of the audit is likely to be.

TABLE 10			
TYPICAL COST PROPOSAL AUDIT OF XYZ CORP.			
Labor	**Hours**	**Rate**	**Total ($)**
Task 1 Processing/regulatory review			
John Smith, Jr. Auditor	8	40	320
Task 2 On-site Audit			
Bill Murry, Sr. Auditor	16	80	1,280
John Smith, Jr. Auditor	16	40	640
Task 3 Audit Report			
Bill Murry, Sr. Auditor	16	80	1,280
John Smith, Jr. Auditor	8	40	320
Jane Jones, Secretary	16	30	480
Linda Bloom, Technical Editor	4	40	160
TOTAL LABOR	**84**		**4,380**
Expenses			
Two roundtrip airfare Washington D.C. to Philadelphia			500
Two nights 2 staff lodging @ $110/night			440
Meal per diem $45/day/person			180
Rental car 2 days @ $70/day			140
Courier - Federal Express			50
Phone/Fax			50
Copying/Report Reproduction			50
Pictures			35
TOTAL EXPENSES			**1,445**
TOTAL AUDIT COST			**5,925**

The RFP should require the consultant to identify all its terms and conditions for conducting an audit. As so much of the audit contract is related to the consultant's terms and conditions, additional explanation of typical terms and conditions is necessary. Terms and conditions usually cover two or three single-spaced typed pages. The first term often states that the proposal is only valid for sixty to ninety days from the date of receipt. Most terms also include a statement that the consultant will submit invoices to the company twice a month and that the consultant will be paid promptly. Another key term is one that states that should the client cause unforeseen delays, the consultant has the right to increase its fee. Most consultants are adding a term that indicates that a ten percent handling and administrative charge will be added to all expense items. Many companies refuse to pay this ten percent fee and this term is often negotiated during contract finalization. Termination by the client is allowed in another standard term; however, the client must inform the consultant in writing seven days prior to the intended termination date. As most audits are completed quickly, a company is likely to pay for a significant portion of an audit even if it issues an intent-to-terminate letter to a consultant.

By far the most important standard term relates to limits on the consultant's liability. Liability terms that limit consultant liability to the client for any loss related to the performance of the audit, professional negligence, or errors or omissions will not exceed the value of the contract. If an audit costs $5000, a consultant's liability is limited to that value. Many companies feel this is too low and seek to negotiate higher limits of liability.

A selection team, similar to the one formed to rank the SOQ, should be established to pick the best qualified contractor.

The team should develop a scoring system that weighs technical expertise against costly considerations. Contractors should not be selected solely on cost. The consultant chosen has to be trustworthy, reliable, and cost-efficient. One should remember that it is not possible to go back and "fix" an audit; no consultant will be willing to upgrade another consultant's work.

Remember: The two key criteria in contractor selection should be the number of similar audits the consultant has done over the last six to twelve months and the experience of its staff.

After the company has selected a consultant, negotiations begin. These negotiations often involve fine tuning RFP technical requirements (such as what issues will be studies), but more often involve issues such as payment schedules, the ability to secure insurance, and issues of confidentiality. A company should make it clear to the selected consultant that it has only been selected, *not* awarded the contract. This interim period between selection and award also allows a company to dismiss a selected consultant it cannot come to terms with, without being contractually tied to the consultant.

As noted earlier, most consultants seek to limit their liability in terms and conditions to the value of contract. Consultants have been able to secure errors and omissions (E&O) insurance in the commercial insurance market since late 1987. This insurance is usually issued with a face value of $1 million and is very expensive. A company should require a consultant to submit copies of the cover sheet of all its insurance policies. If the consultant has an errors and omissions policy, the company can negotiate a change in terms to provide an increased degree of coverage. In most cases the consultant will feel that the company should pay some fee for the increased protection it receives from the consultant's coverage. Many consultants charge a flat fee for E&O coverage that ranges from one to five percent of the contract value to arbitrary flat fees of $500 to $750 per audit.

Another key issue that must be negotiated is how the company wishes to deal with environmental sampling during an audit. Current professional practice among consultants is that no environmental sampling should be undertaken during a compliance or risk audit. The audit should, however, indicate whether sampling is necessary to uncover risks fully.

Concern over corporate liability due to the presence of asbestos-bearing materials has become a major element in auditing. An auditor should carefully inspect the building and identify materials that could be classified as potential asbestos-bearing materials (PABMs). No consultant can definitively state that a facility is asbestos-free based on visual inspection. Samples must be taken and analyzed in a laboratory. Companies must clearly state, that the inspection during an audit will only identify PABMs that can be visually noted *without* using intrusive methods. This simply means that the consultant will not punch holes in the walls, rip out ceilings, or otherwise damage the site during the initial audit. If documents picked up during the audit or the visual inspection indicate that asbestos may be present, sampling should be conducted. The audit report, however, is not going to be able to state conclusively whether there is asbestos in a building until the second phase of the audit, the sampling, is undertaken.

The possibility of soil and groundwater contamination at sites is also a major concern of companies. As noted earlier, sampling of soils and groundwater should not be conducted until the initial audit is complete.

The audit should identify areas where there is soil staining and discuss the reasons why groundwater contamination may be present. The audit report will not, however, be able to quantify the degree or extent of contamination. Such data can only be developed based on sampling results. Therefore, the contract must be explicit about the level of detail to be discussed in the report.

It is not unusual to develop a two-part contract with a consultant. The first part includes all the technical requirements, costs, and terms and conditions to conduct an environmental audit. The second part of the contract includes language that requires the consultant to develop a sampling plan, including a full cost proposal, based on audit results. This plan is to be submitted with the final audit report.

A contractual requirement for the consultant to develop a sampling plan will eliminate the time delay between the issuance of a final audit report and the beginning of sampling. Furthermore, the contract should specify that costs noted in the sampling plan are not fixed and are negotiable at the time of their submittal. By mixing an audit and sampling plan under a single contract, a company can avoid having to go out to bid a second time. This saves time and cuts down on paperwork. The contract should also state that the company is under no obligation to use the same consultant to perform both the sampling and the audit. This allows a company to split an effort if the audit portion of the contract is not performed adequately.

When all contractual issues are resolved, the company should send a formal notification of contract award, which includes the contract as an attachment, to the consultant.

A company should require that the consultant inform it in writing when the audit is scheduled. The consultant should also be required to inform the company one week prior to submittal of the final report.

Most audits of industrial facilities usually take less than one month to complete after contract award. Due to this short time frame, company monitoring of performance should be quite limited. The company should speak with the consultant informally at least once a week to ensure that no serious problem goes undetected.

Never accept a consultant's audit report only in final form. The consultant should be required to submit an audit report in draft to allow the company the option of requesting clarifications and, where appropriate, typographical corrections. A company must never try to change audit findings.

The company's general counsel should review all audit reports in draft, to spotlight any actions the company must take in order to eliminate liabilities. Audit reports should be marked "Attorney-Client Work Product." Such a label may provide some limited protection from subpena.

Once a final report is issued by the consultant, it cannot be changed. If an error made by the consultant is identified after the fact, a separate letter from the consultant to the company must be issued. The report, however, cannot be amended.

An audit report is valid for one to two years. If a facility's operations do not change, the audit is probably useful for the longer period of time. No audit should be used that is more than two years old. Although no set rule exists regarding how long an audit may be used, most consultants and attorneys agree to the two-year limit. Refer to Tables 11 and 12.

TABLE 11

SAMPLE INVOICE FORMAT

Payment Request for Environmental Audit of _____

Period beginning _____ and ending _____

Labor	**Number of Hours Utilized**	**Hourly Rate**	**Total Charge**
List employee name			
Total Labor Charge			_____

Expenses

Air Fare
Car Rental
Hotel
Meals
Photo
Telephone
Courier
Reproduction
Binding
Other

TABLE 12

DRAFT: FIXED-COST LETTER-STYLE CONTRACT FOR AN ENVIRONMENTAL AUDIT

Mr. Tom Smith
XYZ Corporation
111 Crescent Drive
Philadelphia, PA

Re: Environmental Assessment of _____

Dear Mr. Smith,

ABC Consultants is pleased to submit this letter proposal to conduct an environmental assessment of _____.

ABC will conduct the assessment in three phases. Phase One involves contacting local, state, and federal officials to identify previous land use and to determine if the property has a history of environmental noncompliance. ABC will also collect relevant information about properties immediately adjacent to the site.

Phase Two involves an inspection of the property. ABC will identify the sue and storage of hazardous materials, the storage of and disposal of hazardous wastes, the presence of PCBs, the presence of potential asbestos building materials, and identify any underground storage tanks. ABC will also identify the presence of absence of soil staining, drums, and other environmental risks seen during the inspection.

Phase Three is the submittal of findings and recommendations in a final report.

This effort does not involve any environmental sampling. No asbestos sampling will be conducted. All findings are based on nonintrusive methods. The assessment report may, however, recommend sampling be conducted. ABC can complete this three-phase effort for a fixed cost of $ _____. This cost includes all expenses.

Please countersign this letter if you accept this proposal. ABC can submit a report to you no later than _____, 19___ assuming ABC receives authorization to proceed no later than _____, 19___. ABC will also give a verbal report of findings to you the day after the field inspection is completed.

If you have any questions regarding this proposal, please call me at _____.

Sincerely,

ABC

Accepted:

_____   _____
                  Name   Date

Note: Consultants always attach their standard terms and conditions to letter contracts.

Many companies wonder if they can reuse an audit if it is more than one year old and facility operations have not changed. Under these circumstances, audit results may be viewed as valid; however, lawyers and financial institutions will not allow the use of such results to reflect current conditions. In short, basic audit findings are acceptable, but a company may have to conduct a reaudit to meet the requirements set by lawyers. This may appear unfair, but remember that an audit only reflects what conditions were at a site on a particular day.

Should a company keep old audit reports in files beyond the two-year limit? If the audit file contains clear records of how follow-up actions were taken based on audit findings, there is no reason not to keep those records. The issue of records retention, however, is best resolved by a company's general counsel.

CONSULTANT LIABILITIES

Introduction

Hazardous waste projects are particularly vulnerable to litigation over consultant liability. A successful lawsuit proving professional negligence can result in the plaintiff losing clients, and considerable income if the courts levy heavy fines, or assign a negligent consultant the financial responsibility of cleaning up a hazardous waste site. The following considerations must therefore pervade all consultant business decisions:

- A consultant must thoroughly understand his or her professional duties, responsibilities, and contractual limitations.
- Sound business practices will reduce the risk of liability claims.
- Basic liability exposures must be recognized and reduced.
- The consultant can reduce liability risk, but it can never be totally eliminated. Therefore, professional liability insurance is mandatory.
- All contract amendments must be properly worded and reviewed by a corporate lawyer or contracts administrator.
- The environmental report must be written in a concise and unambiguous manner.

Proposals

The proposal sets the contractual tone between the consultant and his or her perspective client. The typical proposal consists of the following elements:

- Cover letter.
- Project summary.
- Proposal outline.
- Introduction.
- Scope of work.
- Experience on similar projects.
- Costs.
- Schedule.
- Confidentiality agreement.
- Resumes of key personnel.
- Project team.

The client should always accept the terms of the proposal formally in a written contract. While verbal contracts are legally binding, practically speaking, they are unenforceable. If the client does not accept all elements of a contract's standard terms and conditions, the conflict should be resolved prior to starting the project.

If the client requires immediate project startup, notice to proceed can be given by a signed letter of authorization. Whether using a notice to proceed or a formal contract, the two parties must always address and correct major concerns before starting the project.

If a client prefers his or her own contract, the consultant must work out mutually agreeable contract terms. Of course, if agreement cannot be reached on legally sensitive issues, the consultant should decline to bid on the proposal.

Elements of the Contract

A contract is a legally binding agreement between two or more parties; a formal dispute over any aspect of a contract can be cause for litigation. Consultants reduce their liability risk by using a properly worded contract.   The following checklist of form and content will help the consultant assess the strength of his or her contract:

- A detailed scope of work is provided.
- The duties and responsibilities of both parties are clearly defined.
- Provision is made for payments.
- If the client's credit history is a concern, provision is made for a retainer.
- The effective dates of applicable local, state, and federal regulations are listed.
- Provision is made for a formal notice to proceed.
- Court remedies are defined; those who are to pay legal costs are identified.
- If the client requests a change in the scope of work, provision is made for the requisite financial adjustment.
- If the client terminates the contract, the consultant should be given the opportunity to negotiate a new contract.

Contract Issues

Scope of Work

The scope of work defines the project elements to be performed by the consultant; the level of detail is a function of the project's complexity. The consultant should attempt to limit the scope of work to low risk tasks whenever possible, and when undertaking high risk tasks, to use the proper caveats.   Performance standards, also included in the discussion of the scope of work, must promise that level of professional performance considered satisfactory within the industry.   Finally, quality control methods must be defined within this section.   The consultant should schedule project reviews with the client at major milestones; the cost of these meetings should be included in the estimated fee.

Schedule

Clients expect services to be performed on time.   Some contracts may require the consultant to assume the responsibility for delays, the costs of which can be significant.   Therefore, the consultant should clearly define schedule requirements in the contract.   Strong client relations and communication are two of the most important elements in controlling the schedule, and eliciting the client's support if the schedule cannot be met.

Budget

The project budget must contain the consultant's profit on labor and other direct costs (ODCs). The budget must also contain funds set aside for contingencies such as delays due to inclement weather, equipment failure, and/or subcontractor failure to perform.

All consultant firms face competition. Fortunately, for the client, competition frequently forces consultants to discount labor rates, and to remove contingency fees from their proposed budgets. Ideally for both the consultant and the client, the budgeted fee will be an equitable exchange for services rendered.

Report Preparation

The environmental report is usually the final product to be delivered to the client. It must contain a summary of findings, recommendations, and conclusions. Since it is a formal document that can be distributed to third parties, the consultant must write a legally defensible report. The report must be clear, concise, and unambiguous, otherwise the client or a third party may later claim that they received misleading information. If such a charge is proven in a court of law, the consultant can be held liable for damages. The following guidelines can help the consultant write a report containing minimal disputable information:

- Define the scope of work.
- State all limitations placed on the consultant regarding the scope of work, schedule, and budget.
- State that the information presented in the report is limited to a clearly defined area of the site.
- Provide detailed site maps of the study area, and always document with photos.
- The reports should be written objectively; the consultant is responsible for reporting only facts.
- All conclusions stated in the report must be supportable. Do not state opinions.

It is practically impossible to write the perfect report. A report favoring the consultant would be full of caveats and disclaimers rendering it useless to the prospective client, and vice versa. Therefore, the content of a workable report is a compromise between the needs of both parties. The well written report must detail the limitations inherent in performing an environmental audit.

The draft and final reports are prepared using the scope of work defined in the proposal and contract. Furthermore, the consultant warrants that he or she will conform to accepted professional standards in effect at the time of the investigation. It is advisable to declare that the statements and conclusions contained in the report are merely estimates of the environmental conditions of the site. The report's statements and conclusions are not a guarantee of the site's environmental conditions. Therefore, the consultant should always recommend a Phase II environmental investigation to corroborate Phase I findings. The statements and conclusions contained in the report are valid for a specified period of time from the date of the report.

Third Party Use Disclaimers

Clients routinely request that third parties be allowed to use environmental assessment reports. The release of a report to a third party without disclaimers on its use is potentially the greatest source of consultant exposure to liability. Since the client's attorney, bank, seller, or a third party did not sign the contract with the consultant, they are not bound by the terms of the consultant's contract. Use of the report should therefore be strictly limited to the client. The contract should clearly state this limitation, and should also apprise the client that use of the report by third parties is the client's sole responsibility.

Realistically, environmental reports are routinely given to third parties, and consultants are legally responsible for their content. Also, though third parties are not signatories to a contract, they may still sue the consultant if they incur loss or damage due to the report's content. Therefore, the consultant should be aware that, regardless of disclaimers, his or her report may be given to third parties such as lenders, buyers, or sellers.

Contract Terminology

A consultant must understand the terminology used by the client, attorney, insurance broker, and third parties when discussing professional liability claims and insurance. The following is a discussion of basic litigation terms.

Hold Harmless and Indemnity Provisions

Term such as "**hold harmless**," or "indemnity" sometimes appear in contracts and may be unacceptable conditions for the consultant's insurance carrier, who may look for more conservative language. The insurance carrier may not cover the consultant who agrees to hold harmless and indemnify others from losses arising from the consultant's negligence, since the law stipulates that the consultant can be held liable regardless of such a clause. An acceptable and insurable hold harmless and indemnification clause might read as follows:

> The consultant agrees to defend, indemnify, and hold harmless, the property owner, its officers, agents, and employees, from all claims arising out of the negligent acts, errors, or omissions of the consultant.

The following hold harmless and indemnity provisions are examples of legally risky language since they assign the consultant liabilities that are not legally the consultant's responsibility. Signing such a contract could negate the consultant's insurance coverage, and expose him or her to legal liability. For example:

- The consultant holds harmless and indemnifies his or her client for liability arising out of the consultant's and the client's negligent acts, errors, or omissions.
- The consultant holds harmless and indemnifies his or her client against losses resulting from the consultant's firm's acts, errors, or omissions or those of its subcontractors.
- The consultant holds harmless and indemnifies his or her client for all loss, injury, or damage arising from the project,

regardless of fault or cause--in effect, assigning the consultant strict liability for the entire project.

Warranties and Guarantees

Some clients require warranties and guarantees of performance in their contracts. The following is an example of a typical warranty clause:

> The consultant warrants that all work performed on this project will conform to the previously submitted written scope of work. The consultant further warrants that all professional services will be performed to the professional standards current at the time the work is performed.

Terms and conditions form the bulk of contract language, but are often given little attention during contract negotiations. The consultant should be sure that the firm's contract administrator reviews the contract's terms and conditions before the firm signs the contract.

Insurance

A consultant's insurance coverage should include the following:

- Workmen's compensation coverage.
- Automobile insurance.
- Comprehensive general liability insurance.
- Environmental impairment coverage.

Having insurance coverage does not justify taking risks. Numerous or financially large claims against a firm will result in increased premium costs, which will increase the consultant's cost of doing business. The consequences could also include cancellation of the firm's coverage.

Liability

Liability refers to absolute, contingent, or probable responsibility legally assumed by a consultant when performing environmental work. A consultant's liability can encompass all legal damages.

Damages

Damages refers to loss or harm to a person or property; it also describes the amount of money payable for loss or injury. Damages can be incurred by the consultant, the client, or by third parties. Damage to the consultant's client and to third parties can include property damage, personal injuries, and economic loss. Generally, property damage and personal injuries suffered by a consultant's acts or omissions are covered by the consultant's liability insurance.

Exposure to Client and Third Party Claims

There are four reasons for which a consultant can be held liable for damages; these are:

1. Breach of Contract.

2. Breach of Warranty or Fraud.

3. Negligence.

4. Willful Misconduct.

Liability for Breach of Contract

A consultant is liable for damages caused by a breach of contract. Formal contracts must contain the fee, scope of work, and schedule. Legal obligations toward the client include warranty, indemnities, and waivers of consequential damages. To be considered in breach of contract, a consultant must fail to comply with a provision(s) of the contract. For example, if a consultant issues a report to a third party without the client's permission, the consultant is violating the terms and conditions of the contract. The consultant's breach of contract could therefore cause the client to incur additional liability from the third party. Another example of breach of contract, called failure to perform, would occur if the consultant did not disclose all contamination located on a property. In this case, the consultant could be held liable for the cost of site cleanup.

Liability for Breach of Warranty and Fraud

A warranty is a contractual promise that professional services will attain a particular level of quality. Liability exists for breach of warranty. The consultant should consequently limit comments about the level of service only to attainable performance. An example would be a promise to drill to a certain depth barring "changing conditions." Breach of warranty would also occur, for instance, if a consultant were foolish enough to promise an unattainable level of groundwater remediation.

Consultant fraud is also easily defined. For example, if a consultant were to submit a report for a site he or she had never visited, the act would constitute fraud. The cost of correcting a warranty may exceed a contract's value. A charge of fraud levied against a firm could also ruin that firm's reputation, and could result in bankruptcy.

Liability for Negligent Acts or Omissions

To establish liability for negligent acts or omissions, the plaintiff must legally prove that professional negligence has occurred. Negligent acts or omissions could be as simple as allowing contaminated drilling fluids to flow onto private property adjacent to a hazardous waste site. The client or his or her subcontractor would then be responsible for cleanup. Unfortunately, hazardous wastes projects involve high risk and exposure to such liability. Perhaps the simplest way to reduce liability is to pursue project work selectively, determining site characteristics, and evaluating the client before bidding for the job. Such analysis should include potential for release of hazardous substances, proximity to sensitive receptors, potential off-site pathways, and the volume and type of hazardous substance(s) involved.

Liability for Willful Misconduct

Willful misconduct can include a variety of acts such as using drugs or drinking while performing work for the client. Committing assault while on the job would also qualify. However, not all willful misconduct involves overt criminal acts. Refusing to abide by a designated health and safety plan would also qualify as willful misconduct.

Extent of a Consultant's Duty

In determining a consultant's duty, courts will evaluate the nature of the risk involved in performing professional work.

Defining the Duty

One duty imposed by common law involves exercising reasonable care. Legally, a party must exercise that degree of care exercised by a "reasonable, prudent man."

Liability for Breach of Duty

Once a duty to a party is established by contract or by law, a party is considered negligent when it breaches that duty. If a consultant has breached a duty to the client or to a third party, the consultant is legally liable for all damages suffered by the injured party or parties caused by the breach of duty. If a consultant is found guilty of gross negligence, he or she may be assessed punitive damages, in addition to other damages.

Contract Negotiations

The ideal outcome of any contract negotiation is the mutual satisfaction of both the consultant and the client. A typical win-win scenario occurs when the consultant negotiates a few that provides his or her firm with a reasonable profit, while staying within the client's budget, and meeting the client's scope and schedule.

The following is a summary of basic negotiating rules for consultants:

- Prepare for negotiations; list every major issue.
- Know your minimum and maximum goals before negotiating.
- Never give a concession without getting one in return.
- When the client makes a concession, take it.
- Do not make the first concession. Negotiation losers are usually those who make the first major concessions.

- Inform the client that concessions on individual issues are based on reaching a satisfactory overall agreement.
- You may ultimately work with the people with whom you negotiate. Therefore, allow the prospective client to feel satisfied with the outcome of the negotiation.

The successful project depends largely on a well planned proposal and contract. Every project is governed by three major considerations: scope, schedule, and budget. The clearly defined goals stated in a properly presented proposal and contract will ensure that each of the above considerations is met.

INSURANCE INDUSTRY'S LIABILITY ISSUES

Introduction

The insurance industry becomes involved with Superfund-mandated cleanup costs when PRPs attempt to have their insurance companies pay for these costs. Given the enormity of the potential costs for hazardous waste cleanup, it is not surprising that PRPs would seek other sources to assist in financing the costs. Similarly it is not surprising that insurance companies would resist paying these costs.

The type of insurance policy that is typically involved in a superfund litigation is known in the industry as a **comprehensive general liability** (often referred to as just general liability) insurance policy. Hazardous waste claims typically come under a coverage section in the policy known as premises and operations. Policyholders, especially larger corporations and organizations, often obtain several layers (increased amounts) of coverage by purchasing additional policies. These policies are known as **excess or umbrella liability** policies and typically provide the same or similar coverage as the primary policy. Consequently these "following form" policies are also brought into the insurance coverage litigation. While property insurance policies, particularly all risks policies, have also been brought into the debate, most of the litigation has focused on general liability and related policies.

The affected policies are older general liability policies, purchased before 1986 when a blanket exclusion for all hazardous waste, pollution and related claims was placed in liability policies. PRPs have conducted and are conducting extensive searches for older policies as the actual policies are the most clear and convincing evidence that coverage was in effect. Because many policies have been destroyed as part of regular records attrition management programs, other evidence of coverage, e.g., correspondence, check stubs, etc., are sought to prove that coverage was in effect. This whole area of reconstructing a PRP's past insurance program has given rise to a new consulting area called insurance archeology.

Insurance Coverage Litigation

The insurance coverage litigation between PRPs, as policyholders, and insurance companies has produced several specific coverage issues. These issues will be discussed briefly in the next section. For interested readers, the author has written a number of articles that discuss these coverage issues in greater depth. Before discussing the specific coverage issues, some general comments are in order.

For each of the specific coverage issues, judgements (often several) can be found to support both sides of the issue, that is, some court decisions support the side of the PRPs and some support the side of the insurance companies. Because insurance is regulated by the states, is covered by contract law, and rarely involves constitutional questions, cases do not come before the United States Supreme Court to produce a single definitive decision. Consequently a single coverage issue often comes before several state and federal district jurisdictions, including appellate courts. Until a preponderance of jurisdictions reach the same decision on the same coverage issue (which may never happen), the issue remains unresolved.

The above situation is both fluid and dynamic. An attempt will not be made to report completely on all the decisions of various jurisdictions on the specific issues. Not only would that be beyond the scope and purpose of this chapter, it would quickly become outdated, because new decisions are continually being written. Interested readers can consult various litigation reporters, or specific law firms which deal with the area of hazardous waste insurance coverage litigation.

The interpretation of insurance policies is governed by specific rules or doctrines. One of the more interesting rules which has substantial applicability to hazardous waste coverage litigation is the rule of **adhesion**. This rule states that ambiguity in insurance policy language should be interpreted in favor of policyholders and against insurers. Anyone who has ever read or attempted to read an insurance policy knows it is a complicated document and subject to claims of ambiguity. Not surprisingly, lawyers for PRP policyholders have made liberal use of the doctrine of adhesion in arguing that insurance policies should cover Superfund-mandated hazardous waste cleanup costs.

Insurance Coverage Issues

Seven specific insurance coverage issues have been identified involving hazardous waste claims. Each will be briefly discussed from the viewpoints of both the PRP policyholders and the insurance companies. Where possible, the author will opine as to which viewpoint, if any, is dominating. While all issues are important, the first four have probably received the most attention. Finally, each of the issues can be considered a potential defense to be used by insurance companies to deny coverage. While all the issues/defenses are rarely applicable to, and thus used in, a particular dispute, in theory if insurance companies are successful in any one of the seven areas, coverage would be denied. The number of potential issues/defenses helps to explain why the litigation often becomes so involved, drawn out, and expensive.

Pollution Exclusion

Prior to the early 1970s, no mention was made of pollution or hazardous waste in general liability policies. Without any other defense, coverage would presumably exist under the premises and operations section of the policy. In the early 1970s (1973 for all standard forms) insurers began to include an exclusion which limited coverage for pollution claims. An example of this standard partial pollution exclusion is included below:

> This insurance does not apply to bodily injury or property damage arising out of the discharge, dispersal, release, or escape of smoke, vapors, soot, fumes, acids, alkalis, toxic chemicals, liquids or gases, waste materials or other irritants, contaminants or pollutants into or upon land, the atmosphere, or any water course or body of water; but this exclusion does not apply if

such discharge, dispersal, release, or escape is sudden and accidental.

This was an attempt to exclude all types of pollution claims except those that resulted from sudden accidental dispersals. In other words, insurers were attempting to exclude the "gradual pollution" claims resulting from, perhaps, a slow leak that nobody had bothered to trace and correct and which, over years, could result in a serious contamination problem, while still maintaining coverage for pollution which might occur from a sudden and accidental event, like an exploding chemical tank.

The term **sudden and accidental** has become the focal point of extensive litigation. Insurers claim that the event had to happen in an instantaneous or very short period of time. Policyholders claim that the term sudden and accidental only meant unintended or unexpected and do not connote any temporal quality. Policyholders also have incorporated the adhesion rule by noting that the term sudden and accidental was not defined in the policy and thus was ambiguous. Some courts resorted to using dictionaries, which (depending on the particular meaning or set of meanings chosen) resulted in findings for insurers in some cases and for policyholders in others. The situation is complicated by the fact that some insurance organizations, in arguing for the pollution exclusion back in the early 1970s, stated that the exclusion was merely a clarification of existing policy wording and coverage. Not surprisingly, these arguments have come back to haunt insurers today, as policy holders use them to argue their cases.

In litigation involving the pollution exclusion, findings for policyholders have the effect of negating the exclusion, thus coverage becomes effective for both sudden and gradual pollution events. While it appears that more recent court decisions have been favoring the interpretation of insurance companies, overall the courts are widely divided and this issue is still largely unresolved.

In 1986, insurers decided to exclude all types of pollution/hazardous waste claims, both sudden and gradual, from general liability policies. Included below is an example of the 1986 exclusion:

This insurance does not apply to:

1. "Bodily injury" or "property damage" arising out of the actual, alleged or threatened discharge, dispersal, release, or escape of pollutants:
   a. At or from premises you own, rent, or occupy.
   b. At or from any site or location used by or for you or others for the handling, storage, disposal, processing, or treatment of waste.
   c. Which are at any time transported, handled, stored, treated, disposed of, or processed as waste by or for you or any person or organization for whom you may be legally responsible.
   d. At or from any site or location on which you or any contractors or subcontractors working directly or indirectly on your behalf are performing operations:
      i. If the pollutants are brought on or to the site or location in connection with such operations.
      ii. If the operations are to test for, monitor, clean up, remove, contain, treat, detoxify, or neutralize the pollutants.

2. Any loss, cost, or expense arising out of any governmental direction or request that you test for, monitor, clean up, remove, contain, treat, detoxify, or neutralize pollutants.

It might be noted that pollution coverage is currently available in limited forms in restricted markets under **Environmental Impairment Liability** (EIL) insurance policies. The EIL insurance market developed in the 1970s to provide gradual pollution coverage, which most insurance industry officials thought had been excluded from liability insurance forms. After the 1986 blanket pollution exclusion, the EIL market now provides both sudden and gradual pollution coverage.

Expected and Intended Damages

Liability policies have never been meant to cover intentionally caused or expected injuries and damages. Both public policy considerations and specific policy wording argue against such coverage. An example of policy wording is included below:

The insurer will pay on behalf of the insured all sums which the insured shall become legally obligated to pay as damages because of bodily injury and property neither expected nor intended from the standpoint of the insured.

When applicable, insurers contend that pollution/hazardous waste damages are not covered as they were either intended or at least should have been expected by the policyholders from their actions. Policyholders have countered by saying that either their actions were not intended, that is, they were accidental; or even if their actions were intentional, they did not intend or expect damages to occur.

Court decisions have been widely split on this issue; intent and what the policyholder should reasonably have expected to happen, as indicated by the specific facts of the case, have largely decided the outcome. In one celebrated case involving the Shell Oil Company as a PRP policyholder, insurers were able to prove that Shell knew its actions were causing damages. Reportedly Shell officials replaced dead ducks, which were apparently dying from Shell's disposal of hazardous waste, with live ducks. Insurers successfully argued that this action constituted evidence that Shell knew its actions were causing damages, hence no coverage. This has become known in litigation circles as the "dead duck" defense.

Trigger of Coverage

The term **trigger** is used to denote which insurance policy responds to a loss, that is, which policy is triggered. Typical wording in liability insurance policies requires that the bodily injury and/or property damage occur "during the policy period"--the period of time the policy is in effect--typically one year. In acute loss situations like an automobile accident or a fire, the time and date of the injury or damage are clear and will fall within a particular insurer's policy period. When losses occur over time, such as the seeping of hazardous waste into an aquifer, the exact timing of the loss and determining which policy or policies are triggered is more difficult.

The trigger of coverage issue for losses which occur over time first gained notoriety in the asbestos injury cases. The question arose: If a worker is exposed to asbestos in 1945 but is not diagnosed for asbestosis until 1975, which insurer(s) must respond? The resulting litigation initially produced two triggers of coverage:

1. The **exposure trigger**--those insurance companies with policies in effect when the worker was exposed to the asbestos must respond.

2. The **manifestation trigger**--those insurers with policies in effect when the worker was diagnosed with asbestosis must respond.

Both triggers produced reasonably short and definable loss periods and consequently short and definable policy periods. Since in any particular case only one trigger would be held to be the applicable trigger, the insurance industry's liability for asbestos injuries was, while not trivial, within reasonable bounds.

This situation was changed dramatically in 1981 by the famous, or infamous depending on your viewpoint, decision in the Keene case. Partly due to the ambiguity of the trigger issue (recall adhesion), and the fact that different insurers did not necessarily agree on a single trigger, the judge in the Keene case reached the startling conclusion that not only should both the exposure and manifestation triggers apply but a third trigger--the entire time period between exposure and manifestation, the so-called latency or residency period--should also apply. This became known as the **triple trigger** and was devastating to the insurance industry, as it meant all the insurance policies in effect from the time of exposure to the time of manifestation must respond. In an event like asbestos injuries, this can mean a period of 30 to 40 or even to 50 years.

While the triple trigger rule has established a strong foothold in asbestos cases, the situation in hazardous waste litigation is not clear. Not surprisingly, PRP policyholders argue for triple trigger as it maximizes their coverage. For instance, in the Shell Oil Company case referred to earlier, Shell had sued 260 liability insurers (primary, excess and umbrella) that provided coverage from 1947 to 1983. Shell sued for one billion dollars in cleanup costs it was responsible to pay for cleanup at two hazardous waste sites in Colorado and California. Insurers, in contrast, argue for either an exposure trigger (when the hazardous waste was disposed of) or a manifestation trigger (when a cleanup order or suit arises). A fourth trigger of coverage, the **injury in fact trigger**, which lies somewhere between the manifestation, exposure and triple triggers may also be claimed by either insurers or policyholders, depending on the circumstances. As is readily apparent, the final determination of the trigger issue will have enormous significance for both PRP policyholders and insurance companies.

The stakes for the insurance policy regarding the trigger issue go far beyond hazardous waste risks. If triple trigger became the established precedent for continuing types of injuries and damages over time, then insurers would be subject to the "stacking" of their limits in any number of risk situations. Pharmaceutical drugs, inside building exposures (light, noise, toxic gases, radiation), and products like silicone gel breast implants, are examples of risk situations where it can be argued that injuries are occurring in a cumulative fashion over time. The application of the triple trigger doctrine in these situations could result in the triggering of numerous liability insurance policies over extended periods of time. For insurance policies written on the predominant occurrence form rather than claims-made form, insurers would never be able to close their books. **Occurrence policies** require only that injuries or damages occur during the policy period, even though the actual claim may be made long after the policy period expired. **Claims-made policies** require that the claim be made during the policy period for coverage to be in effect; thus once the policy period has expired, the insurer is not responsible for future claims. Whenever a new type of injury is discovered, and where it can be established that latent damage had been occurring over an extended time, all older occurrence insurance policies over this period could be triggered. Thus it is not surprising that insurers are putting up a particularly staunch defense on the trigger issue. Universal application of the triple trigger would have an enormously adverse financial impact on the insurance industry. The industry, which has as one of its principal objectives the reduction of uncertainty, would find itself facing a very uncertain financial future.

Covered Damages

The general liability insurance policy typically states that the insurer will pay on behalf of the insured all sums which the insured shall become legally obligated to pay as damages because of bodily injury or property damages.

Insurers contend that government-mandated hazardous waste cleanup costs under Superfund are not damages covered under liability policies.

They argue that such costs are economic losses, and constitute equitable monetary relief rather than legal relief, i.e., monetary amounts awarded by a court of law. PRP policyholders counter that policy wording is unclear, i.e., ambiguous, and they favor a common everyday definition of damages that would include monies they are having to pay for cleanup costs.

On this particular issue, as of July 1992, six state supreme courts (California, Iowa, Massachusetts, Minnesota, North Carolina, and Washington) have held for policyholders that these costs are covered damages; and two state supreme courts (Maine and New Hampshire) have held for insurance companies that these costs are not covered damages. At the federal appellate court level, four U.S. Circuit Courts of Appeals (2nd, 3rd, 9th, D.C.) have held for policyholders while two U.S. Circuits (4th, 8th) have held for insurers. Of the more critical coverage issues, the issue of covered damages has produced the most discernible trend in court decisions, and this trend favors the policyholders.

Duty to Defend

A standard feature of liability insurance policies is that the insurance company has the right and duty to defend the policyholder in any litigation resulting from actions covered under the policy. Defense costs would include attorney fees, court costs, expert testimony, investigations, studies, and other costs associated with defending the policyholder. Obviously these costs can be substantial.

When it is determined that the insurer's policy must respond (that is, the insurer has been unsuccessful in applying other defenses/issues in this section), there is little debate that the insurer must provide defense cost coverage. There are other cases, however, where the insurer may be asked to provide defense cost coverage. Generally the insurer's duty to defend is broader than the duty to indemnify for bodily injuries and property damages. Typical policy wording is included below:

> The insurer shall have the right and duty to defend any suit against the insured seeking damages on account of such bodily injury or property damage, even if any of the allegations of the suit are groundless, false or fraudulent, and may make such

investigation and settlement of any claim or suit as it deems expedient.

PRP policyholders argue that insurers have a duty to defend and pay defense costs for actions brought against them by EPA under Superfund, even if cleanup costs may not be held to be covered property damages. In addition, policyholders argue that a letter from EPA naming them as PRPs for cleanup costs has the same practical effect as a lawsuit and triggers an insurer's duty to defend.

Another more troublesome situation involves the amount or the limit on defense cost coverage. Typically, in a liability insurance policy, no specific dollar limit applies to defense costs. Policy limits apply to bodily injuries and property damages to establish maximum amounts payable by the insurance companies for injuries and damages. Since 1966, insurers have had an explicit clause in policies which states that defense cost coverage ceases when policy limits for bodily injuries and property damages are exceeded. A typical clause is included below:

> The insurer shall not be obligated to pay any claim or judgement or to defend any suit after the applicable limit of the insurer's liability has been exhausted by payment of judgements or settlements.

Earlier policies did not contain such a clause. Consequently substantial litigation has involved the issue of whether insurers have to pay defense costs without limit on these earlier policies. For instance, an insurer may have written a $20,000 policy in 1947, the limit of which has been exhausted by covered property damages, but the insurer is being asked to pay for defense costs at today's rates without limit.

Multiple Occurrences

Policy limits in liability policies are usually expressed as X dollars per occurrence (there may also be per injured person limits). Older policies had per accident rather than per occurrence limits. The basic idea is to establish a maximum amount of money that the insurance company will pay for a particular event or loss.

If a hazardous waste/pollution situation involves multiple claims (e.g., EPA, state governments, private parties), questions have arisen as to whether this should be considered one occurrence or multiple occurrences. For instance, in the Jackson Township case, the court held that each of the 200 plus claimants was a separate individual occurrence. The result was a huge increase in the potential liability of the insurance company as the full policy limit became available for each claimant or occurrence.

Since 1986, the premises and operations portion of the general liability policy has been subject to an annual aggregate limit. This type of limit stipulates the maximum liability of the insurance company for a particular year, regardless of the number of occurrences. Earlier policies, at least the primary policies (excess and umbrella policies usually have aggregate limits), did not have an annual aggregate limit on this portion of the coverage from which pollution/hazardous waste claims arise. On these earlier policies, insurers may incur liabilities in substantial excess of their per occurrence limits. In addition, as new claimants come forward, policy limits may never be exhausted, and could also lead to additional defense cost coverage.

Care, Custody and Control Exclusion

Liability insurance policies have an exclusion called the **care, custody and control exclusion**, as shown below.

This insurance does not apply to property damage to:

1. Property owned or occupied by or rented to the insured.

2. Property used by the insured.

3. Property in the care, custody, or control of the insured or as to which the insured is for any purpose exercising physical control.

The effect is to exclude coverage for damage to property that is in the care, custody, and control of the policyholder. Coverage for such damage is more appropriately provided by property insurance policies.

In the context of hazardous waste claims, coverage for hazardous waste cleanups for waste that was disposed of on the policyholder's property would be excluded. If the waste is shipped off the insured's property to a waste site handled by another party, then the exclusion does not apply. Insurance companies have been generally successful in upholding this exclusion, with at least one notable exception. In the Summit case, the judge held the exclusion invalid, arguing that the public policy of cleaning up the environment overrode the clear wording of the policy exclusion.

10 MANAGING ENVIRONMENTAL COMPLIANCE

INTRODUCTION AND OVERVIEW

Today's environmental manager must maneuver through 12 major arenas of environmental compliance. Without comprehension of each of these arenas, one can never hope to achieve compliance. Much of the available information about compliance is unfortunately written in either legal or highly technical terminology, making it difficult for straight business managers to comprehend. For this reason, corporations and their facilities require specialists who can read, understand, and relate their company's operations to the legal boundary conditions established by the federal programs.

This final chapter provides an overview of the remaining environmental regulations you must work within. The specific subjects covered are transportation, storage tanks, pesticides, water quality, discharges to sewers, and air pollution.

TRANSPORTING HAZARDOUS MATERIALS AND WASTES

The transportation of hazardous materials and hazardous wastes is regulated under both federal and state laws. The U.S. Department of Transportation (DOT) administers the **Hazardous Materials Transportation Act (HMTA**; enacted in 1981), under which it defines "hazardous materials" (wastes are a subset) and issues inspection, training, and transport requirements. HMTA requirements apply to most forms of transportation including rail, motor vehicles, aircraft, and vessels (pipeline transport is regulated separately). These requirements apply regardless of the destination or origin of the shipment (including transportation between two facilities owned and operated by the same company).

The U.S. Environmental Protection Agency (EPA) regulates transportation of hazardous wastes as part of EPA's administration of the **Resource Conservation and Recovery Act**. In addition, DOT jointly administers with EPA and the Federal Emergency Management Agency (FEMA) a set of incident/spill reporting and response requirements.

States are required to adopt laws and regulations that are consistent with the federal system. Many states have done this by incorporating the relevant federal regulations into their own rules. Some states set additional vehicle and driver safety requirements, and restrict the routes and timing for hazardous materials transportation between states (i.e., interstate carriers), while state agencies regulate intra-state transport of hazardous materials.

In most states, more than one agency regulates transportation of hazardous materials; and in many cases their authority overlaps. Typically, these agencies include, at a minimum, the state police or highway patrol and the state environmental agency. Other agencies may separately regulate rail transportation within that state.

DOT is changing its transportation regulations to make them more straightforward and consistent with international shipping requirements (following UN recommendations). The new shipping regulations may be followed immediately or phased in accordance with DOT's fixed schedule over several years. Most requirements were to have been met by October 1, 1993, although certain packaging requirements can be deferred until 1996.

By congressional mandate, all other federal and state laws must conform generally to the HMTA. As a result, all hazardous materials transportation nationwide must meet minimal safety standards and common labeling formats. States and localities cannot use overly strict regulations to stifle transportation of these materials.

Under HMTA, **hazardous materials** are defined as those which might create an "unreasonable" risk to health and safety or to property when being transported. Hazardous materials include, but are not limited to, hazardous substances as defined by the federal Comprehensive Environmental Response, Compensation, and Liability Act of 1980 (CERCLA) and hazardous wastes defined by RCRA.

DOT regulations apply both to the **shipper**, who is the person who offers a hazardous waste for transportation, and to the **carrier**, who is the person who transports a hazardous material by air, highway, rail or water. As a shipper or carrier of any material appearing on DOT's Hazardous Materials Table (49 CFR, parts 100 to 177) in any quantity or any material listed in DOT's Hazardous Substances Table in a quantity in a single package that equals or exceeds its reportable quantity (RQ) or any RCRA hazardous waste, you are subject to DOT regulations in the following areas:

- Packing and repacking.
- Labeling, marking, and placarding.
- Handling.
- Vehicle routing.
- Manufacturing of packaging and transportation containers.

Each of the hazardous materials identified by DOT falls into one or more of the following categories. In 1991, DOT realigned its hazard classes to give them numbers in accordance with international standards:

- Class 1 -- explosives.
- Class 2 -- hazardous gases and cryogenic liquids.
- Class 3 -- flammable and combustible liquids.
- Class 4 -- flammable and combustible solids.
- Class 5 -- oxidizers.
- Class 6 -- poisons.
- Class 7 -- radioactive materials.
- Class 8 -- corrosives.
- Class 9 -- miscellaneous hazardous materials.

DOT has extensively revised its Hazardous Materials Table to incorporate the new information requirements. In cases where more than one type of hazard is present, DOT has established a "precedence of hazard" system so that more severe hazards are addressed adequately.

DOT regulations define documents that must be prepared for each shipment of hazardous materials, as well as labels and placards that must be placed on each shipment. For each material, shipping papers must include:

- DOT-prescribed shipping name.
- Quantity.
- Identification number.
- The hazard class.

Each shipping paper or manifest must contain an emergency response telephone number provided by the shipper, which can be called for additional information in the event of a spill or release.

Shippers must certify that the materials are properly described, packaged, and labeled. For hazardous wastes, the "manifest" (described below) serves as the required shipping document. Manifest forms may be obtained from EPA or state hazardous waste regulatory agencies. Copies of shipping papers must be kept by shippers and carriers for at least six months and may be subject to review by DOT.

The shipper is required to use the most descriptive name available in the Hazardous Materials Table, avoiding use of "not otherwise specified" (n.o.s.) whenever possible for supplementing the n.o.s. description with additional information. In some cases, additional descriptions may be mandatory. For example, if a shipment is made under an exemption, the shipping paper must contain the notation "DOT-E" followed by the exemption number assigned. The words "limited quantity" or phrase "Ltd Qty" must be added for a material offered for transportation under this category; that is, below the maximum amount of a hazardous material for which there is a specific labeling and packaging exemption.

Shippers must ensure that each package is properly marked and labeled and that the transportation vehicle is properly placarded. The carrier cannot accept hazardous materials for transport unless the materials are properly marked and labeled, and the carrier's vehicle properly placarded. There are numerous requirements prescribed for specific hazardous materials, and certain exceptions apply to specific materials or quantities, and may vary by the mode of transport used. The shipper must certify by signature that the materials offered for transportation are "properly classified, packaged, marked, and labeled, and are in proper condition for transportation according to the applicable regulations of the Department of Transportation."

Each package with a rated capacity of 110 gallons or less, and each portable tank and tank car must be marked with the proper shipping name and United Nations (UN) Number or North American (NA) Number as shown in the Hazardous Materials Table. This marking must be on the package surface or on a label, tag or sign. For hazardous substances, the letters "RQ" must also appear on the marking. Cargo tanks and bulk packagings must be marked with the proper identification number on each side and on each end.

Also, each container of hazardous materials must be labeled as specified in the Hazardous Materials Table or the Optional Hazardous Materials Table, unless specifically excepted by the regulations. Generally, the label is a distinctive diamond shape and indicates the hazard class of the materials (e.g., "Nonflammable gas and Poison"). DOT specifies the size, color, and text of each label.

Each vehicle, freight container or cargo tank must be placarded unless excepted. Many hazardous materials have specific placarding requirements. DOT has specified the size and design of these placards, which generally are the same as those of the package labels, and must be prominently displayed on the vehicle. In general, placards are not required for infectious substances, for materials in the D class of Other Regulated Materials (ORMs), for limited quantity shipments, or for properly packaged small quantities of flammable liquids or solids, oxidizers, organic peroxides, corrosives, Poison B, and ORM-A, B, C, and radioactive materials.

ORMs are hazardous materials that either do not meet the definitions of the other hazard classes or have been reclassified as ORM. ORMs are divided into classes: ORM-A materials can cause extreme discomfort to passengers and crew in the event of a leakage; ORM-B materials are capable of causing significant damage to a transport vehicle from leakage; ORM-C materials are specifically named in the regulations and are unsuitable for shipment unless properly identified and prepared; ORM-D materials are ones such as consumer commodities that present limited hazard during transportation; and ORM-E materials are those not included under other classes, such as hazardous wastes or other hazardous substances.

Certain ORMs may be subject to less stringent shipping requirements (including papers, marking, labeling, and placarding) in the following circumstances:

- ORM-A, B, or C liquid, not over one pint in one package.
- ORM-A or B solid, not over five pounds in one package.
- ORM-C solid, not over 25 pounds in one package.

The DOT regulations also include technical requirements for the construction of containers and transport equipment for each mode of transportation--highway, rail, air, and water. In addition, the regulations outline loading, unloading, and general handling procedures for each mode. For example, liquid hazardous materials must be packed with their closure upward, with markings (orientation arrows, for example) clearly indicating which side is up.

DOT is replacing its cumbersome material-specific packaging standards with general guidance and performance standards based on the characteristics of the hazardous materials being packaged. These standards address such matters as compatibility between the contained materials and the packaging, prohibitions against leaks or emissions, ability to withstand vibrations, temperature variations, etc. The new regulations define **Packing Groups**, provide varying levels of protection, and assign materials to each group as required to provide safe shipping. There are three **Packing Group (PG)** designations that the reader should be familiar with: PG I - High Danger, PG II - Medium Danger, and PG III - Minimal Danger. Materials classified as inhalation poisons must comply with new packaging requirements. Other packaging manufactured after October 1, 1994, must comply with new requirements, and non-complying packaging may not be used after October 1, 1996.

DOT regulates hazardous materials carriers as an extension of its regulation of motor vehicles. The regulations address qualifications for drivers based on physical examination, driving record, and their written and practical driving tests. In **New Jersey**, for example, the Department of Motor Vehicles implements DOT and state requirements covering drivers of vehicles transporting hazardous materials. These drivers must first secure licenses appropriate to the vehicles they drive, and then pass a special examination to obtain approval for hazardous material transport (called a Certified Driver License or CDL). DOT also requires vehicles to meet minimum safety requirements covering lights, brakes, other operating equipment, and maintenance schedules. The 1990 HMTA amendments require certain motor carriers to acquire "safety permits," and require certain transporters of hazardous substances (as defined by EPA) to maintain specified financial responsibility levels.

In some cases DOT regulations may restrict routing of hazardous materials carriers to avoid populated areas, tunnels, and narrow streets when possible. These regulations also require the driver to stop regularly to check tires--every two hours or 100 miles. Also, each state, often in consultation with city and county governments, establishes route and time restrictions on the transportation of hazardous wastes (and in certain cases, hazardous materials). In general, transportation of hazardous materials must minimize transit time and exposure to congested or dangerous routes.

Effective as of December 31, 1990, whenever there is a shipment of hazardous materials, emergency response information must accompany the shipment. Emergency response information useful in cleaning up a spill must accompany each shipment. The following information must be included for each hazardous material being transported:

- A description and technical name of the hazardous material, as required on the shipping papers.
- Immediate hazards to health.
- Risks of fire or explosion.
- Immediate precautions to be taken in response to an accident.
- Immediate methods to handle fires.
- Initial methods to handle spills that do not involve a fire or explosion.
- Preliminary first aid measures.

This information must be available on the transport vehicle, but kept away from the materials themselves; it must be immediately accessible to the transporter and regulatory or fire agency personnel (e.g., in the cab of a truck) for use in responding to a release. Additionally, each "person who offers a hazardous material for transportation" must provide a **24-hour emergency response** telephone number, monitored by personnel able to provide callers with additional information regarding the hazardous materials being transported, and proper emergency response procedures. DOT's regulations provide that this number be monitored while materials are actually being transported, so that those who ship only at certain times need monitor the telephone only during those times.

Carriers must notify appropriate agencies of any transportation incidents (including those which occur during loading, unloading, or temporary storage). Generally, immediate notification is required if an incident:

- Causes death or serious injury.
- Involves more than $50,000 in damage.
- Involves the release of radioactive materials or etiologic agents.
- Requires public evacuation.
- Causes a major transportation artery or facility to close for more than an hour.

In such cases one must make an immediate call to the National Response Center at **1-800-424-8802**. Notification should also be made at the same time to the state highway patrol or police.

Notification must include the following information:

- Name of person reporting.
- Name and address of carrier represented by person reporting.
- Phone number where person reporting can be contacted.
- Date, time, and location of incident.
- Extent of injuries (if any).
- Proper shipping name, hazard class, UN Number, name, and quantity of hazardous materials involved (if known).
- Type of incident, nature of hazardous materials involvement, and whether a continuing danger exists at the scene.

The requirements for submitting a written report are more stringent than for making verbal notification. That is, a written report is required even for some incidents that were not sufficiently serious to warrant a call to the National Response Center. DOT requires that its Detailed Hazardous Materials Incident Report (**Form F5800.1**) be submitted within 30 days for each incident where any of the following apply:

- A verbal notification was made to the National Response Center.
- Any quantity of hazardous materials has been released unintentionally from a package, including from storage tanks.
- Any quantity of hazardous waste has been released.

Generally, you must also send a copy at the same time to the state highway patrol or police. The carrier must retain a copy of the report for at least two years.

These reporting requirements are in addition to similar EPA-administered requirements under RCRA and CERCLA. CERCLA also requires the carrier to report to the National Response Center all spills larger than a specific RQ. In order to alert drivers and emergency responders to this requirement, the letters "RQ" must appear on all shipping papers if the transporter is carrying in a single container a substance on DOT's Hazardous Substances Table equal to or in excess of its RQ.

If hazardous wastes are released, three additional requirements apply:

- State and local agencies must be verbally informed.
- The written report must be made within 10 days.
- A copy of the hazardous waste manifest must be attached to the report.
- An estimate of the quantity of waste removed from the scene, the name and address of the facility to which it was taken, and the manner of disposition must be included in the report.

DOT has incorporated information necessary to comply with these requirements into its *Emergency Response Guidebook*, which can be obtained from the Government Printing Office.

Employees who directly affect hazardous materials transportation safety must receive training in the following:

- Familiarization with the regulatory requirements that apply to transportation of hazardous materials.
- Recognition and identification of hazardous materials.
- Specific requirements of their job functions involving hazardous materials transportation.

Employees who handle or transport packaging containing hazardous materials and those who have the potential for exposure to hazardous materials as a result of a transportation accident must also receive safety training in the following:

- Accident prevention.
- Exposure protection measures.
- Limited emergency response actions.

Drivers must receive training on the safe operation of the motor vehicles they operate. This training must cover the following:

- Pre-trip safety inspection.
- Use of vehicle controls, including emergency equipment.
- Vehicle operation.
- Procedures for navigating tunnels, bridges, and rail crossings.
- Requirements for vehicle attendance, parking, smoking, routing, and incident reporting.
- Materials loading and unloading procedures.

Operators of cargo tanks and vehicles with a portable tank of at least a 1000 gallon capacity must also receive further specialized training, including retest and inspection requirements for cargo tanks. These driver training requirements may be met through compliance with an appropriate state-issued commercial driver's license that has a tank vehicle or hazardous materials endorsement.

To implement HMTA, DOT is authorized to inspect transport facilities and vehicles as well as records relating to hazardous materials transportation. DOT also can invoke administrative enforcement procedures, including subpoenas and hearings, and issue compliance orders. When necessary, enforcement can be pursued in federal court with assistance from the U.S. Department of Justice. HMTA imposes civil and criminal penalties for violations of its provisions. Amendments to HMTA in 1990 increased civil penalties for violations of its provisions. Amendments to HMTA in 1990 increased civil penalties for violations of HMTA or DOT regulations to $25,000 per violation, with a minimum fine of $250. These amendments also eliminate the defense for an employee who acts without knowledge. Criminal penalties exist for knowingly tampering unlawfully with any document, packaging, label, marking or similar items. Other willful violations can result in criminal penalties of up to $25,000 and/or five years imprisonment. DOT can also seek federal court orders suspending or restricting specific transport activities if the activities pose an "imminent hazard."

RCRA imposes a comprehensive "cradle-to-grave" regulatory framework on all activities related to hazardous wastes. This framework includes important controls on the transportation of hazardous wastes. Nationally, EPA has coordinated these requirements with DOT's broader regulation of transport of all hazardous materials. If you ship hazardous waste, you are required by EPA to apply for an **EPA Identification Number (ID Number)** using EPA **Form 8700-12**. This ID Number must then be used on all documents, allowing federal and state agencies to track each transporter's activities.

All your hazardous waste shipments must be accompanied by a **hazardous waste transportation manifest**. Carriers are to refuse to carry any hazardous wastes that are not documented by manifests. When hazardous wastes are transported across state lines, usually the manifest for the state that will be receiving the wastes must be used. Some states allow the generator-state manifest to be used. A carrier must retain its copy of the manifest for at least three years from the date the waste was first accepted. Carriers must comply with the terms of the manifest. If the shipment cannot be delivered according to the terms of the manifest, the carrier must contact the generator for further instructions and revise the manifest accordingly.

RCRA requires carriers to prepare contingency plans and train personnel in spill response procedures. In response to hazardous waste discharges during transport, transporters must:

- Inform appropriate federal, state, and local agencies of the spill.
- Carry out immediate containment actions (e.g., diking a spill area), although emergency response agencies typically assume site command once they reach the scene of the spill.

Carriers retain legal and financial responsibility for cleanup; each responsible party, including the original waste generator, shipper, and/or site or facility operator, also remains potentially liable for the ultimate costs of the cleanup. RCRA's general inspection, recordkeeping and enforcement provisions apply to carriers. Thus, a violation of a legal requirement may result in criminal or civil penalties, or both.

Transporters of hazardous materials must turn to regulations for more detailed information on all applicable requirements. These can be found in Volume 49 of the Code of Federal Regulations (49 CFR). The primary source of information specific to chemicals being shipped and their mode of transportation is 49 CFR, Parts 171-177. Part 172 provides lists of hazardous materials and covers communication regulations while Part 173 covers the shipment and packaging regulations. Part 174 is specific to rail transport, 175 to aircraft, 176 to vessels, and 177 to public highways.

PESTICIDES

Federal law requires comprehensive regulation of the manufacture, handling, and use of pesticides. The U.S. Environmental Protection Agency (EPA), in cooperation with state and local agencies, implements the basic federal regulatory framework governing pesticides known as the **Federal Insecticide, Fungicide, and Rodenticide Act (FIFRA)**. This law was initially enacted in 1947 and has been amended several times, most recently in 1988.

FIFRA provides for the registration and classification of pesticides and prescribes controls over their application and use. It allows EPA to delegate primary enforcement responsibility to state agencies. Most states supplement FIFRA's regulatory program with additional state requirements. These requirements fall into four general areas:

- Licensing of pesticide applicators and/or dealers.
- Registration of pesticides with a state agency.
- Notice and recordkeeping procedures.
- Pesticide storage and use restrictions.

Although FIFRA establishes a national basis for the regulation of pesticides, the implementation of FIFRA's regulatory scheme varies from state to state. All states must enforce requirements at least as stringent as those in FIFRA, and many states also have additional requirements. States designate various agencies to regulate pesticides. Depending on the state, the implementing agency could be the Department of Agriculture (as in **Washington** state), the state environmental protection agency (as in **New York** state), or some other agency.

FIFRA requires that each pesticide be registered by use, prior to its distribution or sale. Upon registration, a pesticide is **classified** by EPA according to its use.

Under FIFRA there are three use categories:

- Pesticides classified for **general use** are those found not to have unreasonable adverse effects on the environment when used according to directions.
- Pesticides classified for **restricted use** are those that may have unreasonable adverse effects, including injury to the applicator, if additional regulatory restrictions are not imposed.
- The third classification, **mixed use**, is "use" dependent; a pesticide in this category may be classified for general use for some applications and classified for restricted use for other applications.

Use classifications may trigger specific federal and state requirements. FIFRA and state laws regulate the following:

- Pest control activities, which are defined to include the use and application of pesticides.
- Certification of applicators of restricted materials, based on demonstrated competency in the use of pesticides.
- Worker safety precautions, including requirements to warn field workers before pesticide applications and to prevent exposure of workers not actually involved in pest control activities. EPA and state agencies establish time periods after application during which workers may not return to a treated field.

FIFRA requires registration of each pesticide prior to its distribution or sale. Anyone can register a pesticide, but the pesticide's manufacturer is the most common registrant, since the registration process can take years and be very costly. The 1988 amendments to FIFRA required the **reregistration** of any pesticide containing an active ingredient first registered before November 1, 1984. This reregistration process is still underway. You can ensure that any pesticide product you distribute, sell or use is properly registered by checking the label. The phrase "EPA Registration No." or "EPA Reg. No." and the registration number must appear on the label of any properly registered pesticide.

The registration process focuses on the proposed use of a particular pesticide, and involves submission of an application which must include:

- Name, address, and other applicant information.
- Complete labeling information and instructions for use.
- Pesticide name and complete formula.
- Use classification request.
- Complete testing data.

In addition to this information and other testing information that EPA may require, the application must be accompanied by a registration fee. The amount of the fee depends on the type of pesticide, and ranges from $50,000 to $150,000.

Registration of the following pesticides may be expedited:

- Those that would be identical or substantially similar in composition and labeling to a currently registered pesticide ("me too" applications).
- Those that would differ in composition and labeling from a currently registered pesticide only in ways that would not significantly increase the risk of adverse environmental effects.

EPA must notify an applicant within 45 days whether the application is complete, and then grant or deny the application within 90 days of receipt. The expedited review process also applies to applications to amend existing registrations that do not require scientific review of data.

An **experimental use permit** may be obtained if additional data are needed before a pesticide can be registered. Any pesticide containing an active ingredient first registered before November 1, 1984, must be reregistered to meet current scientific and regulatory standards. EPA issued four lists in 1989 of those pesticides that require reregistration. **All updated reregistration application information should have been submitted to EPA by October 24, 1990.** If an application for a certain pesticide was not received by that date, EPA initiated procedures to suspend the pesticide's registration.

EPA has 18 to 33 months after its lists are issued to review data submissions to determine whether any gaps in ingredient data exist. Once the data are deemed complete, EPA must decide within one more year whether the pesticide is eligible for reregistration. The registrant then has eight months from EPA's eligibility determination to submit any additional product-specific data required by EPA. EPA then has 90 days to review all data and a total of six months to finalize registration. **Reregistration must be completed by December 1997.**

Two conditions must be met to keep a pesticide registration current:

- The registrant must remain in compliance with FIFRA, applicable EPA regulations, and state law and regulations.
- The registrant must pay an annual **maintenance fee** to EPA. The fee depends on the number of pesticide registrations a single registrant maintains, and begins at $650.

FIFRA grants EPA the authority to cancel or suspend registrations and to change use classifications. These actions may be taken if new information indicates that the pesticide violates FIFRA provisions, causes unreasonable adverse effects or presents an imminent hazard or emergency. The registrant bears the burden of proof that its registration should not be canceled. Any person who distributes, sells or possesses a pesticide that becomes cancelled or suspended must notify EPA and the state pesticide agency of the quantity of pesticide owned and the location of each storage site. Any person owning pesticide stocks who suffers financial loss as a result of the pesticide's suspension or cancellation may seek an indemnity payment from EPA. This indemnity is not available to a person who, with knowledge of facts indicating that registration requirements were not met, continued to produce the pesticide without notifying EPA.

The 1988 FIFRA Amendments restrict **automatic indemnity** to certain end users such as farmers. With limited exceptions, indemnification for all other persons is not authorized unless congress approves a specific line-item appropriation.

Purchasers of suspended or canceled pesticides who are not end users and who cannot use or resell the pesticides generally can look to the seller for reimbursement. A seller can avoid this reimbursement obligation if, at the time of sale, it notifies the buyer in writing that the pesticide is not subject to reimbursement by the seller in the event of suspension or cancellation.

A **pesticide applicator** is any person who is authorized to use or supervise the use of any restricted use pesticide. Applicators of pesticides classified for restricted use must be certified. Although the general rules for this certification were established by EPA, the states are charged with implementing certification programs. Because of this, the programs vary widely depending on the state. Applicators can be certified as **private** or **commercial** applicators. Private applicators are those who apply restricted use pesticides on:

- Their own property.
- Their employer's property.
- Another's property, if done in exchange for agricultural services.

Commercial applicators are those who apply restricted use pesticides under any circumstances other than those listed for private applicators. They can be certified in one or more categories. EPA's regulations specify 10 categories, but states are free to adopt some, all or additional categories as needed.

EPA's categories for commercial certification are:

- Agricultural pest control--further subdivided into plant and animal categories.
- Forest pest control.
- Ornamental and turf pest control.
- Seed treatment.
- Aquatic pest control.
- Right-of-way pest control.
- Industrial, institutional, structural, and health-related pest control.
- Public health pest control.
- Demonstration and research pest control.

All commercial applicators must demonstrate knowledge in a number of pertinent areas. This knowledge is demonstrated through an examination administered by the state. EPA's regulations define certain basic elements this examination must include; states are free to augment (but not reduce) these basic requirements.

At a minimum, the examination must test the following areas of competency:

- Label and labeling comprehension.
- Knowledge of safety precautions and procedures.

- Environmental awareness.
- Pest recognition and relevant biology.
- Pesticide types, formulations, compatibility, hazards and other relevant knowledge.
- Equipment types, maintenance, usage and calibration.
- Application techniques.
- Applicable state and federal laws and regulations.

As with commercial applicators, private applicators must demonstrate knowledge and ability to safely and correctly use restricted use pesticides. This knowledge must be demonstrated by a written or oral examination. At a minimum, an applicant for certification as a private applicator must demonstrate the ability to:

- Recognize common pests.
- Read and understand label information.
- Apply pesticides according to label instructions.
- Recognize local environmental concerns that must be considered during application to avoid contamination.
- Recognize poisoning symptoms and know procedures to follow in the event of mishap.

The EPA may require a registrant or applicant to submit information regarding methods for safe storage and disposal of excess quantities of pesticides. EPA may also require that container labels outline directions and procedures for safe transportation, storage and disposal of the pesticide, its container, rinsates, and any other material used to contain or collect excess or spilled quantities of the pesticide. EPA's regulations for FIFRA specify that pesticide transport must comply with requirements set by DOT for hazardous material transport. Also, pesticide dealers are subject to licensing requirements aimed at ensuring safe storage and adequate notice and recordkeeping procedures.

FIFRA endows EPA and the courts with broad enforcement powers, including the right to inspect products and records and the authority to prosecute violations and to issue stop-sale orders and injunctions. EPA may order a "**voluntary**" or "**mandatory**" **recall** of a pesticide that has been suspended or cancelled. Voluntary recalls are ordered when EPA determines that such actions will be as safe and effective as a mandatory recall. Under voluntary recalls, EPA asks the registrant to design and submit a recall plan within 60 days. EPA may then approve the voluntary plan and order the registrant to conduct the recall, or may find the plan inadequate to protect health or the environment and order a mandatory recall. Under a mandatory recall, EPA prescribes the terms of the recall which the registrant must implement. EPA can order a mandatory recall plan at the outset or if a proposed voluntary plan is inadequate. The registrant of a pesticide may be required to provide evidence of sufficient financial and other resources to carry out a recall plan in the event of suspension or cancellation of a pesticide. Storage costs incurred as a result of a recall are shared by EPA and the registrant according to a percentage formula.

UNDERGROUND STORAGE TANKS

To protect the nation's groundwater resources, the federal government and most state governments regulate the installation and use of underground storage tanks (USTs). This regulation is implemented on the national level by the U.S. Environmental Protection Agency (EPA), and has three main components:

- Registration.
- Technical standards for construction and operation.
- Financial responsibility (to ensure funding for cleanup).

State agencies implement state UST laws with varying degrees of coordination with EPA. Most state laws include the three components mentioned above; some states augment the federal program with more detailed requirements.

UST owners register their USTs with a state-designated agency, which in turn reports these registrations to EPA. This requirement is designed to provide EPA and the states with an inventory of USTs, as well as information about their location and size. In New Jersey, for example, registration is through the Bureau of Underground Storage Tanks, a division of DEP (Department of Environmental Protection).

Technical standards are designated to minimize the likelihood that USTs will deteriorate and leak. Existing USTs must upgrade to federal standards on a certain schedule, depending on when they were first installed. New USTs are required to be constructed and equipped to meet strict requirements. State laws often impose stricter standards and tighter deadlines.   Also, tank owners must demonstrate financial responsibility to cover the costs of cleanup should a leak or spill occur. EPA has defined (and subsequently revised) schedules of amounts and deadlines to meet this requirement.

State UST laws increasingly follow the federal regulations, often imposing additional provisions covering permitting and use of USTs, more stringent technical standards, and mechanisms to assist owners in meeting financial responsibility obligations. Some states also regulate aboveground storage tanks (ASTs).

Federal law defines a "UST" as any one or a combination of tanks used to store hazardous substances or petroleum products which, including the capacity of connecting pipes, is 10% or more beneath the surface of the ground. The federal UST law defines the following as "hazardous substances":

- Gasoline and other petroleum products.
- Substances for which reportable quantities (RQs) are listed in federal Superfund regulations.
- Any other substance designated by EPA (none as of July 1992).

Most states adopt this list into their state UST law, but some states add additional substances. For example, **New York**'s list of regulated substances contains nearly five times as many substances as the federal list.

Tanks exempt from federal UST regulations include:

- USTs, including their piping systems, that are more than 90% above grade.
- Farm or residential tanks smaller than 1100 gallons storing motor fuel for on-site use.
- USTs storing heating oil for on-site use.
- Septic tanks.
- Oil field gathering lines or refinery pipelines.
- Surface impoundments, lagoons, pits, and ponds.
- Storm drains, catch basins, and wastewater collection tanks.
- Separation pumps or flow-through process tanks.
- Lined and unlined pits.
- Well cellars.

- USTs storing hazardous wastes [these are regulated by the Resource Conservation and Recovery Act (RCRA)].

The EPA required all UST owners to register with their state by May 8, 1986. Since then, new USTs must be registered as they are put into service. Many states have additional registration and/or permitting requirements.

If you removed an UST from service AFTER January 1, 1974, you were required to register these USTs by May 8, 1986. For these out-of-service USTs, you were required to provide the following information:

- The date the UST was taken out of service.
- Age of the UST on the date it was taken out of service.
- Size, type, and location of the UST.
- Type and quantity of any substances remaining in the UST on the date it was taken out of service.

USTs removed from service since January 1, 1984, have also been subject to the formal **closure requirements**. EPA requires that you register your UST with your state agency. Some states also require additional steps, which may include the following:

- Separate state registration.
- Permits to install, modify or remove your UST.
- Payment of registration or permit fees.

All the construction and operating standards described earlier apply to USTs defined by federal regulations as "new" or "existing," depending on when they were installed. **In the discussions below, the word "existing" refers to a UST installed before December 22, 1988; the word "new" refers to a UST installed after December 22, 1988.**

Construction standards for "new" USTs include requirements covering corrosion protection, spill and overfill protection, and (for hazardous substance USTs) secondary containment. To comply with "new" UST construction standards, you must meet the requirements in all three areas described below.

The underground portions of all "new" USTs must be designed and constructed to resist corrosion. EPA's regulations define certain options for "new" USTs to meet this corrosion protection requirement:

- The UST is constructed of fiberglass-reinforced plastic.
- The UST is steel, and is cathodically protected.
- The UST is constructed of a steel-fiberglass-reinforced-plastic composite.
- The UST is installed at a "noncorrosive site," as determined by a corrosion expert.

"Existing" USTs also have certain options to meet the corrosion protection requirement. One of the following options must be installed:

- A cathodic protection system.
- Interior lining (such as that described in the American Petroleum Institute's Publication 1631, Recommended Practice for the Interior Lining of Existing Steel Underground Storage Tanks).
- Interior lining and cathodic protection.

Some of these options for "existing" USTs also require monthly monitoring and/or tightness testing.

Unprotected steel USTs often act like a battery. Part of the UST becomes negatively charged and another part positively charged. Moisture in the soil connects these parts and the negatively charged portion of the UST--where the current exits the "battery"--begins to deteriorate. Cathodic protection reverses the electric current. It can come in two forms:

- "Sacrificial anodes" are pieces of metal more electrically active than the UST. Because of this, the electric current exits through them, not the UST, and the UST is protected from corrosion while the anode is sacrificed.
- An "impressed current" system produces an electric current in the ground from anodes not attached to the UST. This outside current is greater than the current in the UST, protecting the UST from corrosion.

Spill and overfill protection involves both technical and procedural measures. You must both **monitor transfers** into your UST, and **equip your UST with certain hardware**, to prevent spilling and overfilling. Spill prevention equipment is a device to prevent the release of product into the environment when the transfer hose is detached from the fill pipe; for example, a spill catchment basin.

Overfill prevention equipment:

- Automatically shuts off flow of product when the UST is no more than 95% full.
- Alerts the operator when the UST is no more than 90% full by restricting the flow or triggering an alarm.

UST systems that store hazardous substances **other than petroleum** must include secondary containment, such as double-walled USTs or external liners, to prevent releases to the environment if the primary container leaks.

Secondary containment systems must:

- Contain released product until it is detected and removed.
- Prevent the release of product to the environment at any time during the operational life of the UST system.
- Be checked for evidence of release at least every 30 days.

Double-walled USTs must:

- Contain a release from the inner UST within the outer wall.
- Detect the failure of the inner wall.

External liners must:

- Contain 100% of the largest UST within its boundary.
- Prevent precipitation or groundwater from interfering with its ability to detect or contain a release.
- Surround the UST completely.

Federal law does not require secondary containment for USTs storing petroleum, but many state laws do.

All USTs and piping must be properly installed according to the manufacturer's instructions and codes of practice outlined by nationally recognized associations or independent testing laboratories. For example, one such code is the Petroleum Equipment Institute's Publication RP100, *Recommended Practices for Installation of Underground Liquid*

Storage Systems. Many states [including **New Jersey, Pennsylvania, California** and **Ohio**] also require that persons who install or modify UST systems be certified by the state. These certification programs typically involve training in UST installation and repair, and passing an examination.

Whether "new" or "existing," your UST system must provide a method for detecting a release from any portion of a UST that routinely contains product. Your release detection system must have a 95% probability of correctly detecting a leak (and of not falsely indicating a nonexisting leak). Release detection methods that were permanently installed **before** December 22, 1990, are exempt from this 95% standard.

If the UST was installed after December 22, 1988, you are required to implement one of the following two options for leak detection:

1. **Monthly monitoring**--Most UST owners/operators select this option. Any one of the following five methods satisfies the requirements for this option:

   ● Automatic UST gauging continuously monitors the UST level.
   ● Vapor monitoring detects any vapor from product in the soil above the water table (the "vadose zone").
   ● Groundwater monitoring detects the presence of any product in or on the groundwater.
   ● Interstitial monitoring (for USTs with secondary containment) detects any product in the space between the primary and secondary containment.
   ● Any other method approved by your state UST agency and EPA can be used.

2. **Monthly inventory control *and* UST tightness testing every five years**--This option can only be used for the first 10 years after installation, and these two procedures **must** be used together. Beginning the first day of the eleventh year after installation, you must switch to monthly monitoring:

   ● Monthly inventory control compares the amount of product that should be in the UST (based on your inventory records) with the amount you find from a monthly test.

- UST tightness testing uses equipment capable of detecting a 0.1 gallon per hour leak rate to ensure the integrity of a UST.

UST piping can be either **pressurized** or **suction** piping. Standards are generally stricter for pressurized piping, in which a leak would push product out of the piping. A leak in suction piping does not necessarily involve contamination of the surrounding soil. The standards discussed below apply to all USTs. As with other technical standards, some states have different requirements for piping. These differences could include more extensive installation requirements or earlier upgrade schedules.

UST piping must be equipped with leak detection. The differing requirements for pressurized piping and suction piping are discussed below.

If the UST system contains pressurized piping, regardless of whether it is "new" or "existing," you must equip it with leak detection. To meet this requirement, you must chose one method from Group A and one method from Group B, as follows:

Group A methods include:

- An automatic flow restrictor that slows or shuts down the flow of product through the pipe in the event of a release.
- An automatic line leak detector that triggers an audible or visual alarm in the event of a release.

Group B methods include:

- Annual line tightness testing that can detect a 0.1 gallon per hour leak rate at one and one-half times operating pressure.
- Monthly monitoring as described earlier--any of the UST methods may be used **EXCEPT** automatic UST gauging.

If your UST system has suction piping, regardless of whether the UST is "new" or "existing," you must equip it with leak detection. There are two options you can choose from to meet this requirement:

- Line tightness testing conducted every three years.
- Monthly monitoring--any of the UST methods may be used **EXCEPT** automatic UST gauging.

The facility is **exempt** from this requirement and does not need to install leak detection if the suction piping:

- Is sloped so that the piping's contents will drain back into the storage UST if the suction is released.
- Includes one check valve in each suction line that is located below the suction pump.

As with USTs, piping may also be subject to corrosion. Whether "new" or "existing," the UST piping must be protected against corrosion, using one of two choices to meet this requirement:

- Coated and cathodically protected steel piping.
- Fiberglass piping.

"Existing" UST systems also have an additional choice: upgrading the existing steel piping by adding cathodic protection.

If you have a UST that was installed before December 22, 1988 (that is, an "existing" UST), you must **upgrade** it to meet "new" UST standards for leak detection and monitoring, corrosion protection, and spill and overfill protection. Otherwise you must close and remove the UST. Some states have earlier deadlines; check with your state UST agency. Important deadlines for upgrading "existing" UST depends on the date it was installed, as follows:

--UST installed before 1965 or age unknown: December 22, 1989.
--1965-1969 installation: December 22, 1990.
--1970-1974 installation: December 22, 1991.
--1975-1979 installation: December 22, 1992.
--1980-December 21, 1988 installation: December 22, 1993.

Suction piping must be upgraded as follows:

- Leak detection: same schedule as USTs.
- Corrosion protection: by December 22, 1998.

Pressurized piping must be upgraded as follows:

- Leak detection: by December 22, 1990.

- Corrosion protection: by December 22, 1998.
- Secondary containment: by December 22, 1998.

Existing USTs must be equipped to prevent spills and overfills from the primary container by **December 22, 1998**.

It is important to understand the response requirements to releases from USTs. A **release** is any spilling, leaking, emitting, discharging, escaping, leaching, or disposing from a UST into groundwater, surface water, or subsurface soils. Unplanned or unauthorized releases (or leaks) of regulated substances from your UST trigger a series of release recording and reporting requirements.

An unauthorized release occurs when:

- You discover free product at your site.
- Your equipment is operating unusually.
- Your leak detection monitoring indicates a release.

If your monitoring or inventory control methods indicate that you may have a release, you are given time to attempt to **confirm** the release before reporting. Your time to confirm or disprove a suspected release depends on the type of monitoring you use. If you use **inventory control**, you can wait until your next monthly check. If a second month of data also indicates a release, you must then report the suspected release **within 24 hours**. From that point on, you have seven days to investigate and determine if there has in fact been a release. If you use some form of **release detection**, you can inspect your equipment to determine if it is working properly. If you find the equipment working properly, or repair or replacement of defective equipment does not solve the problem, you must report the suspected release **within 24 hours**. From that point on, you have seven days to investigate and determine if there has in fact been a release.

You must always clean up any release at your UST site. However, some types of releases need not be reported. You do not have to report spills or overfills if:

- They are completely cleaned up **within 24 hours**.
- They are below the following threshold amounts:
    - For petroleum products, 25 gallons.

-- For other hazardous substances, the RQ established by EPA under the federal Comprehensive Environmental Response, Compensation and Liability Act of 1980 (CERCLA).

Note that specific states may have different reporting thresholds or require that you report any release.

Once you have confirmed that there has been a reportable release, you **MUST**:

- Take immediate action to stop or contain the release.
- **Within 24 hours** notify EPA and the National Response Center **(1-800-424-8802)**.
- Notify your state UST agency.

Within **20 days after the release**, submit in writing to your state UST agency a report about your cleanup progress and any information you have collected about the release. Within **45 days after the release**, submit in writing to your state UST agency a report containing the findings of your investigation of the release. This investigation should be very thorough, and explore all possible damage or potential damage to the environment. Your state agency or EPA might require additional site studies. Based on your reports, your state UST agency may decide to take further action at your site. You may be required to develop and submit a Corrective Action Plan that shows how you will meet the agency's requirements.

EPA's regulations allow facility owners/operators to repair a leaking UST. The UST repair must be performed according to a standard industry code (for example, National Fire Protection Association Standard 30, *Flammable and Combustible Liquids Code*). Within 30 days after the repair, you must prove the integrity of the repair by one of the following methods:

- Have the UST inspected internally **or** tightness tested.
- Use one of the monthly monitoring methods.
- Another method approved by EPA or your state UST agency.

Within 6 months after repair, USTs with cathodic protection must be tested to check that the cathodic protection is working properly. You must also maintain records for each repair as long as the UST is in service.

Piping made of **fiberglass-reinforced plastic** can be repaired according to the manufacturer's instructions. Within 30 days after repair, you must test the piping using one of the methods explained above for USTs. Damaged **metal** piping **cannot be repaired** and must be replaced.

Federal law requires that owners of USTs storing **petroleum** demonstrate "financial responsibility." Financial responsibility means that the owner or operator must ensure that there will be money available to help pay for the costs of cleanup and possible third-party liability in the event of a leak. The basic requirement is $1 million for each UST containing petroleum, although EPA has lowered and/or delayed this requirement for certain classes of owners. This $1 million requirement does not limit total liability for damages caused by a leak from a UST system--it merely sets a minimum that will allow you to operate a UST.

Owners of the following facilities must have $1 million of per-occurrence coverage:

- Facilities with USTs used for petroleum production, refinement or marketing.
- Facilities that handle an average of **more** than 10,000 gallons per month, based on annual throughput for the previous year.

Owners of facilities that handle an average of 10,000 gallons **or less** per month must be able to show $500,000 of per-occurrence coverage. An "occurrence" is a single accident that results in a release from a UST. The term can have other meanings within standard insurance usage, but in general, "per occurrence" means all the costs associated with cleaning up a single release. Owners or operators must also have enough coverage for annual aggregate amounts to cover all the leaks that might occur in one year. The amount of aggregate coverage depends on the number of petroleum USTs owned or operated.

Financial responsibility can be demonstrated through:

- Insurance or risk retention coverage.
- A surety bond.
- A guarantee.
- A letter of credit.

- A financial test of self-insurance.
- A trust fund.
- A state-required mechanism.
- A state fund or assurance mechanism.
- Some combination of the above options.

Many states sponsor a state fund or assurance mechanism to assist UST owners in meeting financial responsibility requirements. This may be a fund that you are required to pay into, or it may be deducted from licensing or registration fees. For example, **Ohio** maintains a Petroleum UST Financial Assurance Fund, supported by an annual fee of $150 on each UST. Participation in the fund satisfies federal financial responsibility requirements.

Closures--USTs may be closed permanently or temporarily. Different requirements apply to each situation. To close any UST permanently, you may have to obtain a UST closure permit from your local UST agency or fire department; in any case, you MUST:

- First, give your state UST agency 30 days notice before emptying the UST of all liquids and accumulated sludges.
- Then, remove the UST from the ground or fill it with an inert solid material as instructed by your state UST agency. Your closure procedures may have to be monitored by a certified inspector.
- Finally, determine if leaks from your UST have contaminated the surrounding environment. You must do this even is you have never had reason to suspect a leak. If your investigation reveals potential environmental damage, you must:
  -- Take immediate action to stop and contain the leak or spill.
  -- Inform regulatory agencies within 24 hours that a leak or spill has occurred (these agencies may vary according to the nature and extent of the release).
  -- Remove any explosive vapors and fire hazards.
  -- Report your progress to regulatory agencies no later than 20 days after confirmation of the leak or spill.

Federal UST law defines temporary closure as closure for 12 months or less. While your UST is closed, you must continue to operate the corrosion protection and leak detection systems. If a leak is found, you must respond just as if the UST was operating. You must also cap all lines attached to your UST, except the vent line.

Recordkeeping--Federal UST regulations impose certain recordkeeping requirements on UST owners. You need to maintain records in the following areas:

Leak detection--Document your leak detection system's performance and upkeep. These records should include:

- Last year's monitoring results, and the most recent tightness test.
- Copies of performance claims provided by leak detection manufacturers.
- Records of recent maintenance, repair, and calibration of leak detection equipment.

Corrosion protection--Show that the last two inspections of your system were carried out by trained professionals.

Repair or upgrade--Show that any repairs or upgrades to your system were properly done.

Closure--Maintain records for at least three years of the site assessment performed after closure, showing what impact your UST had on the surrounding environment.

An owner who fails to register a UST, or who provides false registration information, is subject to a civil penalty of up to $10,000 per UST. Violations of technical standards or requirements of an approved state program are also subject to civil penalties of up to $10,000 per UST per day of violation. Failure to comply with EPA compliance orders subjects a violator to a civil penalty of up to $25,000 for each day noncompliance continues. In some states, like **New Jersey**, the fines are higher. EPA can directly enforce federal requirements and bring civil suits in federal court even when a state has obtained EPA approval for its UST program.

WATER QUALITY STANDARDS

The federal Clean Water Act (CWA) provides the basic national framework for regulating discharges of pollutants into the nation's navigable waters. Through the CWA, federal and state agencies establish standards and goals aimed at protecting the water quality in each state. The U.S. EPA has nationwide authority to implement the CWA. States, however, may apply to EPA for authorization to administer various aspects of CWA's program for permitting discharges into waterways. State water quality laws typically parallel the CWA and are designed to allow a state to qualify for delegation of authority to implement federal requirements. Generally, for a state to obtain EPA authorization to administer all or part of the permitting program, it must match the requirements mandated by the CWA or apply more stringent ones.

The CWA is based on a comprehensive permitting scheme known as the National Pollutant Discharge Elimination System (NPDES). The NPDES program requires a discharger to obtain a permit prior to **discharging** any **pollutant** into **navigable waters** from any **point source**. EPA may issue individual permits to dischargers or issue general permits to groups of dischargers. Each of the terms is broadly defined in the CWA to regulate almost all activities that result in the release of contaminants from a discrete point (e.g., a pipe) and that alter the natural condition of surface water.

NPDES permits establish the level of performance that a discharger must maintain, defined both in terms of the quality and quantity of pollutants. In states that have not received delegation, EPA issues NPDES permits. In states that have received delegation, a state agency issues the permit. Specific states refer to these discharge permits by a state-specific term such as the **New Jersey** Pollutant Discharge Elimination System, or the **New York** State Pollutant Discharge Elimination System.

This regulation defines a **pollutant** as "dredged spoil, solid waste, incinerator residue, sewage, garbage, sewage sludge, munitions, chemical wastes, biological materials, radioactive materials, heat, wrecked or discarded equipment, rock, sand, cellar dirt, and municipal and agricultural waste discharged into water."

As discussed above, EPA may delegate responsibility for administering the NPDES permitting program to individual states that have EPA-approved programs. As of June 1992, 39 states had received EPA approval to conduct state NPDES permit programs. In these 39 states, state water quality agencies issue wastewater discharge permits. In the remaining 11 states, EPA or EPA in cooperation with the state water quality agency issue wastewater discharge permits. Some states have further delegated permitting power to regional or local agencies. For example, in **California** the State Water Resources Control Board (SWRCB) is the EPA-authorized state water quality agency, but NPDES permitting is conducted by nine Regional Water Quality Control Boards located throughout the state.

In order to discharge wastewater from a facility, one must become familiar with the wastewater permitting requirements in the state. Requirements sometimes differ between EPA and the states. Federal regulations mandate minimum nationwide permit application requirements, to which some states have added further requirements. Specifically, federal regulations require the following information be submitted on the permit application at least 180 days (fewer in some states) prior to discharging wastewater from the facility (for new discharges as well as permit renewals):

- The specific pollutants contained in the facility's effluent stream.
- The amount or concentration of these pollutants that are discharged into a receiving water.
- The location of the outfall (the point source).

Other information that individual states may require in their permit application include:

- The name, address, and telephone number of all current owners and operators of the facility, as well as the name of any parent corporation.
- Information concerning any administrative action (e.g., consent orders, notices of violation or other corrective enforcement actions) relating to operation of the facility.
- A description of the nature of the business at the facility.

Wastewater discharge permits must be reviewed and if necessary revised by the facility and the permitting agency.

The CWA describes three broad categories of pollutants:

- **Conventional pollutants** are defined as biochemical oxygen demand, total suspended solids, pH, fecal coliform, and oil and grease. The **best conventional pollutant control technology** (BCT) is required for treatment of conventional pollutants prior to their discharge.
- **Toxic pollutants** include 65 different pollutants identified by EPA. The **best available technology economically achievable** (BAT) is required for treatment of toxic substances prior to their discharge.
- **Nonconventional pollutants** are all other pollutants not classified by EPA and/or your state permitting agency as either toxic or conventional (e.g., thermal discharges). BAT is also required for nonconventional pollutants.

The permitting agency determines the concentration and/or amount of pollutants that may be discharged from your facility on the basis of the established water quality standards for the receiving water and any technology-based effluent limitations or categorical standards or toxic pollutant standards that apply to the discharger.

State **water quality standards** are determined by dividing a state's water bodies into segments, determining the most appropriate use for each segment (e.g., recreational, industrial), setting water quality goals for each segment, and determining what permit conditions to attach to each discharger in order to protect the uses designated for each segment. In addition to water quality standards, permits can incorporate **technology-based effluent limitations** or **categorical standards** established by EPA for specific industries. Categorical standards are established by EPA to provide a national level of pollution control for industrial discharges.

Because it costs less for new facilities to install more advanced technologies than it does to retrofit existing facilities, and because technical alternatives available for new plants may not be suitable for existing sources, Congress requires tighter effluent limits for new sources. EPA has therefore established **national standards of performance** for dozens of specific categories of new sources (electroplaters, textile mills, and manufacturers of organic and inorganic chemicals, among others). These performance standards are more stringent than the technical requirements for conventional, nonconventional, and toxic pollutants, and provide for controls on discharges, both from point sources and into publicly owned treatment works (POTWs). These standards reflect the greatest degree of effluent reduction which EPA determines to be achievable through application of the "**best available demonstrated control technology**, processes, operating methods, or other alternatives, including, where practicable, a standard permitting no discharge of pollutants."

In addition to the national standards of performance, states have the power to provide stricter standards under individual state water quality laws. **New Jersey**, for example, under the New Jersey Water Pollution Control Act, provides its state permitting agency [Department of Environmental Protection (DEP)] with authority to establish New Jersey Pollutant Discharge Elimination System (NJPDES) permit conditions that allow for New Jersey-specific new source performance standards.

In order to identify and control discharges of toxic pollutants, EPA requires states to develop **four lists** describing waters affected by toxic pollutants. By June 4, 1992, states were to have prepared and submitted to EPA an **Individual Control Strategy (ICS)** designed to reduce discharges to toxic pollutants from each point source on the list of point sources preventing improvement of water quality in classified waters (the so-called "C" list). **This ICS is incorporated into the NPDES permit**, the primary control mechanism for reducing point source discharges. Revised permits may therefore be more stringent than previous permits since an ICS must set forth effluent reductions sufficient to attain relevant water quality standards for toxic pollutants. The surface water toxics control program also applies to POTWs. For onsite response actions under Superfund, the ICS will incorporate the decision document prepared for response actions in that area.

There are a number of states that augment EPA's surface water program with their own, often with more restrictive, regulatory efforts. For example, in **Florida**, the Surface Water Improvement and Management Act, enacted in 1987, requires local water districts to prepare a list of water bodies of special significance. The water districts were then to develop surface water improvement and management (SWIM) plans or water bodies. These plans include lists of point and nonpoint dischargers into the priority water bodies, and strategies for restoring and protecting the water bodies.

For owners or operators of an onshore or offshore facility that might discharge harmful quantities of oil into navigable waters, one is required to prepare a Spill Prevention Control and Countermeasure (SPCC) plan. SPCC plans are designed to prevent oil spills, and outline measures that will be taken in the event of a spill. State laws often incorporate SPCC requirements, expanding them to a broad range of facilities. **Pennsylvania's** storage tank law requires that owners of aboveground storage tanks (ASTs) of more than 21,000 gallons prepare spill prevention response plans that are similar in form and content to SPCC Plans. An SPCC plan must be prepared within six months after the facility begins operations, and be fully implemented within one year after it begins operations. The SPCC plan must be amended any time there is a major change in facility design, construction, operation, or maintenance which materially affects the facility's potential to discharge oil. In the event of an oil discharge, a copy of your SPCC plan and supporting documentation must be sent to the EPA office in your region and state wastewater permitting agency.

An SPCC plan must be reviewed and certified by a Registered Professional Engineer. It must include a complete discussion of the facility's compliance with construction and operating guidelines, and other effective spill prevention containment procedures including:

- Prediction of the nature and extent of an oil discharge that would result from equipment failure.
- Description of structures and equipment designed to prevent discharged oil from reaching water.
- Discussion of the facility's compliance with applicable guidelines relating to facility drainage, bulk storage, piping, loading and unloading, oil drillings, facility security, inspections, record-keeping, and personnel training.

It is necessary to review and evaluate the SPCC plan at least every three years and amend it (within six months) as necessary, to include more effective, field-proven, and control technologies if they will significantly reduce the likelihood of a spill. In addition, the SPCC plan must be amended whenever there is a change in facility design, construction, operation, or maintenance that affects the potential for a discharge.

The CWA contains specific provisions regulating the handling of oil and hazardous substances, and establish specific penalties and rules of liability for the unauthorized release of these materials. These provisions focus on reporting unauthorized (i.e., those without a permit) leaks, spills, and discharges to water. These reporting requirements are similar to those under the federal Comprehensive Environmental Response, Compensation, and Liability Act of 1980 (CERCLA). Any persons in charge of an onshore or offshore facility or vessel or who has knowledge of an unauthorized release of oil or a hazardous substance in a reportable quantity (RQ) must immediately report the release as required by the **National Response Center: 1-800-424-8802.** If notifying the National Response Center is not practicable, one may notify the Coast Guard or the On-Scene Coordinator designated by EPA for the geographic area where the discharge has occurred. To encourage prompt and accurate reporting of spills and leaks, information received through this notification process may not be used against the informant in any criminal case (except prosecution for perjury or for giving false statement). However, failure to provide immediate notification may result in a fine and/or imprisonment.

Groundwater--The CWA does not regulate the discharge of pollutants into underground waters that do not flow into (are not hydrostatically connected to) surface water. To cover groundwater that is not connected to surface water, the federal Safe Drinking Water Act (SDWA) addresses selected aspects of groundwater protection. Underground injection wells used for waste disposal may pollute groundwater. Underground injection of hazardous waste is generally being phased out under the federal hazardous waste program's "land ban" provisions. To correct contamination of underground sources of drinking water, SDWA regulates underground injection wells. SDWA requires states to develop programs to prevent contamination of drinking water sources by underground injection, and to establish injection well permit programs.

SDWA also establishes a program to protect aquifers (water-bearing strata of permeable rock or soil) from contamination. The Sole Source Aquifer Program provides that any citizen or group can petition that an aquifer provides at least 50% of the water to a community and should therefore be designated as a sole source aquifer. This special protection prevents any land use that receives **any** federal money from proceeding unless a land use plan is submitted to EPA and approved. The Sole Source Aquifer Program does not affect private (non-federally funded) land use. There are currently approximately 50 sole source aquifers in the U.S.

In addition to the Sole Source Aquifer Program, SDWA requires states to develop programs to protect wellhead areas. "Wellhead protection areas" are defined as surface and subsurface areas surrounding water wells or well fields that supply public water systems. States are to implement their wellhead protection programs within two years of submission. Such wellhead protection programs attempt to reduce contamination in the vicinity of the wellhead by controlling the land uses or by requiring liners under certain land uses.

Stormwater permitting requirements--Traditionally, the NPDES program has focused on reducing pollutants in the discharges of industrial process wastewater and municipal sewage. However, the NPDES program has been expanded to regulate a more diffuse source of water pollution: storm water discharges. New federal regulations require industrial facilities that discharge storm water, as well as municipalities with a population of 100,000 or more, to control discharges of storm water and surface runoff and drainage related to storms or snow melt that contain such pollutants as oil, grease, and pesticide residue, and flow into storm water drains discharging into the nation's rivers, lakes, and streams. These federal regulations apply only to discharges of storm water from point sources; nonpoint sources of storm water are not covered under the new regulatory program.

If a facility is engaged in one of the following "industrial" activities, it must obtain an NPDES storm water permit:

- The manufacturing of certain lumber and wood products, paper, chemicals, petroleum refining, leather tanning and finishing, stone, clay, concrete and glass activities, as well as tobacco, textile, furniture, and printing activities. In short, almost all manufacturing activities may be subject to the new storm water requirements. Facilities involved in these activities are identified in the regulations by Standard Industrial Classification (SIC) code.
- Construction activities, including clearing, grading, and excavation. (EPA's regulations provided that operations that result in disturbance of less than five acres of total land area and that are not part of a larger common plan of development or sale are

exempt from regulation.) This position was recently invalidated in *NRDC v. EPA* and is being reviewed by EPA.

- Operation of a hazardous waste treatment, storage, or disposal (TSD) facility.
- Operation of landfills.
- Operation of a steam electric power generating facility.

The federal regulations establish three distinct types of storm water permits: **individual, general,** and **group**. Applying for an individual permit requires submission of **EPA   Form 1   and Form 2F**. These forms require dischargers to provide a comprehensive set of information, including:

- A site map.
- An estimate of impervious areas.
- The identification of significant materials treated or stored on site together with associated materials management and disposal practices.
- The location and description of existing structural and nonstructural controls to reduce pollutants in storm water runoff.
- A certification that all storm water outfalls have been evaluated for any unpermitted nonstorm water discharges.
- Any existing information regarding significant leaks or spills of toxic or hazardous pollutants within the three-year period prior to the permit application.
- Sampling reports of the facility's storm water taken during "storm events."

The general permit process stipulates regulatory conditions in advance, which dischargers must then meet. This approach allows regulatory agencies to issue a single permit regulating all dischargers of a particular type (which is a more cost-effective approach for permitting than issuing individual permits). If the facility is regulated under a general permit, it is not required to obtain an individual permit.

The general permit is the procedure most likely to be employed in those states which have general permitting authority. Under the CWA, EPA cannot issue general permits in those states that are authorized to administer the base NPDES program (of the 39 states and territories that have baseline NPDES permitting authority, 11 do not have general permitting authority). Therefore, if the facility is located in one of the 11 states with baseline NPDES permitting authority but without general permitting authority, it has no option except to obtain an individual storm water permit.

In order to apply for a general permit, the facility must submit a **Notice of Intent (NOI)**. The federal regulations provide for minimum NOI requirements. At a minimum, however, the NOI will require the legal name and address of the facility's owner and operator, the facility name and address, the type of facility or discharges, and the receiving stream.

The group permitting option provides facilities involved in the same or similar types of operations with the opportunity to file a single two-part permit application. Part 1 of these applications was due on September 30, 1991. Facilities that missed this deadline are precluded from using the group permitting option and must obtain an individual permit or comply with general permitting requirements (if available).

For purposes of data collection, NPDES permits are required for discharges of storm water runoff from:

- Municipal separate storm sewer systems located in incorporated areas with a population of 250,000 or more ("large systems").
- Municipal separate storm sewer systems located in incorporated areas with a population of 100,000 or more but less than 250,000 ("medium systems").

Under the federal program, EPA may issue either one system-wide permit that covers all discharges from storm sewers within a large or medium system or distinct permits for categories within a system. Federal municipal storm water permits consist of two parts:

- Part 1 includes source identification information, discharge characterizations, a description of the legal authority to control discharges, and existing management programs to control pollutants and identify illicit connections to the sewer system.
- Part 2 requires additional data--a proposed management program, an assessment of storm water controls, and fiscal analysis.

Publicly-Owned Treatment Works

The federal Clean Water Act (CWA) regulates all direct discharges into
navigable waters. All such discharges, including treated sewage discharged by sewer systems and sanitation agencies (publicly-owned
treatment works--POTWs), require permits issued under the **National
Pollutant Discharge Elimination System (NPDES)**. These NPDES
permits restrict the quantity and quality of a POTW's discharge. To
meet these permit requirements, individual POTWs place a range of
detailed restrictions on their own industrial users, subject to selective
federal, state, and regional oversight of such "pretreatment programs."

EPA has overall responsibility for the CWA, including development
of selected national pretreatment standards for pollutant discharges into
POTWs. States that are authorized by EPA to administer the federal
pretreatment requirements, will administer this program through its state
water quality agency. POTWs obtain permits for their discharges from
EPA or the state water quality agency by following essentially the same
process required of direct industrial dischargers. POTWs then regulate
industry discharges into their sewer system to be sure of meeting their
own discharge limits. This regulation by POTWs of "indirect dischargers" to the treatment works results in environmental regulation that is
uniquely tailored to local conditions. The federal government, however,
has not completely abdicated its regulatory role. EPA sets general
pretreatment standards and national categorical pretreatment standards for
different industry categories, defining necessary effluent reductions.
EPA or authorized states may also place more stringent requirements on
POTWs if necessary to meet water quality standards for the receiving
waters. To meet these standards, POTWs may in turn regulate indirect
dischargers.

Industrial users of POTWs are not required to obtain NPDES permits
and are ordinarily not required to obtain a state-issued discharge permit.
Instead, their individual POTW imposes restrictions--"pretreatment
standards"--on all its industrial users. Typically, a POTW discharge
permit is required from the POTW, although POTWs differ in their
terminologies.

Regulation of industrial discharges into POTWs serves four
objectives:

- Prevents introduction of pollutants that would prevent the POTW from complying with its own NPDES permit and discharge limitations.
- Prevents introduction of pollutants into POTWs which would interfere with equipment or operations, or endanger POTW personnel.
- Prevents introduction of pollutants that would "pass through" (i.e., would not be treated adequately before discharge to the receiving water body) or would be incompatible with the POTW.
- Improves opportunities to recycle and reclaim municipal and industrial wastes and sludges.

In order to meet these objectives, each POTW imposes discharge limitations on some or all of its industrial users; POTWs also enforce relevant federal or state pretreatment standards.

EPA's general pretreatment regulations provide for the following controls over POTW dischargers:

- POTW evaluation of "significant industrial users" to determine if additional controls are required to prevent "slug" discharges.
- Inspection of the discharger by the POTW, at least annually.
- Submission of sampling and enforcement response plans by dischargers without categorical standards.
- Required reporting of any discharges to a POTW of any RCRA hazardous wastes not subject to self-monitoring requirements.

EPA has established a variety of national pretreatment standards that all facilities in a particular industrial category must achieve before discharging their effluent to a POTW. POTW's pretreatment programs enforce these national standards, as well as their own additional local requirements. An industrial user in any of the designated categories, must conform to national **categorical pretreatment standards** for existing and new sources (or seek a variance from these standards). These pretreatment standards specify quantities and concentrations of pollutants that may be discharged into POTWs and require dischargers to use **"best available technology economically achievable" (BAT)** prior to discharging.

The CWA requires POTWs with a design flow **over five million gallons per day** (mgd) which receive from industrial users pollutants that pass through or interfere with the operation of the POTW to establish industrial pretreatment programs to control industrial discharges into their sanitary sewer systems. POTWs with a design flow of five mgd or less may also be required to establish pretreatment programs if the nature or volume of industrial discharges to the POTW, upsets, violations of POTW effluent limitations, contamination of municipal sludge or other circumstances pose a danger of interference or pass through. EPA or the state water quality agency enforces this federal program statewide. Other POTWs may also choose to establish local pretreatment programs.

Local standards are often more stringent than the national standards, or cover industry groups not covered by the EPA standards. These local requirements are often imposed in response to unique concentrations of point or nonpoint discharges into the receiving waters, or to provide additional protection to receiving waters. In many industrial areas, pretreatment programs provide important elements of the area's overall toxics management effort.

POTWs with approved pretreatment programs are required to issue permits or equivalent individual control mechanisms, and to inspect and sample each significant industrial user at least once per year. Other POTWs issue permits on their own, separate from the federal requirements.

Pretreatment controls may include individual industrial use permits (as discussed above), permits-by-rule, or the prohibition of certain discharges to the sewers. Permits issued by the POTW or permit-by-rule can establish flow restrictions, pollutant concentrations, or other pretreatment thresholds and set forth specific monitoring and reporting requirements.

POTWs with pretreatment programs are required to implement an enforcement response plan with procedures to investigate and respond to industrial user noncompliance. Many POTWs conduct regular inspections of their industrial sewer users.

RCRA excludes solid or dissolved material in domestic sewage (i.e., POTWs serving residences) from its definition of "hazardous wastes." As a result, if you operate an industrial facility that is permitted by your POTW to discharge "hazardous" liquids into sewers, you are not subject to RCRA manifest requirements even though these discharges would otherwise be considered hazardous waste (this is known as the **"domestic**

sewage exclusion"). However, you **must still comply** with any other applicable hazardous waste management requirements (e.g., obtain an EPA Identification Number and comply with both RCRA and state hazardous waste law standards for treatment and storage).

As noted, such discharges must also meet the POTW's specific pretreatment requirements. POTWs which receive pretreated hazardous liquids along with domestic sewage are not considered to have received hazardous waste, and therefore are not required to meet any of the treatment, storage, and disposal requirements mandated by RCRA.

An industrial user, is required to submit written notice to POTWs, EPA, and state hazardous waste authorities of any discharge that would be considered hazardous if not made under the domestic sewage exclusion. These notifications must contain the name and EPA hazardous waste number of each hazardous waste, and the type of discharge (continuous, batch, or other). If the discharge is more than 100 kilograms per month, the notification must also include: identification of the hazardous constituents; estimates of the mass and concentrations of these constituents discharged monthly; and estimates of the mass of discharges in the coming 12 months. Notifications are due within 180 days after the discharge. Notifications need only be filed once with EPA and the state for each hazardous waste being discharged, but must provide the POTW with a follow-up notification whenever there is a change in the discharge. Federal regulations provide for specific exemptions from this reporting requirement.

AIR QUALITY STANDARDS

The federal government and individual states regulate literally thousands of commercial, industrial, and other activities to clean up the nation's air. Strong amendments were made to early versions of the **Clean Air Act (CAA)** in 1970, again in 1977, and most recently in 1990. The law requires the EPA to set health-based national standards for several different pollutants.

Within the framework provided, air quality regulatory compliance varies greatly from one state to another, and even within a state. While all areas are responsible for meeting the same basic national air quality standards, facilities in certain locations may have to comply with more stringent requirements if their air quality is significantly worse (or even significantly better) than the national standards. State and local agencies are given a great deal of discretion in implementing the nation's air quality laws, as well as additional state and local requirements.

In nearly every state, three layers of governmental agencies are involved in regulating air pollution. EPA sets national standards and oversees selected state and local agency actions. The **state air quality agency** continually revises each **state implementation plan (SIP)**, enforces federal standards, controls auto emissions, and sets guidelines for local air pollution control agencies (if any). Many states set and enforce standards for some pollutants which are more stringent than the national standards. Actual permitting is typically performed by **local** agencies--typically state agency field offices, or in some states, regional or county air quality control authorities. These agencies prepare local plans, establish local rules, issue permits, and enforce the relevant requirements through inspections and other procedures.

These programs distinguish in important ways between **existing sources** of air pollution, and **new sources**. New source requirements are often more stringent, on the theory that pollution control technology can be built into new plants from the ground up, and that new sources add to existing air pollution levels. This focus on stricter technology controls for new sources applies to both stationary and mobile sources of air pollution.

Conventional and toxic pollutants are addressed differently. While conventional air pollutants are present in greater quantities, they are also less dangerous, pound for pound, than are air toxics. Both kinds of air pollutants can harm public health.

The 1990 CAA amendments require more emissions inventories from smaller sources than ever before, and also target these polluters for accelerated emissions reductions depending on their area's non-attainment status for the different pollutants--extreme, severe, serious or moderate. Many of these sources will fall under a new federal air quality permit program that began in the mid-1990s. The 1990 CAA amendments also tightened transportation control measures, and imposed the stiffest civil and criminal penalties yet seen for environmental crimes.

Existing sources of air pollutants are regulated according to their emission, sources, and quantities. Provisions vary based on location. Some local agencies often impose additional requirements.

Six "criteria" or "conventional" pollutants are regulated under CAA:

- Carbon monoxide (CO).
- Lead.
- Nitrogen oxides (NO_x).
- Ozone (smog).
- Particulates (10 microns or less in size, or "PM-10").
- Sulfur oxides (SO_x).

Several of the conventional air pollutants are familiar as the building blocks of urban smog. Others pose additional problems for health and visibility.

Some facilities emit smog precursors even though they don't have smokestacks. For example, large bakeries in many areas must install control devices on their ovens to reduce the amount of ethanol--formed by yeast when bread dough rises--which they emit into the air. Ethanol is a **Volatile Organic Compound (VOC)**. Many VOCs are also **Reactive Organic Gases (ROGs)** which can contribute to smog formation. Thus, conventional air pollutants are not just as the six listed "criteria" pollutants, but also as a broader array of pollutants that interact with (or form) these six gases to produce smog, haze, and acid rain, for example.

EPA has set **National Ambient Air Quality Standards (NAAQS)** for the six criteria pollutants, defining air quality targets. When a region or air basin has not achieved these standards in its background (or "ambient") air quality for one or more of these six pollutants, that region is said to be a "**non-attainment area**." Air quality agencies in these areas must impose additional and more stringent or innovative measures to improve air quality toward NAAQS compliance, especially under the requirements of the 1990 CAA amendments.

PSD (Prevention of Significant Deterioration) standards focus on maintaining air quality in especially pristine areas. PSD standards apply to SO_x, particulates, and NO_x, but not to ozone or smog itself. These standards were mandated by Congress in the 1977 CAA amendments to reflect the fact that some areas with pristine air quality should be afforded an extra measure of protection. PSD standards fall into three categories: Class I, Class II, and Class III.

PSD Class I areas automatically include:

- International Parks.
- National Wilderness Areas and National Memorial Parks over 5,000 acres in size.
- National Parks over 6,000 acres in size.

These standards allow very little additional emission of certain air pollutants from certain kinds of large new (or modified) stationary sources (such as power plants) proposed in or near such areas. As a result, it is very difficult to obtain approval for such large new sources of air contaminants in PSD Class I areas, or nearby.

PSD Class II and Class III standards apply in areas designated primarily by the states; usually scenic rivers and some wilderness or agricultural areas are Class II, whereas urban areas are usually Class III. The air in Class II and Class III areas is generally cleaner than the air in areas that are barely in attainment with NAAQS.

Most states have had their own air emission permit requirements for many years. The 1990 CAA amendments introduced for the first time **a new nationwide air quality permit program**, to take effect by the middle of the decade. In most areas, federal air quality permits will be issued by today's state or local permitting agencies; however, these permits will contain many new features mandated by the 1990 CAA amendments.

Many **stationary sources of air pollutants** need a permit issued by the appropriate air quality agency; details on which facilities need a permit vary considerably from one area to another. Most air quality agencies also require that facilities obtain a permit to **install and use any air pollution control device**.

The thresholds for emissions that require permits are being lowered in order to bring many small sources under air quality rules and regulations that previously applied only to large sources of air pollutants.

The following are categories of equipment and processes for which permits are often required by air quality agencies in non-attainment areas:

- Combustion.
  - -- Boilers, heaters, furnaces, and turbines.

-- Project-associated combustion (cogeneration, resource recovery).
-- Internal combustion engines.
- Open-air processes and operations.
- Metal melting.
- Incineration.
- Material processing (and equipment).
  -- Coating.
  -- Cleaning.
  -- Food processing.
  -- Drying.
  -- Curing.
  -- Printing.
- Production of chemicals, petrochemicals, and petroleum products.
- Handling of grains, feed, and food products.
- Handling of minerals.
- Storage and handling of natural organic materials.

The term "**control technology**" simply refers to any piece of equipment added to another piece of equipment or machine solely for the purpose of controlling (i.e., reducing) emissions of air contaminants. Some control technologies, such as the catalytic converters used on automobiles, are built into equipment right from the design stage; others are added after a machine or industrial process has been operating. Other typical control technologies address the type of equipment or methods that must be used to control emissions.

Air districts often require that boilers and steam generators be designed or retrofitted to emit low levels of NO_x. The most common control technologies for boilers are low-NO_x burners, flue-gas recirculation, selective catalytic reduction, and selective noncatalytic reduction.

Air quality agencies in non-attainment areas typically require a two- or three-step process for permits for equipment and processes that emit air contaminants. The first step usually requires sources to obtain a **permit to construct or install** a piece of equipment, or to build an entire facility. Sources at this stage must obtain a "Permit to Install," "Authority to Construct," or "Permit to Construct." These different terms are used interchangeably in the various state air pollution control programs. In the second step, following construction, a source must obtain a **"Permit to Operate"** or **"Authority to Operate."** Occasionally, the permit to construct may serve as a temporary operating permit, whose duration is limited to one or two years. The third step involves **periodic permit renewals**.

Most air districts assess annual permit fees to cover the costs of their review of facilities and permit applications, and their enforcement activities. Fees are assessed on a per-equipment basis. They vary with the size and complexity of the equipment. Typically, air districts send permit holders invoices for their air quality fees, either on the anniversary of the date their permit took effect or on a date triggered by that air district's rules and regulations.

At a minimum, permits under the federal permit program must require:

- Enforceable emissions limitations and standards.
- A compliance schedule for each individual facility.
- Progress reports (including air monitoring results) to be submitted every six months.
- Reports of any deviations from permit requirements (to be submitted immediately).
- Facility recordkeeping.

State and local air quality agencies develop administrative and source- or industry-specific rules. The complexity and comprehensiveness of these rules vary a great deal. Generally, agencies located in areas with poor air quality tend to be more comprehensive in their rulemaking, as they struggle to meet NAAQS. Air quality agencies write two kinds of rules in both attainment and non-attainment areas: administrative rules and source-specific rules. **Administrative rules** govern the interaction and paperwork between the air quality agency and the operator of a controlled source of air pollutants. Administrative rules describe who needs a permit, how the permit should be applied for and obtained, how long it will be valid, and where it should be posted. These rules also outline the agencies' responsibilities toward source operators in their jurisdictions, including how many days they have to respond to petitions by sources, when they can inspect a source and issue penalties, and grievance procedures. **Source-specific rules** cover certain industrial or commercial processes and equipment. These rules are usually written narrowly enough to target emissions of one kind of pollutant from one set of similar processes or equipment. They often include a mix of equipment standards, materials standards, and procedures (or "process standards") describing how equipment or a procedure can be used to minimize emissions. A rule typically defines how many pounds or tons of a certain pollutant can be emitted, per day or per manufacturing process. Alternatively, the rule may specify that a business must use a specified process or products which do not contain more than specified amounts of pollution-causing chemicals. Rules may become more stringent over time.

All air districts define **limits** for various emissions as required to meet SIP goals and the targets set out in regional or local air quality plans. A limit is the maximum emission allowed from a particular kind of source. It is usually developed balancing health and other air quality concerns with technological and economic feasibility. In heavily polluted non-attainment areas like southern **California**, limits tend to be more stringent and exist for a greater number of pollutants. These limits are typically set out in numerous rules, each covering a particular industry category, pollutant, or type of activity.

Air districts often go beyond merely mandating daily or annual emissions totals to dictate **performance standards** for equipment used in the processes that generate air emissions. For example, a number of agencies now require businesses that offer spray coating services to use an HVLP spray gun. By using less paint, the operators may save money over the life of the equipment. As new regulations are passed, some businesses occasionally find that previously unregulated equipment now falls under a new inspection process. A permit may even specify in detail what equipment should be used.

Air quality laws provide a number of "technology forcing" mandates to agencies to devise technical performance standards for emission controls. The laws have coined a number of terms for these various sets of standards. Air districts may be required to apply **Reasonably Available Control Technology (RACT)** in some cases, and **Best Available Retrofit Control Technology (BARCT)** in others. Occasionally, state air agencies or air districts may have to apply even more stringent **Best Available Control Technology (BACT)** or mandate use of the **Lowest Achievable Emissions Rate (LAER)**. The principal differences among these terms have to do with the level of air pollution control technology required, the air quality designation of the jurisdiction for which the control measure is chosen (moderate, serious, severe non-attainment), and the "impact" of controls--a combination of environmental, economic, and energy considerations. Thus, some control measures are more purely "technology forcing" (requiring firms to use state-of-the-art, expensive control equipment), while others balance gains in air quality against equipment costs. Energy requirements are also factored in because most control equipment adds further energy demands to industrial or heat-producing processes. Extra energy, of course, often means more emissions from fossil fuel combustion.

Sources may obtain variances from certain air district rules and regulations. Variances are temporary exceptions from administrative laws and rules intended to give firms the necessary "breathing space" to continue operating while they take the steps needed to meet specified air pollution control requirements. **Variances are granted only for permits to use or operate a source, and for rule compliance actions. Variances cannot be granted for permits to build, erect, alter, or replace a source.** Persons seeking a variance usually must petition their air quality agency. Often, these petitions are reviewed by a Hearing Board made up of appointed citizens. This Hearing Board has authority to deny or revoke permits, approve or reject new or proposed rules, and grant variances from district rules. Each Hearing Board has its own quasi-judicial administrative procedures (i.e., courtroom-like) for conducting reviews and hearings.

Federal rules define "**new**" or "**modified**" sources as those facilities that **either** add further air pollutant emissions to a site, **or** undergo construction costs equal to 50% or more of the costs of building comparable facilities from the ground up. Thus, it takes relatively little modification for a facility covered by these rules to come under federal new source requirements. Generally, the term "new sources" is used to refer to both new and modified sources. New sources of air pollutants typically face more stringent controls than do existing sources, on the presumption that it is easier to incorporate new control technologies into the design stage of a new facility than to add them later to an existing facility. Details vary greatly depending on the local area, the type of facility, and the size of the facility.

New sources may have to go through a permitting process called **New Source Review (NSR).** In this process a state or local air quality agency determines whether a proposed or modified facility can begin operations, and under what conditions. Since state and local requirements are often more stringent than federal standards, you will need to check with your local air quality agency to determine your requirements and status. The NSR process generally applies to major new sources. Most smaller new sources can proceed through the normal permitting procedures without triggering the NSR process. EPA generally defines a major source as one that emits 100 tons or more per year of any air contaminant or mix of air contaminants. Where local or state rules do not specify their own definition of "major sources," sources defined as "Major Polluting Facilities," or Major Modifications of Major Polluting Facilities must meet federal requirements. However, many state and local agencies apply lower thresholds. Different emission thresholds trigger NSR in different non-attainment areas, based on how far the areas exceed NAAQS. The most stringent federal NSR requirements apply to new sources that emit 10 tons or more per year of any air contaminant. This threshold applies only in southern **California**'s South Coast Air Basin. NSR thresholds for other non-attainment areas around the country range from just over 10 tons to 100 tons per year of any pollutant. Some air districts covering non-attainment areas enforce local rules more stringent than federal regulations. For example, most air districts covering non-attainment areas in EPA's Region 9 have NSR thresholds for some pollutants or processes that are less than the 10 ton federal threshold.

The CAA requires EPA to develop NSPSs for certain categories of new sources located anywhere in the country. These standards require the best technologically feasible system of emissions reduction for each category, provided those controls are economically viable. A typical NSPS sets out the date upon which it becomes effective, the kind of process or equipment covered, and requirements that must be observed by the new source operator. For example, the NSPS for new bulk gasoline storage tanks describes storage tank and tank roof designs that all your new gasoline tanks larger than 40,000 gallons (and containing certain liquids) must meet. The basic idea of an offset system is to allow one source to "compensate" for its own emissions--especially its new or increased emissions--by reducing the pollution levels of other sources in its same airshed. This compensation could involve buying some businesses that are heavy polluters, and closing them while retaining their right (under earlier permits) to emit pollutants. It could also mean installing control devices on other sources that result in emissions reductions at least as great as your increase. By letting sources "shop" for offsets, these programs are intended to reduce the costs of compliance. Some offset programs require new or modified sources in polluted areas to reduce emissions by producing offsets larger than the emissions they will create. For example, under the 1990 CAA amendments new sources in extremely polluted areas such as Houston or Baton Rouge will be required to eliminate 1.5 tons of pollutants for each ton they expect to emit. This system ensures that introduction of a new source into the area will actually result in a **net reduction** in emissions, instead of an increase or even no net change.

Many state environmental quality acts require new projects to be evaluated through an **Environmental Impact Report (EIR)** or similar environmental assessment process if the project may present significant environmental effects. EIRs can be required even if only one environmental medium (e.g., air) would be affected by a proposed project. Completion of an EIR is not a substitute for obtaining an air quality permit; the EIR process operates **in addition** to required state agency or local air district permit and rule requirements, including NSR requirements.

"**Market incentive**" tools in the battle against dirty air include emissions credits, marketable permits or emissions trading systems. These systems allow companies to meet emission reductions as they see fit--either by installing air pollution control equipment or by purchasing "credits or permits to pollute" from other companies. In an emissions trading program, companies that exceed their emissions targets or emit less than their allowance may sell their surplus "emission credits" in an open market. These programs make broader use of the ideas and mechanisms developed for so-called offset programs. Title IV of the 1990 CAA amendments sets out a massive new Acid Rain Control Program using a marketable allowance system. The aim is to reduce and control SO_2 and NO_x emissions from fossil fuel-fired electric utility plants. SO_2 and NO_x emissions from power plants react with atmospheric gases to form acid rain, which can be deposited hundreds of miles away from the original source of the emissions. This program covers utilities except those with $\leq$ 25 MW capacities, qualifying cogeneration units, and small power producers. Under Title IV, EPA is required to allocate tradeable SO_2 emissions allowances among major sources. Each allowance is equal to one ton of SO_2. EPA will issue to each particular source a number of allowances equal to the annual tonnage emission limitation set out in its permit. If a source reduces its emissions below

---

TABLE 1

VOC CALCULATION FOR METAL COATING OPERATION

A metal coating operation applied 0.53 gallons of coating in one day. The make-up of this coating is as follows:

> amount of VOC in the coating
> > (less water and exempt compounds)   3.5   lbs/gal
> amount of VOC in the reducer (hardener)
> > (less water and exempt compounds)   6.625 lbs/gal
>
> mix ratio of coating to reducer   2:1

To determine the VOC value of the mixture:

Multiply the amounts of coating and reducer by their proportions in the mix:

> coating 3.5   x   2   =   7   lbs/gal
> reducer 6.625   x   1   =   6.625   lbs/gal
> ───
> sum   13.625 lbs/gal

Add the number of parts in the mix to determine their total:

> 2 + 1   =   3

Divide the sum of the amounts of coating and reducer by the number of parts in the mixture:

> 13.625 ÷   3 =   4.54   lbs/gal

The VOC value of the mixture is 4.54 pounds per gallon.

To determine the VOC applied that day:

Multiply the VOC value of the mixture by the amount of mixture used:

> 4.54 lbs/gal x 0.53 gal =   2.4   lbs

The total daily VOC emission is 2.4 pounds.

its allocation, it may sell its extra allowances to other sources. New sources will need to purchase allowances for their SO_2 emissions from existing sources. Existing sources may also choose to do so if these purchases cost less than reducing their own emissions directly. An example of a simplified calculation sheet for recording the VOC (called VOM in **Illinois**) emitted in a typical metal coating operation is given on the previous page (Table 1).

Most of the air quality regulations discussed thus far concern stationary sources of pollutants. These sources have traditionally borne the brunt of air quality control measures because, unlike individual mobile sources, they are relatively large and identifiable, and they don't move from one jurisdiction to another on a daily basis. Mobile sources cannot be ignored, however, because they represent a very large cause of air pollution. In the urban areas of **California** that are still out of NAAQS attainment for ozone and smog, for example, on-road motor vehicles are responsible for 37% of ROG emissions from all sources, 51% of all NO_x emissions, and 69% of all CO emissions. There are two major ways to control emissions from mobile sources. The first is to change the technologies we use for transportation; the second is to control the indirect sources associated with automobiles. Changes in transportation technology involve improvements in fuel efficiency and emission controls on motor vehicles, or the use of "cleaner" transportation energy such as electricity. Several states, including **Colorado, Massachusetts,** and **California,** have taken steps to mandate cleaner fuels and technologies. The **Federal Motor Vehicle Control Program (FMVCP)** consists of a series of regulations applicable to mobile sources such as automobiles, trucks, aircraft, and other mobile equipment that emit air contaminants. Mobile source emissions are covered by emissions standards (defining how much of any given pollutant may be emitted), equipment standards (e.g., catalytic converters), and materials standards (composition of motor fuels). Some states (especially those with particularly poor air quality in metropolitan areas) implement emission testing programs. Metropolitan areas with more than 200,000 people were required by EPA to incorporate smog testing into their SIPs.

The 1990 CAA amendments mandated a number of new mobile source programs. For example, **California**'s existing clean fuels requirements have been redefined as a pilot program for the rest of the country. In 1988 the city of Los Angeles decided it would purchase

15,000 "clean-fuel" vehicles for use by various city agencies. By 1996, 150,000 clean-fuel vehicles must be made available for sale in California; by 1999, this figure will increase to 300,000 vehicles per year. **Colorado** offers a $200 incentive to alternative fueled vehicle owners who certify their cars with the state.

In 1992, EPA proposed rules regarding on-board emissions diagnostic systems that may replace some existing inspection and maintenance tests. In 1994, all new transit buses are required to use clean fuels like methanol, natural gas or electricity. By 1995, cleaner, reformulated gasoline must be produced for sale in the nine areas with most severe ozone pollution across the country. Also by 1995, all new vehicles will be required to have on-board vapor recovery canisters.

Concern over emissions from mobile sources has also spawned some controls on stationary sources associated with automobiles. For example, gas stations in non-attainment areas must equip their pump nozzles with vapor lock and recovery systems that reduce VOC emissions from filling operations.

CAA and state air laws all give air quality agencies authority to enforce their requirements. Most state laws empower air districts to send inspectors into any site, facility, building, or equipment at any time to determine compliance with air pollution control laws. In **New Jersey** this also includes the right to test or sample any materials at the facility, to sketch or photograph any portion of the facility, to copy or photograph any document or records necessary to determine compliance, or to interview any employees or representatives of the owner, operator, or registrant. The New Jersey Department of Environmental Protection (DEP) may enter and inspect premises without any prior warning. DEP need only present appropriate credentials and follow applicable safety procedures. Owners or operators, or any of their employees, must assist with all aspects of any inspection. Assistance includes making available sampling equipment and sampling facilities for DEPE to determine the nature and quantity of any air contaminant emitted. If a business refuses access to an air quality inspector, that inspector may be able to obtain a warrant to enter and inspect that facility.

Inspection schedules vary depending on the emission levels and complexity of the equipment being inspected and on the compliance history of each firm. Inspections are also made in response to complaints. Air districts often conduct unannounced inspections.

An air quality agency may suspend or revoke a permit for any of the following reasons:

- Failure or refusal to provide information, analyses, plans, or specifications requested.
- Violations of rules, orders, or regulations.
- Failure to correct any condition when required by the air district.
- Fraud or deceit in obtaining the permit.

Generally, the easiest reinstatements are available when information, analyses, plans, or specifications are provided by the source operator. Instances of fraud or violations having to do with equipment changes may require re-permitting.

Air districts generally prefer to encourage voluntary compliance. Therefore, districts use outreach and education rather than fines. Consequently, many air pollution inspectors are encouraged to be as informative and helpful as possible while they are on-site. In the event of a procedural or recordkeeping violation, most air districts issue a **Notice to Comply** or a **Notification of Violation**. Every inspector who issues either of these notices is required to follow up to ensure that compliance has indeed been achieved. The first question you should ask an inspector after receiving a Notification of Violation or a Notice to Comply is: "How long do I have to correct this violation?" Violations are usually required to be corrected within a set time--typically from 14 to 30 days. **Texas** inspectors allow 30 days for compliance before issuing an Order for Compliance. Once issued, an Order for Compliance requires compliance within 180 days after receiving a Notification of Violation. If you do not comply by the date given on the Notice to Comply, a district may issue a **Notice of Violation**. This typically means that you are in violation of either district rules or state air pollution laws. If nothing is done to correct the violation, the problem may be handled as a civil case with potentially substantial fines, through a nonjudicial settlement process, or as a criminal misdemeanor.

To the extent possible, businesses and other air pollution sources should **take immediate action to correct the violation**. In some states, many violations can be resolved by mail through administrative provisions similar to SCAQMD's Mutual Settlement Agreement Program.

EPA received several new or strengthened enforcement powers under the 1990 CAA amendments. Under the new amendments, EPA may issue **administrative compliance order** following a Notice of Violation, and can set out a compliance schedule for up to one year (instead of just 30 days, as was previously the case). Knowing violations of a compliance order are now made felonies under the 1990 CAA amendments. In addition, EPA now has authority to issue administrative penalty orders without going to court (where EPA must rely on the U.S. Department of Justice). The practical effect is that only the largest cases go to court; the others proceed through an administrative law judge. EPA can also issue field citations ("clean air traffic tickets") in amounts up to $5000 per day of violation.

The harshest penalty is 15 years' imprisonment for "knowing" releases of HAPs which may cause imminent danger of death or serious bodily injury. For lesser but "knowing" violations of the act, Congress increased former misdemeanor penalties to felonies, leading to up to five years in prison. Obstructing any of the act's regulatory provisions may be punished by up to two years in jail (as opposed to six months under prior law). Failure to pay fees may result in up to one year in jail. Additionally, Congress doubled all sanctions for repeat offenders. Responsible corporate officials must certify the accuracy of their company's compliance status report. Congress has recognized that employees who violate the act in the normal course of their employment, and at their employer's direction, should not be subject to criminal prosecution. Moreover, since audit, compliance, and monitoring programs are seen as desirable, the CAA amendments do not allow the government to resort to criminal sanctions when such programs are carried out in good faith.

Conduct which **contravenes** the CAA's regulatory scheme includes:

- Knowing violation of a SIP requirement continued more than 30 days after the violator receives a Notice of Violation.
- Violation of an administrative compliance order or administrative penalty requirement.
- Knowing violations of any aspects of CAA, including:
  -- NSPSs.
  -- Permit requirements.
  -- Recordkeeping, inspection, and monitoring requirements.
  -- Emergency orders.

-- NESHAPs.
-- Acid rain or stratospheric ozone controls.
● Knowing failure to pay fees.

The government need only show that the accused was aware that he or she was committing the particular action, not that he or she knew the action was unlawful.

Air Toxics: Emissions

Until recently, the federal Clean Air Act (CAA) has done relatively little to regulate the wide variety of air toxics. The CAA amendments of 1970 required EPA to establish and maintain a list of air toxics which are named as **hazardous air pollutants (HAPs)**, and to set emission standards for several specific sources of such pollutants. The federal regulatory process over the following 20 years was very cumbersome, however, essentially requiring EPA to amass conclusive evidence that a particular air toxic was indeed "hazardous" before regulation could take hold. As a result, in two decades only eight HAPs were designated, and their emissions subjected to **National Emission Standards for Hazardous Air Pollutants (NESHAPs)**.

Once promulgated, NESHAPs were quite strict. No new source of a designated air toxic compound could be built, and no existing source could be modified, except in compliance with the appropriate standard. But the program's overall scale paled in comparison with the scope of the nation's air toxics problem. A few states, like **New Jersey**, developed extensive air toxics regulatory programs of their own. Then in 1990 Congress decided to act decisively.

The 1990 CAA amendments, represent an important new national effort in air pollution control. Air toxics requirements are distinct from the new federal programs described; many sources are required to comply with rules adopted for both "conventional" and toxic air contaminants. In most states, a single air quality agency administers both conventional and toxic air contaminant programs.

The 1990 CAA amendments made major changes in the federal approach to regulating toxic air pollutants. In these amendments, Congress listed 189 substances as HAPs, while giving EPA authority to review the HAP list periodically and add or delete substances. EPA must develop standards for **Maximum Achievable Control Technology**

(MACT) to control air toxics emissions from thousands of sources across the country.

Tens of thousands of sources of air toxics nationwide--existing as well as new, small as well as large--are now required to apply specific technologies to control their ongoing emissions of hazardous air pollutants. Thousands of facilities also have to obtain and comply with air quality permits under a new national permitting program. These new federal regulatory requirements will become increasingly significant during the 1990s as EPA develops its detailed regulations and guidelines. Existing state air toxics control programs will have to be integrated into these new national regulatory mandates.

Through 1990, EPA had set NESHAPs for only eight air toxics:

- Asbestos.
- Benzene.
- Beryllium.
- Coke oven emissions.
- Inorganic arsenic.
- Mercury.
- Radionuclides.
- Vinyl chloride.

In general, each NESHAPs applies to an air toxic emitted in a specific context.

The reader should review Part 61 of 40 Code of Federal Regulations (CFR) to ensure that the facility is in compliance with NESHAPs standards. These requirements cover many pages of federal regulations, so you should refer to specific CFR parts applicable to your operation.

In the 1990 CAA amendments, Congress listed 189 additional HAPs to be controlled by EPA (which can add or subtract from this list). EPA is to develop air toxics emission standards for all of these HAPs over the next several years. EPA has developed a lengthy list of nearly 750 different categories of sources that emit HAPs. This list was published by EPA in the *Federal Register* of June 21, 1991. The list of 750 source categories is divided into the following 20 general industry groups:

- Agricultural chemicals: production and use.
- Chemical production and use activities.
- Fibers: production and use.

- Food and agriculture.
- Fuel combustion.
- Inorganic chemicals: production and use (two groups).
- Metallurgical industry (two groups of nonferrous metals).
- Metallurgical industry: ferrous metals.
- Mineral products processing and use (two groups).
- Miscellaneous industries.
- Petroleum refineries.
- Petroleum and gasoline: production and marketing.
- Pharmaceutical production processes.
- Polymers and resins production.
- Surface coating and processing.
- Synthetic organic chemicals: production.
- Waste treatment and disposal.

The category for production of synthetic organic chemicals is the largest single source category.

EPA has divided its 20 general industry groups into four "bins" for which it will develop **Maximum Achievable Control Technology (MACT)** standards. EPA will issue standards for these bins in sequential order at two to three year intervals. Over the next decade, therefore, EPA will specify technological control standards for all of the source categories in each bin. As required by the amendments, these standards will require all sources to use MACT to reduce their emission of HAPs.

- **Bin One**--The first bin is to cover 40 industry source categories, for which standards were developed by November 15, 1992. These 40 categories include dry cleaners and a new hazardous organic standard (covering production of 400 chemicals). Coke oven standards covering charging, topside, and door leaks were issued by December 31, 1992.
- **Bin Two**--By November 15, 1994, EPA is to develop emission control standards for 25% of all the industry categories.
- **Bin Three**--By November 15, 1997, EPA must set standards for 50% of all listed categories.
- **Bin Four**--By November 15, 2000, EPA must promulgate air toxics emission standards for all listed categories and subcategories.

Existing sources have three years to comply with the MACT standard once it has been defined for that source category. However, a one-year extension may be granted if the permitting authority (EPA or a state agency with an approved permit program) finds the need for additional time to install necessary controls. New sources may be granted a three-year extension under certain conditions.

EPA's first proposed emission control rule under the 1990 CAA amendments is aimed at reducing emissions of perchloroethylene (also called PCE or "perc") from industrial and commercial dry cleaning establishments. As set forth, EPA's rule affects firms emitting more than 10 tons of perc per year--some 3700 of the country's 25,200 dry cleaning facilities. This requires cleaners--depending on their size, and on whether they are new or existing--to control perc emissions by adopting either MACT or less-stringent **Generally Available Control Technology (GACT)**. The EPA rule includes some pollution control measures that require a combination of equipment changes, operating practices, and maintenance procedures. Operators are required to conduct a weekly inspection to prevent solvent emissions from broken or improperly operating equipment. Also periodic recordkeeping to track the amount of perc used at that facility is required. EPA estimates that dry cleaners currently emit over 92,000 tons of perc annually.

Sources emitting 10 tons or more of perc are required to install MACT--in this case, carbon adsorbers, refrigerated condensers, or equivalent devices (approved by the administering air quality control agency). Sources emitting less than 10 tons are only required to implement GACT, which, in the case of dry cleaners, consists of improved recordkeeping and "housekeeping" efforts.

INDEX

A

Printed and bound by CPI Group (UK) Ltd, Croydon, CR0 4YY

25/10/2024

01779162-0001